ZEW Economic Studies

Publication Series of the Centre for European Economic Research (ZEW), Mannheim, Germany

ZEW Economic Studies

Further volumes of this series can be found at our homepage:
http://www.springer.de/cgi/svcat/search_book.pl?series=4041

Vol. 6: B. Fitzenberger
Wages and Employment Across Skill Groups
An Analysis for West Germany
1999. XII, 251 pp. ISBN 3-7908-1235-8

Vol. 7: K. Rennings et al. (Eds.)
Social Costs and Sustainable Mobility
Strategies and Experiences in Europe
and the United States
1999. VI, 212 pp. ISBN 3-7908-1260-9

Vol. 8: H. Legler et al.
Germany's Technological Performance
2000. X, 191 pp. ISBN 3-7908-1281-1

Vol. 9: O. Bürgel
The Internationalization of British Start-up Companies in High-Technology Industries
2000. XIV, 230 pp. ISBN 3-7908-1292-7

Vol. 10: J. Hemmelskamp et al. (Eds.)
Innovation-Oriented Environmental Regulation
Theoretical Approaches and Empirical Analysis
2000. VI, 348 pp. ISBN 3-7908-1313-3

Vol. 11: K.L. Brockmann, M. Stronzik (Eds.)
Flexible Mechanisms for an Efficient Climate Policy
Cost Saving Policies and Business Opportunities
2000. VIII, 130 pp. ISBN 3-7908-1314-1

Vol. 12: W. Smolny
Endogenous Innovations and Knowledge Spillovers
A Theoretical and Empirical Analysis
2000. VIII, 277 pp. ISBN 3-7908-1320-6

Vol. 13: M. Lechner, F. Pfeiffer (Eds.)
Econometric Evaluation of Labour Market Policies
2001. X, 242 pp. ISBN 3-7908-1372-9

Vol. 14: M. Beise
Lead Markets
Country-Specific Success Factors of the Global Diffusion of Innovations
2001. XVI, 305 pp. ISBN 3-7908-1430-X

Vol. 15: O.H. Jacobs, C. Spengel
Effective Tax Burden in Europe
Current Situation, Past Developments and Simulations of Reforms
2002. X, 131 pp. ISBN 3-7908-1470-9

Vol. 16: U. Kaiser
Innovation, Employment, and Firm Performance in the German Service Sector
2002. X, 164 pp. ISBN 3-7908-1481-4

Vol. 17: J. Köke
Corporate Governance in Germany
An Empirical Investigation
2002. VIII, 184 pp. ISBN 3-7908-1511-X

Vol. 18: E. Wolf
What Hampers Part-Time Work?
An Empirical Analysis of Wages, Hours Restrictions and Employment from a Dutch-German Perspective
2003. X, 174 pp. ISBN 3-7908-0006-6

Vol. 19: P. Cecchini et al. (Eds.)
The Incomplete European Market for Financial Services
2003. XII, 255 pp. ISBN 3-7908-0013-9

C. Böhringer · A. Löschel (Eds.)

Empirical Modeling of the Economy and the Environment

With 27 Figures
and 42 Tables

Physica-Verlag
A Springer-Verlag Company

ZEW
Zentrum für Europäische
Wirtschaftsforschung GmbH
Centre for European
Economic Research

Series Editor
Prof. Dr. Wolfgang Franz

Editors

Dr. Christoph Böhringer
Dipl. Vw. Andreas Löschel
Centre for European Economic Research (ZEW)
L7, 1
68161 Mannheim, Germany

ISBN 3-7908-0078-3 Physica-Verlag Heidelberg New York

Cataloging-in-Publication Data applied for
Bibliographic information published by Die Deutsche Bibliothek
Die Deutsche Bibliothek lists this publication in the Deutsche Nationalbibliografie; detailed bibliographic data is available in the Internet at <http://dnb.ddb.de>.

Physica-Verlag Heidelberg New York
a member of BertelsmannSpringer Science+Business Media GmbH

Cover design: Erich Dichiser, ZEW, Mannheim

SPIN 10934371 88/3130-5 4 3 2 1 0 – Printed on acid-free paper

Contents

Introduction[*]

Christoph Böhringer and Andreas Löschel

Centre for European Economic Research, P.O.Box 103443, 68034 Mannheim, Germany, *boehringer@zew.de*, *loeschel@zew.de*

This book contains the proceedings of an international workshop "Empirical Modeling of the Economy and the Environment" held at the Centre for European Economic Research (ZEW, Mannheim) in June 2001. The workshop was organized on occasion of ZEW's 10th anniversary and in honor of the 60th birthday of Klaus Conrad (University of Mannheim) who has been affiliated with the Centre since its foundation. The papers presented by internationally reputed experts in the field of environmental economics cover a wide spectrum of issues in environmental regulation.

Daniel L. Millimet and **Thomas Osang** (Southern Methodist University, Dallas) investigate the implications of environmental regulation for productivity growth. For their econometric analysis they make use of the productivity measurement approach by Conrad and Morrison[1] and apply it to industry-level panel data of U.S. manufacturing industries. They find empirical evidence that environmental regulation has a negative impact on productivity growth providing a potentially important input to the public debate about cost and benefits of stricter environmental standards.

Abay Mulatu, **Raymond J.G.M. Florax**, and **Cees A. Withagen** (Vrije Universiteit, Amsterdam) review the empirical literature on the relationship between domestic environmental regulation and international competitiveness. The motivation for their meta-analysis stems from conflicting views. On the one hand, stringent environmental policy may harm the competitiveness of domestic firms, inducing phenomena such as ecological dumping, ecological capital flight, and regulatory 'chill' in environmental standards. On the other hand, strict environmental regulation may trigger industry's innovation potential and reduce X-inefficiency, thus subsequently increasing its competitiveness (Porter hypothesis). The authors employ a statistically supported meta-analysis to assess contradictory evidence of various studies. Based on their evaluation, the empirical

---

* We thank Christoph Skupnik for assistance in putting together this book. Thanks to Noelle Christ-See for cross-reading parts of the book.

1 Conrad, K. and C.J. Morrison (1989), The Impact of Pollution Abatement Investment on Productivity Change: An Empirical Comparison of the U.S., Germany, and Canada, *Southern Economic Journal* 55 (3), 684-698.

literature does not strongly support the hypothesis of environmental regulation as being detrimental for competitiveness. However, they point out problems in the empirical analyses, particularly stringency measure, type of data, methods of investigation and sectoral disaggregation, which warrant future research.

Heinz Welsch (University of Oldenburg) identifies major determinants of the substantial decline of West German production-related carbon intensity in the second half of the 1980s given the substantial drop of world energy prices over that period. Using a computable general equilibrium model, he first simulates the effects of observed changes of world energy prices and domestic energy policy on the economic structure. His key finding is that the contribution of rest of the world productivity and demand patterns to emission-relevant structural change unaccounted for by energy prices and energy policy is just under 30%. The remainder is split almost equally among patterns of domestic autonomous energy efficiency improvement and domestic labor efficiency patterns. This result suggests that a substantial fraction of emission-relevant structural change in West Germany between 1985 and 1990 has been trade-induced.

Michael Rauscher (University of Rostock) analyses the problem of pollution leakage, i.e. the increase in foreign emissions after a reduction in domestic emissions. This is an important issue when countries unilaterally address international environmental problems. Leakage results from the interdependence of international commodity and factor markets. Therefore, the optimal solution to the leakage problem requires international co-operation. In a second-best world, where broader international co-operation cannot be achieved, trade interventions can be justified to cope with the leakage problem. Using a partial equilibrium model of energy trade and international capital mobility, Rauscher finds that the most important channel through which leakage effects are generated appears to be the market for primary goods, particularly energy. The other main result is that the existence of leakage aggravates the free-rider problem in a world with transfrontier pollution. In addition to Pigouvian taxes, the social planner should use export taxes or import subsidies on energy as well as the subsidization of the domestic use of mobile capital to mitigate counterproductive leakage effects.

Franz Wirl (University of Vienna) tries to reconcile the divergence between computable general equilibrium (CGE) studies of the consequences of environmental policies that reveal a modest – sometimes even positive – impact on economic activity and studies that find a significant contribution of environmental regulation to the slow-down in economic growth and particularly in productivity growth. He points out that many CGE models assume rather frictionless re-organization of production processes and thus ignore the complication in production management due to the additional task of controlling environmental inputs, which weakens incentives and shows up in the empirical data. Using a principal-agent model that accounts for this managerial friction, he provides a partial explanation for the low growth and high unemployment rates in the late 1980s and early 1990s despite the dramatic energy price cut in 1986.

ZhongXiang Zhang (East-West Center, Honolulu) uses a global model based on marginal abatement cost curves for 12 world regions to estimate the contributions of the three flexibility mechanisms under the Kyoto Protocol, i.e. emissions trading, joint implementation, and the clean development mechanism. He shows how the reduction in compliance costs of industrialized regions depends on the extent to which the flexibility mechanisms will be available. Not surprisingly, the fewer the restrictions on the use of flexibility mechanisms will be, the greater the gains from their use. These gains are unevenly distributed, however, with industrialized regions that have the highest autarkic marginal abatement costs tending to benefit the most. Restrictions on the use of flexibility mechanisms not only reduce the potential of the industrialized regions' efficiency gains, but are also not beneficial to developing countries since they restrict the total financial flows to developing countries under the clean development mechanism.

Christoph Böhringer (ZEW, Mannheim), **Glenn W. Harrison** (University of South Carolina, Columbia), and **Thomas F. Rutherford** (University of Colorado, Boulder) evaluate the welfare implications of alternative ways in which the EU could distribute its aggregate emission reduction commitment under the Kyoto Protocol across member states. Using a large-scale CGE model, they compare a uniform proportional cutback in emissions and the actual EU burden sharing agreement with an equitable allocation scheme derived from an endogenous burden sharing calculation. The latter equalizes the relative welfare cost across member states. The authors show that the equitable burden sharing scheme differs substantially from uniform proportional emission cutbacks as well as the specific burden sharing rule adopted by the EU.

Gunter Stephan and **Georg Müller-Fürstenberger** (University of Bern), analyze trading and banking of carbon emission rights within the framework of an integrated assessment model of global climate change that breaks the world economy into two regions, North and South. They find that increasing flexibility in greenhouse gas abatement through trading and banking of carbon rights has no negative impact on the global climate. The accumulation of the atmospheric carbon stock is almost independent of whether there is trading and banking of carbon rights or not. Furthermore, their simulations reveal conflicting interests between North and South with respect to "when and where" flexibility. While the North is best off in a regime that allows both banking and trade of emission rights, the South is best off in a pure trade scenario without banking. As a compromise, trading of carbon rights provides welfare gains to both regions.

Zeger Degraeve (London Business School), **Stef Proost,** and **Gunther Wuyts** (KULeuven) examine the cost-efficiency methodology to decide on new emission standards for cars and fuel quality standards in the context of the Auto-Oil Programme of the European Commission. They detect two major methodological problems that could lead to a bias against European wide measures: the overachievement problem and the problem of decentralization of local measures. The authors show how the first problem can be solved by attributing benefits to overachievement and how the second problem can be handled by explicitly using the behavior of the local government as constraint in the problem. In numerical simulations, they demonstrate that the extended methodology affects the selection of optimal emission standards for cars in Europe and leads to substantial welfare gains.

Till Requate (University of Heidelberg) surveys recent developments regarding the incentives of environmental policy instruments, particularly, taxes, subsidies, auctioned permits, free permits, and emission standards, to spur both R&D and the adoption of new technology. He finds that if the regulator makes long-term commitments to policy levels and does not anticipate the arrival of new technology, taxes provide stronger incentives than permits, auctioned and free permits offer identical incentives, and standards may give stronger incentives than permits. If the regulator anticipates new technologies, taxes and permits can induce first-best outcomes whenever the regulator moves after firms have invested. On the other hand, this result does not necessarily hold if the regulator moves first. Furthermore, the author develops a model where a polluting downstream industry is regulated either by emission taxes or by tradable permits, while a separate monopolistic or duopolistic upstream industry engages in R&D and – in case of R&D success – sells an advanced abatement technology to the downstream firms. He uses this model to investigate three different timings of environmental policy: (i) ex post taxation (or issuing permits), (ii) ex interim commitment to a tax rate (a quota of permits) after observing R&D success but before adoption, and (iii) ex ante commitment before and independent of R&D success. In the specific analytical framework it can be shown that ex interim commitment always dominates ex post environmental policy, whereas ex interim second best taxation dominates ex interim second best optimal permit policy. However, there is no unique ranking between ex ante and ex interim commitment.

Rüdiger Pethig (University of Siegen) analyses the provision of the public good 'environmental quality' using a benchmark model with a complete set of competitive markets, and investigates various ways to finance that public good, including subsidies, (distortionary) taxes, and resource income from selling assimilative services. He relates the raising of the level of environmental quality to the concept of ecological tax reform discussed in the literature. By looking at such tax reforms as providing more of the costly public good 'environmental quality', the author casts doubts on the double dividend hypothesis, since the increase in environmental quality must be financed by distortionary taxes, which, in turn, augments the inefficiency of taxation and, thus, prevents a second efficiency dividend. In the public good perspective of ecological tax reforms, the double dividend conjecture is tantamount to assuming that more environmental quality can be bought for less money, which is rather unlikely even if general equilibrium interdependencies are accounted for.

Klaus Conrad (University of Mannheim) extends the standard Hotelling model of resource depletion by including stock externalities such as accumulated CO_2 and

SO_2 as well as flow externalities like pollutants that can be abated (SO_2). He employs a micro model of a profit-maximizing industry to derive the optimal design of an energy (CO_2/SO_2) tax – both in level and in time profile – in order to bring about socially desirable paths of fossil fuel consumption and carbon accumulation.

Wolfgang Buchholz (University of Regensburg) retraces the ethical justification of intergenerational discounting, i.e. the desire to avoid a distributional bias in favor of future generations. Using a two-generation framework, the author shows that for productive economies it is well possible to use undiscounted utilitarianism instead of intergenerational discounting in order to make ethically acceptable choices on intergenerational distributions. He then considers the implications of this approach for the time consistency problem that has accompanied the discounting question for a long time. A key insight from his analysis is that there is not so much need to worry about the problem of time-inconsistency, even when intergenerational discounting is used as a pragmatic device.

Environmental Regulation and Productivity Growth: An Analysis of U.S. Manufacturing Industries*

Daniel L. Millimet

Department of Economics, Southern Methodist University, Box 0496, Dallas, TX 75275-0496, USA, *millimet@mail.smu.edu*

Thomas Osang

Department of Economics, Southern Methodist University, Box 0496, Dallas, TX 75275-0496, USA, *tosang@mail.smu.edu*

Summary:
We show that traditional measures of productivity change that ignore the unproductive nature of pollution abatement capital within the production process are likely to underestimate the true productivity gains that most manufacturing industries are able to generate in any given year. While the average bias of traditional measures is not large in absolute terms, the bias can be substantial for industries with relatively large pollution abatement capital expenditures. We also find that environmental regulation has a non-trivial adverse effect on productivity change, lowering productivity growth by roughly 0.3% across all industries, and by more than 1% for some industries.

Keywords: Productivity change, pollution abatement, capital expenditure, capital adjustment bias, environmental regulation bias, U.S. manufacturing industries.

1 Introduction

One of the economic indicators closely watched by the Federal Reserve, policy makers on Capital Hill, stock market investors, as well as a large number of media is the monthly report on U.S. productivity growth. While high levels of

* We would like to thank the discussant, Ralph Winkler, for helpful comments and suggestions.

productivity growth are generally seen as benign, low levels or a "slowdown" in productivity change are interpreted as warning signals about the health of the economy, evoking fears of rising inflation, declining growth, and a loss in international competitiveness. Not surprisingly, given its importance among key economic indicators, productivity change has played a central role as a research topic among academic economists. Two issues have dominated the literature on productivity growth. First, the correct measurement of productivity growth[1]; and, second, the identification of factors that cause productivity to change or are at least significantly correlated with productivity growth.[2] While most papers on productivity growth concentrate on either measurement or identification, some studies, such as Conrad and Morrison (1989), deal with both issues. The novel insight of the Conrad and Morrison paper is the treatment of pollution abatement capital (or PA capital for short) as a costly, but unproductive input in the production process. This departure from the traditional view leads to a theoretical formulation of productivity change which can be used to identify a bias associated with traditional measures of productivity growth. The bias, referred to as capital aggregation or CA bias from here on, measures the difference between the true level of productivity growth and its level according to the traditional measure. Once the correct measure of productivity change has been established, Conrad and Morrison modify it to estimate a second "bias" that measures the impact of environmental regulation on the true measure of productivity growth.[3] We will refer to this second bias as the environmental regulation or ER bias from here on.

In this paper, we revisit the theoretical approach presented by Conrad and Morrison and apply it to an industry-level panel data, comprised of U.S. manufacturing industries. Compared to the study by Conrad and Morrison, who apply their theory to three countries – the United States, Canada, and Germany –, our approach has several advantages. First, choosing industries instead of countries should improve data quality and, consequently, inference, as data quality usually declines with the level of aggregation. Second, observing a cross-section of industries over time allows us to test whether the most heavily regulated industries – the ones that have the highest share of pollution abatement capital

---

1 See Nadhiri (1970) for an early survey of the literature. Other, more recent articles on this topic include Maddison (1987) and Bernard and Jones (1996), among others.

2 The literature on the determinants of productivity change is enormous. For an early survey, see Nelson (1981). The slowdown in U.S. productivity growth during the 1970s and 1980s has been researched extensively. For an overview, see the 1988, volume 2(4) edition of the Journal of Economic Perspectives, which contains a number of articles on the topic. More recent studies on the productivity slowdown include Hulten (1992) and Wolf (1996), among many others.

3 The second "bias" is not a bias in the traditional sense of the measurement literature. Rather, it is a measure of the impact that a particular policy (here environmental regulation) has on productivity change. However, in order to avoid confusion with the terminology used in Conrad and Morrison, we will refer to this impact measure as "bias".

expenditure – exhibit indeed the strongest bias as predicted by theory. With a country-level data set that includes only a handful of countries a cross-section comparison yields few new insights due to the small cross-section dimension of the sample. In addition, there is the problem that environmental regulation differs across countries. Third, we perform a number of sensitivity tests to provide evidence on the robustness of our results.

The main findings of this paper are as follows. First, we find both cross-section and time-series evidence for the existence of a positive capital aggregation bias. A positive CA bias implies that the true measure of productivity growth is greater than the traditional measure. Second, we also find cross-section and time-series evidence for a negative environmental regulation bias. A negative ER bias implies that environmental regulation had an adverse effect on productivity growth. Both biases increase, in absolute value, with the pollution abatement intensity of the industry. Third, we find that alternative measures of the stock of capital, different depreciation rates, and differences in price deflators do not change our results in a qualitative sense.

The paper is organized as follows. In Section 2 we introduce the theoretical model and derive the specifications for both types of biases. We describe the data set in Section 3. Section 4 contains the empirical results, while we discuss the results of the sensitivity analysis in Section 5. Section 6 concludes.

2 Theoretical Model

Our theoretical model is a simplified version of Conrad and Morrison (1989). The representative firm maximizes profits by optimal choice of output, x, labor input, L, and investment in productive capital, I, as well as non-productive pollution abatement capital, IPA, subject to equations of motion for productive capital, K, pollution abatement capital, KPA, an emission production function, and an emission constraint imposed by the government. The firm's intertemporal maximization problem can thus be stated as:

$$\text{Max} \quad \sum e^{-rt}\left\{p_{xt}x_t - G(x_t, K_t, w_t) - p_I\left(I_t + IPA_t\right)\right\} \tag{1}$$

subject to

$$K_{t+1} = K_t + I_t - \delta \cdot K_t \tag{1a}$$

$$KPA_{t+1} = KPA_t + IPA_t - \delta \cdot KPA_t \tag{1b}$$

$$TE_t = \xi \cdot x_t \tag{1c}$$

$$\overline{NE}_t - f(KPA_t)\cdot TE_t \;\geq\; 0 \tag{1d}$$

where p_{x_t} denotes the output price, $G(.)$ denotes the variable cost function, w_t denotes the unit cost of labor, p_{I_t} denotes the common price of both types of investment goods, δ is the fixed depreciation rate for productive and abatement capital, and ξ is a parameter that represents the constant relationship between the level of production and the level of pollutant emissions. The total level of (unrestricted) emissions is denoted by TE_t. Note that *KPA* is not included in the cost function since it is not productive.[4] In contrast to *K*, *KPA* is not in the production function and thus not in its dual, the variable cost function. *KPA* does effect overall cost, but as a fixed cost, not as a variable cost. Reducing emissions requires additional investment in PA capital, captured by the function *f*(*KPA*). The function *f*(*KPA*) exhibits diminishing returns, $f' < 0$, $f'' > 0$, as the marginal productivity of emission reduction falls with each additional unit of installed abatement capital.

Incorporating conditions (1c) and (1d) into the objective function, the constrained optimization problem of the firm can now be written as:

$$\text{Max} \;\sum e^{-rt}\left\{p_{xt}x_t - G(x_t, K_t, w_t) - p_{It}(I_t + IPA_t)\right\}$$

$$-\overline{\tau}\left[f(KPA_t)\cdot\xi x_t - \overline{NE}_t\right] \tag{2}$$

subject to (1a) and (1b).

The first-order optimality conditions for this problem are:

$$p - \tau f(KPA)\xi \;=\; G_x \tag{3a}$$

$$q_K \;=\; -G_K \tag{3b}$$

$$q_K \;=\; -\tau f'(KPA)\xi x \;=\; 0 \tag{3c}$$

where $\tau = \overline{\tau}\cdot(1+r)^t$ is the current shadow value of non-abated emissions, and q_K is the rental price of both productive and PA capital.

Equations (3a)-(3c) can be rewritten to facilitate the definition of the homogeneity properties of the cost functions:

$$p_x x/G - \tau f(KPA)\xi x/G \;=\; p_x x/G - \tau NE/G \;=\; \partial \ln G/\partial \ln x, \tag{4a}$$

4 Whenever possible without causing confusion, time subscripts are omitted from hereon.

$$-q_K K/G = \partial \ln G / \partial \ln K, \tag{4b}$$

$$q_K KPA/G = -\tau f'(KPA) KPA \xi x/G = \tau NE/G, \tag{4c}$$

where the last equality in (4c) requires that $f(KPA)$ is homogeneous of degree (-1). While this assumption is not necessary for our analysis, we impose it as a useful simplification.

The expressions in (4) are useful in two ways. First, they allow us to motivate the definition of the total cost function C. Second, they can be employed to define the homogeneity properties of both C and G. Homogeneity of degree one in output of G implies that $\partial \ln G / \partial \ln x + \partial \ln G / \partial \ln K = 1$. From (4), this implies that $p_x x / G - \tau NE / G - q_K K / G = 1$. Rewriting this equality yields $p_x x = G + \tau NE + q_K K$, which, using (4c) is equivalent to $p_x x = G + q_K KPA + q_K K$, the firm's total costs, C. In addition to variable costs, the firm incurs a capital cost for the use of productive capital as well as for the use of pollution abatement capital.

The specification of the above cost function can be used to derive a cost-side measure of productivity growth, $\partial \ln C / \partial \ln t$, similar to traditional accounting productivity measures but adjusted for the fact that KPA is not a productive input. This adjusted productivity growth measure can be written as[5]:

$$\begin{aligned} \varepsilon_C{}' &= -\partial \ln C / \partial \ln t \\ &= \left[(p_x x - q_K KPA)/C \right] \dot{x}/x - (wL/C)\dot{L}/L - (q_K K/C)\dot{K}/K. \end{aligned} \tag{5}$$

Since productivity change leads to a reduction in cost, $\partial \ln C / \partial \ln t < 0$, which implies that $\varepsilon_C{}'$ is a positive number. Equation (5) differs from the traditional productivity measure in two ways: (1) the KPA component has been removed from the capital stock; (2) the effect of pollution abatement has been purged from the output value measure. Thus, the traditional productivity measure can be written as the sum of the correct measure (5) and a bias term composed of the two adjustments mentioned above:

$$\begin{aligned} \varepsilon_C &= -\partial \ln C' / \partial \ln t \\ &= (p_x x/C)\dot{x}/x - (wL/C)\dot{L}/L \\ &\quad -\left[q_K (K + KPA)/C \right] \left(\dot{K} + \dot{KPA} \right) / (K + KPA), \end{aligned}$$

---

5 For a proof, see the Appendix in Conrad and Morrison (1989).

$$= \left[(p_x x - q_K KPA)/C\right]\dot{x}/x - (wL/C)\dot{L}/L - (q_K K/C)\dot{K}/K$$

$$-q_K (KPA/C)\left[(\dot{KPA}/KPA) - (\dot{x}/x)\right],$$

$$= \varepsilon_C ' - b_1, \tag{6}$$

where b_1 is the "error bias" that distinguishes the true measure (5) from the traditional productivity growth measure (6). The sign of this bias term is ambiguous and depends on the relative size of changes between PA capital spending and changes in output. A positive value of b_1 implies that the true measure is greater than the traditional measure. In this case, the traditional measure underestimates the true change in productivity. Such a situation will arise when pollution abatement growth exceeds output growth, as is to be expected in years when regulation is initially put into place or increases suddenly.

The second bias we are interested in measures the impact of environmental regulation, as manifested by investments in PA capital, on productivity growth. To derive the environmental regulation bias, we use the true productivity measure (6) but then impose the counterfactual condition that no regulation took place, i.e. $KPA = 0$. In addition, the computation of K must be adapted since the return to capital in (5) depends on the existence of pollution abatement capital. Without regulation, the ex post rate of return to capital is higher, which in turn implies that the price of capital paid by the firm is higher as well. We thus multiply the last term in (5) by a factor $1 + KPA / K$. The new productivity growth measure that corresponds to the case of no regulation is thus defined as:

$$\bar{\varepsilon}_C = -\partial \ln C'' / \partial \ln t$$

$$= (p_x x)\dot{x}/x - (wL/C)\dot{L}/L - q_K (1 + KPA/K)(K/C)\dot{K}/K$$

$$= \left[(p_x x - q_K KPA)/C\right]\dot{x}/x - (wL/C)\dot{L}/L - (q_K K/C)\dot{K}/K$$

$$-q_K (KPA/C)\left[(\dot{K}/K) - (\dot{x}/x)\right]$$

$$= \varepsilon_C ' - b_2, \tag{7}$$

where b_2 is the "regulation bias" that distinguishes the correct measure in the presence of regulation (5) from the correct measure of productivity growth measure in the absence of regulation (7). The sign of the ER bias term is ambiguous as well and depends on the change in productive capital spending relative to the change in output. A negative value of b_2 implies that productivity growth without regulation, $\bar{\varepsilon}_C$, is higher than productivity growth in the presence of regulation, $\varepsilon_C{}'$. Such a situation will arise when output growth exceeds growth of productive capital, as is to be expected in years (or industries) with high levels of regulation.

3 Data

Annual data on PA capital expenditures, *IPA*, are taken from the Census Bureau's Pollution Abatement Costs and Expenditures.[6] Since the data for 1987 are not available, we estimate the missing values using an extrapolation method.[7] Three-digit SIC level data on value added, *x*, and total new capital expenditures, *IPA* + I, are taken from the Annual Survey of Manufactures.[8] We construct stocks of productive and pollution abatement capital, *K* and *KPA*, respectively, from their corresponding flow variables in two different ways. First, we follow Acemoglu and Zilibotti (2001) (A-Z from here on) and construct the estimated capital stock at time *t*, K_t^e, as the ratio of the weighted average investment flow at time *t*, $\hat{I}^T_{c,i,t}$ and the time-invariant depreciation rate, *d*. *T* denotes the first year for which investment flow data are available. The weighted average investment flow is defined as:

$$\hat{I}^T_{c,i,t} = \frac{\sum_{s=T_{c,i}}^{t} (1-d)^{t-s} I_{c,i,s}}{\sum_{s=T_{c,i}}^{t} (1-d)^{t-s}}.$$

As an alternative measure, we construct a starting value for total capital using the stylized fact that the value of the capital stock is approximately three times the value of output. Once the initial value of the total capital stock has been determined, we use the perpetual inventory method to derive the changes in the stock of capital over time. We then use the fraction of *I* and *IPA* in total investment in order to decompose the total capital stock into its two parts, *K* and *KPA*. Both methods of constructing the two capital stock data are based on the assumption of an annual depreciation rate of 8%.

---

6 U.S. Department of Commerce, Bureau of the Census, "Pollution Abatement Costs and Expenditures", Current Industrial Reports MA200, Washington, D.C., GPO, various years.

7 See Jaffe and Palmer (1997) for a similar procedure.

8 U.S. Department of Commerce, Bureau of the Census, "Annual Survey of Manufactures", Washington, D.C., GPO, various years.

Annual data on the output price index, p_x, the price index for investment, p_I, the size of the labor force, L, as well as wages, w, are taken from the NBER-CES Manufacturing Database (Bertelsman and Gray, 1996). The price of capital, q_K, was calculated from the zero profit condition as the ex post price of the total capital stock, including both productive and pollution abatement capital. Four-digit SIC level data have been aggregated to the three-digit SIC level, using value added as weight if necessary.

4 Empirical Results

Table 1a contains three different productivity indexes for each manufacturing industry in our sample: the first index was computed using the traditional method (6); the second index was calculated using the correct (*KPA*-adjusted) definition (5); and the third index was computed using the *KPA*-adjusted definition but with the no-regulation conditions imposed (7). In addition, Table 1a contains two bias terms for each industry, the capital aggregation (CA) bias, which we use to rank the industries in the table, and the environmental regulation (ER) bias. The indexes and bias terms in Table 1a are based on the A-Z measurement of the capital stock, with a depreciation rate of 8%. In addition, we have applied a uniform deflator to all nominal variables.

Table 1a shows that the CA bias is positive for most industries. This means that the traditional productivity measure (5) underestimates the true productivity gains for almost all industries. For the few industries for which the reverse is true, the magnitude by which the traditional measure overestimates productivity growth is minimal, except for one industry – primary nonferrous metals (SIC 333). Overall, the magnitude of the CA bias is fairly small for most industries. Close to 60% of all industries exhibit a CA bias of one tenth of one percent or less (in absolute terms), and only 9 industries have a CA bias that exceeds half a percentage point (in absolute terms). More important, however, is that most of the industries with a strong CA bias are heavy investors in pollution abatement equipment, exactly as one would expect from our theoretical model. This can be seen by a comparison of the ranking of industries by CA-bias in Table 1a with the ranking of industries by PA intensity in Table 3. The correlation between the ranking of these two measures is very high, with a correlation coefficient of 0.59 that is statistically significant at the 1% level. Except for one industry, petroleum refining (SIC 291), the ER-bias is negative: Environmental regulation lowers productivity growth, even in industries that are not particularly PA-intensive.[9] The mean loss in

---

9 One possible explanation for the different sign of the ER bias on petroleum refining, as suggested by the discussant, may be the fact that the petroleum industry was, at an early stage, subjected to governmental regulation via tradable permits instead of

productivity because of environmental regulation across all industries is 0.29%, with some industries losing 1% or more. On average, the productivity loss is more substantial for industries that are PA-intensive, as can be seen from the negative correlation coefficient of -0.39, statistically significant at the 1% level.

Using the alternative capital stock measure does not change the results in a qualitative sense (see Table 1b). As before, the PA-intensive industries are more likely to exhibit high values of the CA bias, with a significant correlation coefficient of 0.39. The ER bias is negative for most industries and increases, in absolute terms, with an industry's PA-intensity (the correlation coefficient is -0.48 and significant at the 1% level). However, a few differences emerge. Overall, the results based on the alternative procedure are smaller in magnitude, with a mean true productivity growth rate (6) of 2.73% for all industries, compared to a value of 8.34% in the previous table. Similarly, the mean bias terms for CA and ER bias are now 0.02% and -0.05%, respectively, compared 0.20% and -0.29% in Table 1a.

The time series results for the A-Z capital measure are presented in Table 2a. Similar to the cross-section evidence, we find time-series evidence for the existence of both CA and ER bias. The CA bias is positive except for 2 years, 1987 and 1988, both years of higher than average productivity growth. In these years, the impact of strongly increasing output growth overwhelms the small changes in *KPA*. As noted by Conrad and Morrison, when output growth is strong, lowering the valuation of output to its marginal cost has a large impact relative to the increase in pollution abatement capital. In the first half of our sample, from 1984-88, the average CA bias is close to zero as a result of the two outliers, while CA bias is positive, as most researchers would expect, in each of the remaining years. From 1989-93, the traditional measure of productivity growth underestimates the correct value by 0.37% per year. The ER bias is negative in both sub-periods, as expected. However, the absolute value of the regulation bias trends downward over time. This is surprising since PA capital expenditures by all manufacturing industries (SIC codes 20-39) increased from 2.88% of new capital expenditure in 1984 to 7.0% in 1993 (see Osang and Nandy, 2000). Thus, while firms increased their relative expenditure on pollution abating equipment between 1984 and 1993, the adverse impact on productivity growth that resulted from these investments diminished over time. One explanation for this result may be that early pollution abatement investments were easier to identify for firms since they often involved retrofitting of existing plants, typically with end-of-pipe abatement technologies. Over time, as abatement technologies became an integral part of new technologies, the distinction between regular and pollution abatement capital spending became increasingly difficult and arbitrary for most firms leading to abatement data that are noisier than in the earlier years.

---

environmental standards or taxation, a form of regulation not covered by the theory presented in Section 2.

As Table 2b reveals, there are no qualitative differences in the time-series evidence when the alternative capital measure is used. As before, the main difference to Table 2a lies in the reduced absolute value of all three indexes and the two bias terms.

5 Sensitivity Results

To test the robustness of the previous results, we perform two sets of sensitivity tests.[10] First, we reestimate Tables 1-2 for different depreciation rates (d = 5% and d = 10%). Once again, while raising or lowering depreciation rates alters the absolute values of the indexes, there is no change in results in a qualitative sense. Second, we abandon the assumption of a uniform deflator for all nominal variables, using instead variable-specific deflators if available. In particular, we use the shipment deflator for value added and total labor cost and the investment deflator for both productive and PA capital. Using specific deflators causes only minor changes in the results. In effect, both cross-section and time-series results are nearly identical to our findings reported in Tables 1 and 2. For example, with variable-specific deflators, the average CA bias across all years and industries is 0.20%, while the average ER bias is -0.27%. With a uniform deflator, the corresponding numbers are 0.20% and -0.29% (see Table 1a).

6 Summary and Conclusion

In this paper, we have shown that environmental regulation – requiring firms to invest in unproductive pollution abatement equipment – creates a challenge for researchers as far as the correct measurement of productivity growth is concerned, and, in addition, has an adverse effect on industry-level productivity growth. Unless properly adjusted, traditional measures of productivity change are likely to underestimate the true productivity gains that most industries are able to generate in any given year. While the average bias of traditional measures is not large in absolute terms, it is important to know that the true change in productivity is slightly higher than the official statistics indicate. This is particularly true in years when low (unadjusted) productivity growth rates lead to widespread pessimism about the overall health of the economy. With regard to its economic impact, we find that, for almost all industries and for most years, environmental regulation leads to a non-trivial reduction in productivity growth. Knowledge about the impact of environmental regulation on productivity growth, both in terms of the

[10] All results that pertain to this section are available upon request.

sign of the change and its magnitude, is important and should play a role in the public debate about cost and benefits of stricter environmental standards.

Table 1a: Productivity Growth Indexes, CA and ER Biases: Cross-Sectional Analysis.

	Productivity			Biases	
SIC	Traditional	KPA-Adj.	KPA=0	CA Bias	ER Bias
291	4.27%	6.07%	5.76%	1.80%	0.31%
261	2.51%	3.46%	3.83%	0.95%	-0.36%
281	4.93%	5.78%	6.69%	0.85%	-0.91%
203	10.37%	11.10%	11.53%	0.74%	-0.43%
311	4.24%	4.95%	5.21%	0.71%	-0.26%
341	1.48%	2.12%	2.32%	0.63%	-0.20%
262	3.70%	4.29%	4.52%	0.59%	-0.23%
286	6.53%	7.07%	7.58%	0.54%	-0.51%
282	6.10%	6.55%	6.97%	0.45%	-0.42%
324	4.05%	4.46%	4.82%	0.41%	-0.36%
287	5.55%	5.96%	6.62%	0.41%	-0.66%
386	18.22%	18.50%	18.80%	0.28%	-0.30%
364	8.81%	9.02%	9.24%	0.21%	-0.22%
331	5.87%	6.08%	6.27%	0.21%	-0.19%
284	13.10%	13.30%	13.70%	0.19%	-0.41%
343	8.29%	8.47%	8.70%	0.18%	-0.23%
299	9.34%	9.52%	10.09%	0.18%	-0.57%
295	10.38%	10.55%	10.78%	0.17%	-0.23%
322	9.10%	9.25%	9.52%	0.15%	-0.26%
205	12.38%	12.49%	12.61%	0.10%	-0.12%
342	6.47%	6.56%	6.78%	0.09%	-0.22%
332	2.82%	2.90%	3.12%	0.08%	-0.22%
283	12.01%	12.09%	12.52%	0.07%	-0.43%
201	7.50%	7.56%	7.81%	0.06%	-0.25%
206	11.77%	11.83%	12.11%	0.05%	-0.29%
251	6.25%	6.30%	6.50%	0.05%	-0.19%
355	7.90%	7.94%	7.98%	0.04%	-0.04%
371	7.58%	7.62%	7.86%	0.04%	-0.23%
289	7.53%	7.57%	7.98%	0.04%	-0.41%
352	9.54%	9.57%	9.79%	0.03%	-0.22%
243	7.66%	7.68%	7.87%	0.02%	-0.19%
207	11.51%	11.53%	12.12%	0.02%	-0.59%

Table 1a: continued.

	Productivity			Biases	
SIC	Traditional	KPA-Adj.	KPA=0	CA Bias	ER Bias
345	5.58%	5.60%	5.72%	0.02%	-0.12%
221	5.64%	5.66%	5.86%	0.02%	-0.19%
275	7.27%	7.28%	7.41%	0.02%	-0.13%
351	8.80%	8.82%	8.91%	0.02%	-0.09%
242	10.25%	10.26%	10.50%	0.01%	-0.24%
346	6.21%	6.21%	6.28%	0.01%	-0.06%
349	7.10%	7.11%	7.29%	0.01%	-0.18%
335	9.61%	9.61%	9.91%	0.00%	-0.30%
225	7.91%	7.91%	8.01%	0.00%	-0.10%
344	7.30%	7.29%	7.40%	-0.01%	-0.11%
265	7.09%	7.08%	7.19%	-0.01%	-0.11%
204	15.91%	15.89%	16.28%	-0.01%	-0.39%
367	8.35%	8.33%	8.48%	-0.02%	-0.15%
208	16.05%	16.01%	16.38%	-0.04%	-0.38%
285	11.30%	11.20%	11.75%	-0.10%	-0.55%
333	8.89%	8.05%	9.12%	-0.85%	-1.08%
Mean	8.15%	8.34%	8.64%	0.20%	-0.29%

Notes: A-Z capital measure, d=8%, and uniform price deflator. Industries ranked by CA bias.

Table 1b: Productivity Growth Indexes, CA and ER Biases: Cross-Sectional Analysis.

	Productivity			Biases	
SIC	Traditional	KPA-Adj.	KPA=0	CA Bias	ER Bias
291	2.68%	4.08%	3.91%	1.40%	0.17%
261	3.32%	3.86%	4.08%	0.54%	-0.21%
262	0.44%	0.64%	0.61%	0.20%	0.03%
286	5.22%	5.29%	5.47%	0.07%	-0.17%
282	2.68%	2.74%	2.82%	0.06%	-0.08%
203	2.21%	2.26%	2.31%	0.05%	-0.05%
324	1.25%	1.28%	1.31%	0.03%	-0.04%
341	-0.89%	-0.88%	-0.87%	0.01%	-0.01%
322	3.21%	3.22%	3.23%	0.01%	-0.01%
331	3.83%	3.84%	3.85%	0.01%	-0.02%
205	3.09%	3.09%	3.10%	0.00%	-0.01%
355	0.74%	0.73%	0.74%	0.00%	0.00%
364	1.43%	1.43%	1.44%	0.00%	-0.01%
346	0.49%	0.49%	0.50%	0.00%	0.00%
343	1.63%	1.63%	1.64%	0.00%	-0.01%
344	0.85%	0.84%	0.85%	0.00%	0.00%

Table 1b: continued.

	Productivity			Biases	
SIC	Traditional	KPA-Adj.	KPA=0	CA Bias	ER Bias
295	2.57%	2.56%	2.58%	0.00%	-0.02%
351	2.56%	2.56%	2.56%	0.00%	0.00%
345	-0.09%	-0.10%	-0.09%	0.00%	0.00%
275	0.79%	0.78%	0.79%	0.00%	-0.01%
265	0.46%	0.46%	0.46%	-0.01%	-0.01%
225	2.34%	2.33%	2.34%	-0.01%	-0.01%
386	9.77%	9.77%	9.79%	-0.01%	-0.02%
251	1.34%	1.34%	1.35%	-0.01%	-0.01%
367	2.11%	2.10%	2.11%	-0.01%	-0.01%
349	1.20%	1.19%	1.20%	-0.01%	-0.01%
221	1.33%	1.32%	1.33%	-0.01%	-0.01%
352	2.03%	2.02%	2.03%	-0.01%	-0.01%
342	1.59%	1.58%	1.59%	-0.01%	-0.01%
243	2.31%	2.30%	2.31%	-0.01%	-0.01%
242	4.31%	4.29%	4.31%	-0.01%	-0.02%
332	0.92%	0.91%	0.93%	-0.02%	-0.02%
201	1.52%	1.50%	1.52%	-0.02%	-0.02%
206	2.84%	2.82%	2.85%	-0.02%	-0.03%
335	3.78%	3.76%	3.78%	-0.02%	-0.02%
371	3.20%	3.17%	3.20%	-0.02%	-0.02%
208	5.59%	5.56%	5.59%	-0.03%	-0.04%
284	3.46%	3.43%	3.48%	-0.04%	-0.06%
299	3.62%	3.58%	3.65%	-0.04%	-0.07%
285	2.60%	2.56%	2.60%	-0.04%	-0.04%
289	1.96%	1.92%	1.97%	-0.04%	-0.05%
311	3.18%	3.14%	3.23%	-0.04%	-0.09%
207	4.79%	4.74%	4.80%	-0.05%	-0.06%
204	5.09%	5.04%	5.10%	-0.05%	-0.06%
283	5.08%	5.00%	5.12%	-0.07%	-0.12%
287	4.08%	3.99%	4.24%	-0.10%	-0.26%
281	2.84%	2.74%	2.94%	-0.10%	-0.20%
333	8.48%	8.02%	8.49%	-0.46%	-0.48%
Mean	2.70%	2.73%	2.77%	0.02%	-0.05%

Notes: "Rule of Thumb" capital stock measure, d=8%, and uniform price deflator. Industries ranked by CA bias.

Table 2a: Productivity Growth Indexes, CA and ER Biases: Time Series Analysis.

	Productivity			Biases	
Year	Traditional	KPA-Adj.	KPA=0	CA Bias	ER Bias
1985	7.26%	7.63%	8.00%	0.36%	-0.38%
1986	14.41%	14.60%	15.10%	0.19%	-0.50%
1987	18.09%	17.82%	18.70%	-0.27%	-0.88%
1988	13.30%	12.94%	13.61%	-0.36%	-0.67%
1989	0.77%	0.96%	0.86%	0.19%	0.09%
1990	5.47%	5.86%	5.99%	0.39%	-0.13%
1991	1.87%	2.82%	2.75%	0.95%	0.07%
1992	4.98%	5.16%	5.26%	0.17%	-0.10%
1993	7.16%	7.31%	7.44%	0.15%	-0.13%
1985-1988	13.26%	13.24%	13.85%	-0.02%	-0.61%
1989-1993	4.05%	4.42%	4.46%	0.37%	-0.04%
1985-1993	8.15%	8.34%	8.64%	0.20%	-0.29%

Notes: A-Z capital measure, d=8%, and uniform price deflator.

Table 2b: Productivity Growth Indexes, CA and ER Biases: Time Series Analysis.

	Productivity			Biases	
Year	Traditional	KPA-Adj.	KPA=0	CA Bias	ER Bias
1985	-2.01%	-2.01%	-2.00%	0.00%	-0.01%
1986	4.52%	4.51%	4.53%	-0.02%	-0.03%
1987	10.91%	10.79%	10.93%	-0.12%	-0.15%
1988	7.28%	7.06%	7.33%	-0.21%	-0.27%
1989	-0.39%	-0.39%	-0.32%	0.01%	-0.07%
1990	0.73%	0.82%	0.81%	0.09%	0.01%
1991	-2.52%	-2.29%	-2.38%	0.23%	0.09%
1992	2.54%	2.68%	2.69%	0.14%	-0.01%
1993	3.29%	3.38%	3.38%	0.10%	0.00%
1985-1988	5.17%	5.09%	5.20%	-0.09%	-0.11%
1989-1993	0.73%	0.84%	0.83%	0.11%	0.01%
1985-1993	2.70%	2.73%	2.77%	0.02%	-0.05%

Notes: "Rule of Thumb" capital stock measure, d=8%, and uniform price deflator.

Table 3: Ranking of Industries by PA Intensity: Average Intensity 1984-1993.

SIC	Industry	PA-Intensity
291	Petroleum refining	0.2276
281	Industrial inorganic chemicals	0.1861
333	Primary nonferrous metals	0.1798
311	Leather tanning and finishing	0.1636
261	Pulp mills	0.1630
287	Agricultural chemicals	0.1559
286	Industrial organic chemicals	0.1444
324	Cement, hydraulic	0.1353
262	Paper mills	0.1056
332	Iron and steel foundries	0.0946
282	Plastics materials and synthetics	0.0876
331	Blast furnace and basic steel products	0.0741
299	Misc. petroleum and coal products	0.0708
341	Metal cans and shipping containers	0.0594
289	Misc. chemical products	0.0592
285	Paints and allied products	0.0505
207	Fats and oils	0.0502
283	Drugs	0.0437
342	Cutlery, hand tools, and hardware	0.0408
371	Motor vehicles and equipment	0.0404
203	Preserved fruits and vegetables	0.0397
335	Nonferrous rolling and drawing	0.0392
284	Soaps, cleaners, and toilet goods	0.0372
295	Asphalt paving and roofing materials	0.0346
201	Meat products	0.0346
386	Photographic equipment and supplies	0.0344
251	Household furniture	0.0340
322	Glass & glassware, pressed or blown	0.0324
242	Sawmills and planing mills	0.0306
243	Millwork, plywood, & struct. memb.	0.0297
221	Broadwoven fabrics mills, cotton	0.0288
204	Grain mill products	0.0280
343	Plumbing & heating, except electricity	0.0273
208	Beverages	0.0268
364	Electric lighting \& wiring equip.	0.0248
206	Sugar & confectionery products	0.0235
349	Misc. fabricated metal products	0.0230
352	Farm and garden machinery	0.0225
345	Screw machine products., bolts	0.0202
367	Elec. components & accessories	0.0180
275	Commercial printing	0.0169
344	Fabricated structural metal prod.	0.0153

Table 3: continued.

SIC	Industry	PA-Intensity
265	Paperboard containers \& boxes	0.0151
225	Knitting mills	0.0132
205	Bakery products	0.0102
346	Metal forgings and stampings	0.0095
351	Engines and turbines	0.0088
355	Special industry machinery	0.0055

Note: PA-intensity = IPA/(I+IPA).

References

Acemoglu, D. and F. Zilibotti (2001), Productivity Differences, *Quarterly Journal of Economics* 116 (2), 563-606.

Bernard, A.B. and C.I. Jones (1996), Comparing Apples to Oranges: Productivity Convergence and Measurement Across Industries and Countries, *American Economic Review* 86 (5), 1216-1238.

Bertelsman, E.J. and W. Gray (1996), *NBER Manufacturing Productivity Database*, NBER Technical Working Paper No. 205.

Conrad, K. and C.J. Morrison (1989), The Impact of Pollution Abatement Investment on Productivity Change: An Empirical Comparison of the U.S., Germany, and Canada, *Southern Economic Journal* 55 (3), 684-698.

Hulten, C.R. (1992), Growth Accounting When Technical Change is Embodied in Capital, *American Economic Review* 82 (4), 964-980.

Maddison, A. (1987), Growth and Slowdown in Advanced Capitalist Economies: Techniques of Quantitative Assessment, *Journal of Economic Literature* 25 (2), 649-706.

Nadiri, M.I. (1970), Some Approaches to the Theory and Measurement of Total Factor Productivity: A Survey, *Journal of Economic Literature* 8 (4), 1137-1177.

Nelson, R.R. (1981), Research on Productivity Growth and Productivity Differences: Dead Ends and New Departures, *Journal of Economic Literature* 19 (3), 1029-1064.

Osang, T. and A. Nandy (2000), Impact of U.S. Environmental Regulation on the Competitiveness of Manufacturing Industries, *In Submission*.

Wolf, E.N. (1996), The Productivity Slowdown: The Culprit at Last? Follow up on Hulten and Wolff, *American Economic Review* 86 (5), 1239-1252

Environmental Regulation and Competitiveness: An Exploratory Meta-Analysis

Abay Mulatu

Department of Spatial Economics, Master-point and Tinbergen Institute, Vrije Universiteit Amsterdam, De Boelelaan 1105, 1081 HV Amsterdam, The Netherlands, *amulatu@feweb.vu.nl*

Raymond J.G.M. Florax

Department of Spatial Economics, Master-point, Vrije Universiteit Amsterdam, *rflorax@feweb.vu.nl*

Cees A. Withagen

Department of Spatial Economics, Master-point and Tinbergen Institute, Vrije Universiteit Amsterdam, Department of Economics and CentER, Tilburg University, *cwithagen@feweb.vu.nl*

Summary:
The relationship between domestic environmental regulation and international competitiveness has evoked various speculations. The common neoclassical train of thought is that strict environmental regulation is detrimental to the competitiveness of industry, and that it induces phenomena such as ecological dumping, ecological capital flight, and regulatory 'chill' in environmental standards. A different view is that strict environmental regulation triggers industry's innovation potential, and subsequently increases its competitiveness. The impact of environmental regulation on competitiveness has been analyzed in terms of international capital movements, new firm formation, and international trade. The paper presents a statistically supported evaluation of the literature, in order to assess what the main conclusions regarding the relationship between environmental regulation and competitiveness are when it comes to studies on international trade flows. The synthesis of the literature is subsequently used to present guidelines for future primary research in this area.

Keywords: Environmental policy, strictness, competitiveness, meta-analysis.

1 Introduction

The persistent integration of the world economy has increased the apprehension for potentially negative effects that domestic environmental regulation may exert on a country's position in international trade. International trade and the environment are intertwined in various ways (Ulph, 1997), among other things because the occurrence of international trade implies changing patterns of production and consumption that may have an impact on the level of pollution. Environmental pollution may also be the cause of spatial externalities, for instance, when consumption and production in one country imply non-negligible side effects on the level of pollution in other countries. The regulation of international trade through trade policies may also be used to enforce international agreements on environmental issues.

The economic literature on international trade and the environment centers on the problems that arise from these relationships. It is sometimes argued that, in order to advance international trade and protect domestic firms at the same time, governments should not choose a stringent environmental policy, because a stringent policy may harm the competitiveness of domestic firms. A lenient environmental policy may, however, cause 'too much' pollution. Leniency can nevertheless be attractive as it prevents 'ecological capital flight', firms leaving the country and establishing production plants in places with a less stringent environmental policy (see Rauscher, 1997, on the concept of ecological dumping). A contrasting view is purported in the hypothesis suggested by Porter (1991), who argues that a strict environmental policy may actually increase industries' competitiveness because it triggers innovation efforts of firms and reduces X-inefficiency.

These considerations are obviously interesting in the context of international trade theory, and they have important policy implications. An appealing example of the latter is the introduction of a carbon tax, and the arguments this provoked in the political debate in the Netherlands. Several large energy consumers have, at least temporarily, been exempted from this tax because the legislator was persuaded by the argument that the Netherlands would otherwise become a less attractive location for industry. A close look at the economic literature shows, however, that only a fairly limited number of empirical studies addresses the issue of potentially negative effects of environmental policy on competitiveness measured in terms of trade flows.[1] These studies are mainly concerned with the United States.

---

1 Impacts on competitiveness have also been measured in terms of foreign direct investments and new firm formation (see below). The term 'competitiveness' is actually rather fuzzy, and has been interpreted and operationalized in different ways (for a general discussion, see Ekins and Speck, 1999; Thomson, 1998).

The objective of the present paper is to review the empirical literature on environmental policy and trade and to assess its main findings. The relationship between domestic environmental policy – in particular its 'strictness' – and a country's international competitiveness is investigated in three major strands of the literature. The first set of studies deals with the impact of environmental regulation on international capital movements or foreign direct investment (see, e.g., Bouman, 1998; Hettige et al., 1992; Leonard, 1998). The foreign direct investment literature generally reveals that empirical evidence on the 'pollution haven' hypothesis is fragile and inconclusive (Jeppesen et al., 2000). In a similar vein a second set of studies focuses on the impact of heterogeneous environmental regulation on the flow of capital in terms of domestic new firm formation. Although initially empirical evidence was thought to be rather weak, some recent studies have shown that environmental regulation affects the location behavior of pollution intensive manufacturing firms (Henderson, 1996; Greenstone, 1998; Becker and Henderson, 2000; List and McHone, 2000). In the third strand of literature, there is a rather limited set of studies that deals with the impact of environmental policy on international trade. Literature surveys of the latter (Jaffe et al., 1995; Jeppesen et al., 2000; Jeppesen et al., 2002, point out that the available evidence for negative effects of environmental regulation on international trade is again rather mixed).

In what follows, we will focus on the third strand of the literature. We will use a traditional literature review as well as a statistically based approach, generally referred to as 'meta-analysis', to address the question of whether the literature on environmental regulation and competitiveness is conclusive or not. The meta-analysis also serves as a useful starting-point for further primary research to the link between environmental policy and competitiveness, because insights derived from existing studies are of pivotal importance for the development of guidelines for future research.

The outline of the remainder of the paper is as follows. In Section 2 we review the literature, which constitutes the traditional approach to summarizing the state of the art in research. In this Section we also construct the database employed in subsequent sections. Section 3 introduces meta-analysis as an alternative approach to summarizing and synthesizing research results. Meta-analysis has been developed in the context of experimental sciences, and it constitutes a statistically rigorous approach to the assessment of research results. Various exploratory meta-analytical techniques are applied to the available set of studies on environmental policy and international trade flows. The final Section 4 contains conclusions regarding the conclusiveness of the literature. It also elaborates on the implications of the meta-analysis for environmental policy-making, and presents some guidelines for future primary research into the relation between environmental policy and competitiveness.

2 A Survey of the Empirical Literature

The empirical literature on environmental policy and international trade can be divided into three groups depending on the approach being used: an exploratory approach, the Leontief approach, or an econometric approach. Each of these will be discussed below, using the traditional state of the art review tool to summarize the main findings in a narrative way. Moreover, we set the stage for the meta-analysis in Section 4 by discussing in some detail how the data from each study have been used for the meta-analysis. This necessarily involves some technical details, most of which are treated in the appendix. In particular it is our aim to derive and quantify so-called effect sizes and p-values. Effect sizes can be defined as mean standardized differences between control and experimental groups. In the case at hand the effect size indicates the effect of increased environmental stringency on trade performance. For the standardization there are several options. One might use the pooled standard deviation, or the standard deviation of each separate group. It should be noted that under the null hypothesis of no effect of stringency on trade performance, the effect sizes are Student-distributed. The appendix presents a more formal approach. Alternative definitions of effect size that are used are correlations and differences in proportions. It is fairly straightforward to extract the required probabilistic information from the econometric studies. This is more difficult for exploratory and Leontief type studies. They will therefore be given somewhat more attention below. In the meta-analysis we will also use the concept of p-value. In general this gives for a specific statistical test the probability that the test statistic takes a value that is contradictory to the null hypothesis. For our analysis it gives the probability that the null hypothesis of no effect of environmental policy on competitiveness is accepted. Knowledge of t-statistics and degrees of freedom suffice to figure out p-values from statistical tables.

2.1 Exploratory Studies

One approach to investigating the effect of regulation on international trade patterns is to explore whether a shift in trade patterns of pollution-intensive industries from developed to developing countries has occurred, the underlying assumption being that the shift is due to less restrictive environmental standards in developing countries. In this type of studies the impact of environmental policy is captured by a stringent versus non-stringent dichotomy.

Low and Yeats (1992) consider 6 years (1965, 1975, 1985-1988) and make a distinction between industrial countries and all other countries. In the first category explicit data are given for EEC(10) and North America, for the second category there are separate data for including Eastern Europe, Latin America, South-East Asia and West Asia. In the construction of our database the industrial countries are the experimental group (meaning subject to strict environmental regulation) and the developing countries are the control group.

The first type of data used from this study are those giving the share of environmentally dirty goods originating in different regions. To give an example, in 1975 40% of the total trade in environmentally dirty goods originates in EEC(10). The experiment we have in mind is that the members of the experimental group were subject to treatment (environmental policy), whereas the members of the control group were not. What we observe are yearly average changes in shares over a period under consideration. The effect size (Hedges' *g*) is calculated as follows. For the period 1975-1988 we calculate the average yearly changes in the shares of the two groups, industrial and all other countries, as well as the standard deviation of these changes per group using the changes in the individual regions within the groups. From this we calculate the pooled standard deviation. This yields an estimate of the effect size of -1.972. The effect size multiplied by the expression

$$\sqrt{n_e n_c /(n_e + n_c)}$$

involving the sizes $n_e = 3$ and $n_c = 5$ of the experimental and control group, respectively, has a *t*-distribution. The *t*-value is -1.708, implying that the probability that a *t*-value is smaller than -1.708 or larger than 1.708 equals 0.163. This is the *p*-value used in the database. The interpretation of this figure is that it is safe to state that there is no significant decrease of the share of industrial countries in dirty exports. The same procedure is used for data for the periods 1965-1988 and 1965-1975, in order to allow for the possibility that the results differ according to the period reviewed. No significant negative effects are found.

Another type of data refers to the same years and the same regions, but the data present the regions' dirty exports as a share of total exports from that region. For these data we also construct Hedges' *g* (see Appendix) with the pooled standard deviation. In calculating this standard deviation we use the standard deviations for the regions which are explicitly reported on. The conclusion of no significant negative effect remains valid.

Finally, the third type of data refers to the concept of Revealed Comparative Advantage (RCA), defined as the share of a specific industry in a country's total exports as a fraction of the share of the industry in total world exports. Low and Yeats produce data on the number of countries with RCA indices exceeding unity, for 40 industries over two time periods, 1966-1968 and 1986-1988. We take the pulp and wastepaper industry as an example. In the first period there were 5 industrialized countries with an RCA exceeding unity and 7 other countries with an RCA exceeding unity. Hence 42% of the countries with a share larger than unity was industrialized. For the second period this amounted to 44%. The effect size for this case is therefore 0.44-0.42. The estimated standard deviation is calculated using the total number of countries having an RCA exceeding unity.

The procedure outlined above, assuming a normal distribution of the standardized difference in proportions, yields a test statistic of 0.151, with a p-value of 12%. In this example the null hypothesis that the effect size is zero, is not rejected. Some of the empirical results, however, demonstrate a tendency to developing countries specializing in 'dirty' industries. There has been a disproportionately large increase in the number of developing countries that develop a revealed comparative advantage in 'dirty' industries. Over the period 1966-1988 there was a 14% increase of industrial countries with a revealed comparative advantage in dirty industries. For developing countries the percentage is approximately three times higher.

Sorsa (1994) provides data on the share of environmentally sensitive goods in total exports, for 7 industrial countries as well as for the total developing world, for 1970 and 1990. Since the control group consists of one element only we have used the standard deviations in the experimental group to compose the estimated effect size. The test statistic is -2.218, implying a p-value of 6.8%. Hence it is likely that industrial countries export less dirty goods over time.

Sorsa also provides data on revealed comparative advantage in environmentally sensitive goods (not per industry). Again the same approach is taken, yielding a test statistic of -1.384 with p-value 21.6%.

There are also data on the correlation between the share in world trade of environmentally sensitive goods and the share of environmental expenditures in gross domestic product. These are calculated for private expenditures (three countries) and total expenditures (also three countries). The correlation coefficients can be used for testing because

$$t = r\sqrt{df} / \sqrt{(1-r^2)},$$

where r denotes the correlation coefficient, has a t-distribution with df degrees of freedom. Moreover, the correlation analysis of environmental expenditures and the share of environmentally sensitive goods in world trade show no significant relationship.[2]

Finally, we have used data on the number of industries with a revealed comparative advantage. Of all industries having a revealed comparative advantage 73.7% were in industrial countries in 1970, and 60.5% in 1990. This gives a test statistic of -1.233 and a corresponding p-value of 21.8%. Sorsa reaches the same conclusions as Low and Yeats, of "little systematic relationship between higher environmental standards and competitiveness in environmentally sensitive goods". She also shows, however, that some countries maintain or even increase their comparative advantage.

---

[2] With the exception of Austria, for which Sorsa (1995) finds a significantly positive effect.

In sum, the empirical evidence of exploratory studies is at best mixed. It is obvious, however, that the exploratory studies suffer from various methodological deficiencies and the application of statistical techniques is, of course, hampered by the low number of available observations. Although exploratory studies can be used to attain preliminary insights into the relationship between environmental regulation and international trade, they lack a theoretical basis, and they use a poor specification of the differential effects of environmental regulation. Finally, an important drawback of this type of study is the inability to control for other factors that are potentially relevant to the observed changes in specialization patterns (see Jaffe et al., 1995).

2.2 The Leontief Approach

A second approach towards assessing the regulation-competitiveness issue is in the spirit of Leontief's attempt to measure whether American exports are labor- or capital-intensive relative to imports. On the basis of Leontief's input-output model, the pollution content of products can be assessed, taking into account the pollution related to direct inputs as well as to intermediate inputs from other sectors. Walter (1973) and Robison (1988) have used this approach. Also part of Kalt's (1988) work belongs to this category. The theory behind the Leontief approach is based on Baumol and Oates (1975), who indicate that: "Undertaking pollution abatement will reduce the abating country's comparative advantage in producing high-abatement-cost goods and improve the comparative advantage in low-abatement-cost goods" (Robison, 1988: 188). In the Leontief studies use is made of input-output matrices to calculate the overall abatement-cost contents of imports and exports.

Walter (1973) investigates the pollution content of US trade. Basically, the approach is as follows. A group of 83 goods and services is selected, and it is assumed that the share of environmental control costs in the final value of the goods is a proxy for the costs incurred to meet environmental criteria. Direct environmental production costs as well as indirect environmental costs[3] result in a measure labeled 'Overall Environmental Control Loading' (OECL). Subsequently, the OECL is multiplied by the value of US exports and imports to obtain the total environmental cost content of US trade. The pollution contents of US exports and imports across all goods and services, during 1960-1970 turns out to be 1.75% and 1.52% of total exports and imports, respectively.[4] Walter argues that the difference is negligible, and concludes that US environmental policy is generally trade neutral. It is, however, difficult to cast this conclusion in a probabilistic

3 The latter are calculated multiplying the abatement cost vector by the total requirement matrix.

4 The import figure is obtained assuming that the pollution content of import commodities is the same as for domestically competing products.

framework, because the analysis contains no numerical comparison over time, and hence nothing can be said about increasing stringency.

Walter also presents ratios of abatement contents for imports over exports in trade with Japan and Canada, amounting to 1.11 and 1.29 respectively, whereas the ratio is 0.81 for trade in general. This suggests that the US imports relatively dirty commodities from Canada and Japan and relatively clean commodities from the rest of the world. Treating the data as independent, which strictly speaking they are not, and assuming that Japan and Canada have similar environmental policies, we arrive at a positive impact of stricter environmental policy, but the effect is not significant. Compared to the rest of the world, the US has a relative advantage in dirty products, whereas one would expect the reverse in view of the Baumol and Oates theory. For Japan and Canada the same argument applies, but reverse.

Robison (1988) sets out to determine whether environmental control costs affect US comparative advantage, and what the impact of environmental cost is on US trade with Canada and the rest of the world. Input-output tables are used to determine the abatement cost content of US trade for the years 1973, 1977 and 1982. Prices are endogenously determined through a full-fledged input-output model with 78 sectors. Abatement cost changes are assumed to be reflected in the value added of sectors, which subsequently translate into price changes via the input-output price equation. It is assumed that these price changes actually occur, at least in the long run. Abatement costs are defined in a way similar to Walter (1973), although Robison considers a 'modified' total requirement matrix in order to take account of abatement costs in capital goods. From this study we extract three types of data.

First, Robison provides data for three years on the ratios of average abatement content of imports over exports, for total US trade and Canada-US trade. Over the three years, the means are 1.236 and 1.115 with standard deviations of 0.133 and 0.029, respectively. The pooled standard deviation is 0.10, yielding an effect size of -1.249 (Hedges' g). The value of the test statistic is -1.529 with a p-value of 20.1%. With the same data, one can analyze the differences in average annual change over time. This approach results in a p-value of 41.2%.

Robison also considers the effect of a one percent price increase on trade in three specific years, supposedly through an increase in abatement costs, for individual sectors, keeping everything else unchanged. We consider the average impact over the years, for total US trade as well as for US-Canada trade. We arrive at test statistics of -3.068 and -3.876 (p-values 9.2% and 6.1%, respectively), indicating that on average, over the years, the balance of trade deteriorates as a consequence of stricter environmental regulation.

Finally, Robison provides sectoral data on the impact of a one percent price increase on the balance of trade (in general as well as for Canada only) in the three years. We record the average and perform a test on the total effects over the sectors, per year. This procedure results in effect sizes that are significantly different from zero and negative.

Part of Kalt's (1988) study employs an approach similar to Walter and Robison. For 1967 and 1977, Kalt provides data on the total abatement cost component of exports and imports and of the abatement factor dollars per dollar of exports and imports, for manufacturing industries only as well as for all industries. It is assumed in Kalt's set-up that the same abatement cost structure applies to both years. These data allow us to test, in the usual way, for changes in the abatement cost ratio over time in both types of industries. Kalt concludes that there is an indication that "environmental regulation was a source of shifting comparative advantage" over time. However, our statistical analysis cannot confirm this statement. Our test statistics for the change in the import export ratio of factor dollars of abatement costs give values of -4.622 and -3.742 for all industries and manufacturing industries respectively, which are statistically not significant (*p*-values are 13.6% and 16.6%, respectively). For the case of total abatement costs we obtain a similar result.

Much in the same way as for the exploratory approach, the studies based on the Leontief approach suffer from lack of conclusiveness. Although both Walter (1973) and Robison (1988) cast their conclusions in terms of environmental regulation, there are alternative – equally plausible – explanations for the small observed shifts in trade patterns. One frequently cited example is the process of industrialization in developing countries, which raises the relative importance of manufacturing in the economy (see, e.g., Jaffe et al., 1995).

2.3 Econometric Studies

The econometric studies are based on either the Heckscher-Ohlin framework or on the gravity model. The Heckscher-Ohlin model rests on the following assumptions: factor immobility between countries, perfect factor mobility among industries, identical technologies in all countries, and different endowments of productive factors. It suggests that a country specializes in the production of commodities that require intensive use of resources that are relatively abundant (see, e.g., Helpman and Krugman, 1985). Environmental policy can be incorporated in this kind of analysis: regulations deprive industries of the right to pollute, and can hence be considered a drain on endowments resulting in loss of comparative advantage.

The gravity model is frequently used to model bilateral trade flows (see e.g., Helpman and Krugman, 1985). Trade flows are specified as a function of the potential supply of the exporting country (measured by, e.g., gross domestic product or population), the potential demand of the importing country (usually measured in a similar way), and some measure of friction to trade between the countries (often based on a distance measure). In order to investigate the impact of differing environmental regulations, an operational environmental variable is usually added to the set of variables that reflect the three factors mentioned above.

Kalt's (1988) analysis is based on the Heckscher-Ohlin model. He specifies net exports of an industry as a function of physical capital, human capital, unskilled labor, research and development, and environmental control costs. He analyses a cross-section of 78 industries for 1977, distinguishing three groups of industries: all industries, manufacturing and manufacturing excluding chemicals. For each cross-section two regressions are run, one with and one without correction for heteroscedasticity. A significant negative estimate of the environmental cost variable (costs of regulation based on a survey of 48 firms) is obtained only when the sample is confined to the 52 manufacturing industries. A similar result, but with a lesser degree of significance, shows up for a specification where the change in net exports over the period 1967-1977 is the dependent variable, assuming that environmental costs were negligible in 1967. A third set of regressions involves net export performance in 1977 as the dependent variable and total direct and indirect factor inputs as independent variable. Pollution abatement expenditures again have a negative impact on export performance, but statistically significant for manufacturing excluding chemicals. Kalt concludes that environmental regulation has a significant negative effect on US manufacturing. Finally there are 4 correlations between regulatory costs for 78 industries and trade performance (measured either as a change in contribution of exports in value added or as exports in total industry shipment). Results are mixed and statistically insignificant.

Tobey (1990) is concerned with multi-factor and multi-commodity extensions of the Heckscher-Ohlin model for the US. Net exports of each of five commodity groups, which are all qualified as dirty,[5] are regressed on country characteristics, in particular endowments of land, capital, labor, natural resources (such as coal and oil production), and stringency of environmental regulation. The analysis includes trade flows to and from 21 countries, in 1975. Stringency of environmental regulation is represented by a qualitative index that ranges from 1 to 7 based on Walter and Ugelow (1979). The average of this index for developed countries is 6.1, and for developing countries the average equals 3.1. All estimates for the stringency variable turn out to be statistically insignificant. Subsequently, an omitted variable test is performed. It consists of two regressions; in the first regression strictness is not incorporated as an independent variable, in the second it is. If environmental policy does not play a role, one would expect that half of the residuals is negative, and that the same percentage is obtained for industrial and non-industrial countries. Therefore, a test is performed on proportions. This is

5 Industries are labeled 'dirty' if they show pollution abatement costs greater or equal than 1.85% of total costs, in 1977. The number of industries considered is 34, grouped into five commodity groups (i.e., mining, paper, chemicals, steel, and non-ferrous metals).

done for three groups: industrial versus moderately developed countries, industrial versus less developed countries and industrial versus moderately plus less developed countries. The conclusion of no significant impact of environmental policy is maintained. The differences in proportions resulting from this exercise are used in our meta-analysis.

Next, two extensions of the model, one allowing for non-homothetic preferences and one allowing for scale economies (larger countries have an advantage on the export market, reflected in national income as a fraction of world income as an explanatory variable), are considered. Regarding the first extension, Tobey only mentions that the environmental variable is not significant, and that the omitted variable test does not support the hypothesis of a negative effect. For the second extension no numerical data are reported either, but it is mentioned that the stringency variable is not significant in the regression; the outcome of the omitted variable test is significant but points in the 'wrong' direction.

Tobey performs a second set of regressions, where the dependent variable is the change in exports in 1984 compared to 1970, and where the independent variable is the stringency index. This specification is motivated by the fact that differential stringency across countries in the 1960s may not have been strong enough to show up in a cross-section Heckscher-Ohlin model. Although resource endowments are thus basically assumed constant, this specification does not reveal any significant effect either.

In a recent paper Van Beers and Van den Bergh (2000) perform a gravity analysis with the 1975 data employed by Tobey, for five dirty sectors, and a country sample similar to Tobey's. In their model, stringency does not have a significant effect on the chemicals and steel industries, whereas its effect is significantly negative for mining and non-ferrous metal, and significantly positive for the paper and pulp industry. The estimates are elasticities and are well-suited for the meta-analysis.

Diakosavvas (1994) also follows Tobey's (1990) framework, but his analysis centers on agriculture. Ten agricultural commodities for 23 countries (including five less developed countries) over the period 1984-1986 are considered in a Heckscher-Ohlin based cross-country model. Net exports of a country are regressed on endowments of labor and capital, the environment, government policies, and stringency of environmental regulation (i.e., the Walter and Ugelow measure). The regression results for five out of ten commodities subject to stricter environmental regulation suggest that environmental policy indeed causes net exports to fall. This also holds for another set of regressions where the dependent variable is total exports.

Van Beers and Van den Bergh's (1997) analysis is essentially based on the Tobey (1990) approach as well, although there are three noteworthy differences. First, the gravity model, which considers bilateral instead of multilateral trade flows, is used. Bilateral exports are regressed on land area, gross domestic product and population (as a proxy for potential supply of one country as well as potential demand in the other), and strictness of environmental policy in both countries. The distance between countries as well as dummy variables for membership of the European Union and the European Free Trade Association are added. Second, the measures of environmental regulatory stringency are different. Two variants of 'output oriented' environmental stringency measures are considered: a broadly defined measure using seven environmental indicators (e.g., protected land area, and paper recycling rate), and a narrowly defined measure based on a subset of these indicators considered to better reflect private environmental costs.[6] Third, three types of bilateral trade flows are used as dependent variables: total bilateral trade, 'dirty' bilateral trade, and 'dirty' footloose bilateral trade flows. The sample consists of 14 OECD countries and 9 developing countries in 1975, and 21 OECD countries in 1992.

The regressions for 1975 employ the Walter Ugelow (1979) measure of strictness. For total bilateral trade a statistically significant positive relationship is found between strictness of environmental policy and export performance. This also holds when attention is restricted to non-resource based industries (such as chemical materials, and iron and steel). When only dirty industries are included the impact is insignificant. For 1992 the 'new' stringency measures are used. If the broad stringency measure is employed, it turns out to be insignificant. The narrow measure produces a statistically significant negative effect, except for the case of dirty industries, where it is insignificant.[7] For all estimates t-values are provided.

Xu (2000) replicates the Van Beers and Van den Bergh (1997) analysis, although with a number of modifications. First, a different measure of regulatory stringency is used, based on the work by Dasgupta et al. (1995). Second, the sample comprises 31 UNCED-report countries (which range from highly industrialized to extremely poor), randomly sampled from a total of 145 countries; this differs markedly from the Van Beers and Van den Bergh (1997) sample of 21 OECD countries for 1992. Finally, the effects of macroeconomic and cyclical disturbances are removed from the export flow variables, something that is ignored in Van Beers and Van den Bergh (1997). Regressions are performed for three dependent variables: bilateral exports, bilateral exports of environmentally sensitive goods, and bilateral exports of non-resource-based (footloose) environmentally sensitive goods. Another distinction is between regressions having no import tariff variables included among the independent variables, and others in which they are included. Regressions are also performed using maximum

---

6 'Output oriented' measures reflect tangible outcomes of stringency of regulation. Although Van Beers and Van den Bergh (1997) make substantial efforts to improve the measurement of environmental stringency, Co et al. (1999) point out some caveats in their measures.

7 It should be noted that counterintuitive results, such as positive effects of stringency on exports, were obtained as well.

likelihood to correct for possible heteroscedasticity. Altogether we obtain 12 estimated elasticities with *t*-values.

In part, the results sharply contrast those obtained by Van Beers and Van den Bergh, as Xu finds statistically significant positive coefficients for the environmental policy variable, implying that a strict environmental policy is beneficial to export performance, whereas Van Beers and Van den Bergh find insignificant or significant negative effects for their 1992 data. Xu argues that this departure from the earlier results can be explained by the use of a different set of environmental policy measures, the removal of cyclical fluctuations from the export figures, and/or the inclusion of developing countries in the sample. The latter argument is, however, not fully convincing, as one would expect a negative impact of regulatory action given the likelihood of observing diverging regulatory standards in North-South trade.[8]

Han and Braden (1996) use a cross-commodity regression based on the Heckscher-Ohlin model for 19 manufacturing sectors in the US, for the period 1973-1990. Net export of a sector is expressed as a function of that sector's use of factors of production and pollution abatement cost, all of which show sectoral as well as temporal variation captured by fixed or random effects terms.[9] They distinguish between one way and two way models. In the latter type of models, sectoral attributes are allowed to change over time. Moreover, in some regressions, the expenditures on abatement are multiplied by a time trend, allowing for the impact to vary over time. So, denoting abatement expenditures by AB, the right-hand side of the regression includes $\beta_5 AB_{it} + \beta_6 AB_{it} t$ (in the notation of the authors).

Four panel regressions are executed: one and two way fixed effects and one and two way random effects. It is found that the coefficients β_5 and β_6 defined above are negative and statistically significant in all cases. Therefore, the effect of stringency on export performance is negative. Since *t*-values are given for the individual coefficients only, it is necessary to construct standard errors for the total effect (see Appendix for details). These models generate data for all years.

In the fixed effects model the time invariant intercept captures the sector-specific effects. For four years (1975, 1980, 1985 and 1989) and all 19 sectors under

8 In another paper Xu (1999) considers five 'dirty' sectors (i.e., wood, paper and printing, chemicals, non-metal, and metal) in 30 countries including most of the OECD countries, in 1988, and he employs the same stringency measure as before. The paper reports an insignificant effect of environmental regulation on the share of dirty industries in total value added of the manufacturing industry. This study is, however, not considered in the sequel, as it does not address the trade effect of environmental regulation.

9 Panel techniques also constitute a way to circumvent the problem of unmeasured heterogeneity inherent in OLS regressions. More arguments for the use of panel data in this context are given in Co et al. (1999).

consideration, abatement expenditures elasticities of net exports are presented. Regression coefficients give the change in net exports in terms of dollar values. The *t*-values refer to these coefficients. It is therefore not straightforward to find standard errors corresponding to elasticities (see the Appendix for a discussion of this problem).

A third type of data is produced from a time-series regression per manufacturing industry. This yields sector-specific β_5's as well as β_6's. For about half of the 19 sectors statistically significant and negative estimates are found. Given that the time variable is also involved, we can derive the impact of abatement costs, in terms of changes in exports in dollar values, for each sector and for each year. Again the problem is to find the standard error of effect size (see the Appendix).

The results of Han (1996) partly overlap with those reported in Han and Braden. We focus here on distinguishing features of Han's study. He considers the effect of differential environmental stringency on international competitiveness across countries, using an environmental regulation index based on the "the ratio of the emission reduction due to regulation and the emission in the absence of regulation". Using a Heckscher-Ohlin framework, a panel of 34 countries at every five year interval from 1975 to 1990 is estimated for nine sectors. Of the five environmentally sensitive goods sectors considered, only mining is found to show a statistically negative relationship between stringency and competitiveness. Two other environmentally sensitive goods sectors (paper and pulp, and chemicals) show significant positive estimates for the stringency variable. The effect of stringency on net exports of non-polluting sectors is insignificant. For all estimators *t*-values are provided.

Grossman and Krueger (1993) use cross-commodity regressions data in a reduced form model to assess whether pollution abatement costs in the US explain the pattern of Mexican specialization and trade. In particular, patterns of US imports from Mexico, and US foreign direct investments in Mexico are scrutinized. Three different patterns are considered. The first is the 1987 patterns of US imports from Mexico for 135 manufacturing industries. The explanatory variables are human capital share, physical capital share, and tariff rates as well as a stringency variable (abatement costs). In addition, an injury rate is included in some of the regressions as a proxy for the major costs of US labor protection laws in American manufacturing. It is found, as expected, that higher US abatement costs have a positive effect on imports. But the result is statistically insignificant. Another pattern is the 1987 US imports from Mexico that have entered under the offshore assembly agreement.[10] One would expect a positive relationship between Mexican value added in US imports under the provision, and pollution abatement costs. However, a negative relationship is found. Finally, the sectoral pattern of value

[10] This allows for duty free re-entry of US-made components.

added by 'maquiladora' plants[11] is considered. It concerns the question of whether American firms invest in 'maquiladoras' to avoid environmental regulatory costs. As this pertains to direct foreign investment rather than trade flows, these estimates have been excluded from our analysis.

2.4 Conclusion

Table 1 summarizes the studies discussed in this section. It describes the studies according to the type of study and it gives the type of data we have been able to extract from each study. There are 13 studies. The study by Kalt is partly econometric and partly of the Leontief type. There is a large variety of estimates in the studies: univariate estimates, mainly from the Leontief studies (which are means), bivariate estimates, such as correlations and standardized mean differences, and multivariate estimates from econometric studies, giving dollar value effects and elasticities.

Table 1: A Categorization of the Stringency-Competitiveness Trade Flow Literature According to Type of Approach, Type of Estimate, and Number of Available Estimates.

Study	Type of approach[a]		
	Exploratory	Leontief	Econometric
Low and Yeats (1992)	×		
Sorsa (1994)	×		
Walter (1973)		×	
Robison (1988)		×	
Kalt (1988)		×	×
Tobey (1990)[c]			×
Van Beers and Van den Bergh (2000)			×
Diakosavvas (1994)			×
Van Beers and Van den Bergh (1997)[e]			×
Xu (2000)			×
Han and Braden (1996)			×
Han (1996)			×
Grossman and Krueger (1993)			×
Total	2	3	9

11 'Maquiladoras' are foreign-owned firms (usually somehow based in the US), with most of them located at the Mexican side of the US-Mexican border.

Table 1: continued.

Study	Type and number of estimates[b]							
	Univariate	Bivariate				Multivariate		Total
	μ	r	g	g_p	Δ	β	η	
Low and Yeats (1992)			6	40				46
Sorsa (1994)		6		1	2[d]			9
Walter (1973)	1							1
Robison (1988)	8		2					10
Kalt (1988)	4	4				18		26
Tobey (1990)[c]				9		20		29
Van Beers and Van den Bergh (2000)							5	5
Diakosavvas (1994)						20		20
Van Beers and Van den Bergh (1997)[e]							10	10
Xu (2000)							12	12
Han and Braden (1996)						436	76	512
Han (1996)						9		9
Grossman and Krueger (1993)						2		2
Total	13	10	8	50	2	505	103	691

[a] The categories refer to exploratory, Leontief, and econometric approaches, as in Section 3. One study combines different approaches.

[b] The symbols used are: μ is a sample mean, r a correlation coefficient, g Hedges's difference in means, g_p Hedges's difference in proportions, Δ Glass's difference in means, β a regression coefficient not representing an elasticity, and η an elasticity (see Appendix A for details).

[c] Tobey (1990) presents 13 estimates in a quantitative fashion. The results of two alternative specifications are mentioned in the text, but no numerical details are given. Sufficient information for binary and categorical coding is, however, available.

[d] The standard deviation is based on the experimental group because too few observations are available for the control group (see Appendix A for details).

[e] Not all results are published in the Kyklos version of the paper; estimates are taken from an extended version of the article (Van Beers and Van den Bergh, 1996).

There are substantial differences across studies in terms of number of observations. We will pay attention to this below, when discussing the significance of the results from the studies. There are also major differences in theoretical approaches, which make a comparison rather difficult. Finally, we would like to mention that Leontief and exploratory studies do not control for other dimensions that may have an impact on the effect of environmental policy on competitiveness.

3 Exploratory Meta-Analysis

3.1 Statistical Rigor Through Meta-Analysis

Based on the results obtained in the previous section an exploratory meta-analysis will be performed. Meta-analysis has been developed in the context of the experimental sciences and refers to the statistical analysis of research results of studies performed previously. In economics meta-analysis is gaining ground, for instance, in industrial economics (Button and Weyman-Jones, 1992), labor economics (Jarrell and Stanley, 1990, Card and Krueger, 1995, Ashenfelter et al., 1999), and transport economics (Button and Kerr, 1996). Especially in environmental economics, stimulated by the work of Smith (1989), Smith and Kaoru (1990a,b), and Smith and Osborne (1996), many meta-analyses appeared.

In a nutshell the empirical literature on the effect of stringency of environmental regulation on competitiveness is commonly characterized as one with mixed and rather vague evidence. This has led to divergent views on the issue, and numerous speculations as to the likely reasons for the differences and the lack of robustness of the results. That conclusion is, however, based on a qualitative and narrative summary of what we think is the essence of the quantitative empirical results. This is the general practice in state of the art literature reviews compiled for journals and books. Although this practice is valuable in its own right, there are a number of problems associated with it. Obviously, the selection of the most relevant conclusions is to a certain extent arbitrary (Van den Bergh et al., 1997). Moreover, usually some sort of vote-counting procedure is implicit in a literature review. 'Vote-counting' refers to the practice of counting significantly positive, significantly negative and insignificant results. The statistical inference is that the category representing the majority of cases represents the true underlying relationship (Light and Smith, 1971). Hedges and Olkin (1980) have shown that the vote-counting methodology is inadequate, because it tends to lead to the wrong conclusion more often when the number of studies increases. The basic argument is that the Type-II errors of the underlying studies do not cancel one another (see also Hedges and Olkin, 1985). Notwithstanding this basic flaw of the vote-counting methodology, we will show below to what conclusion it leads when applied to the environmental regulation and international trade literature.

The rather crude comparison that is being made in vote-counting techniques (i.e., a categorical classification into (significantly) positive, zero, and negative effects) is also unsatisfactory because it is insufficient to determine whether the results of different studies agree (Hedges, 1997). In order to address that question we perform an analysis more sophisticated than simple vote counting. In the analysis p-levels are combined. This meta-analytical technique tests the null hypothesis of no significantly negative effect of environmental stringency on competitiveness by considering p-values from the individual studies. Another problem associated with a narrative survey of the literature is the following. The results of an empirical study may provide a relatively good estimate of the sampling uncertainty of results, but non-sampling issues such as research design, model specification, and estimation techniques, are usually relatively constant within a study (Hedges, 1997). Meta-analysis taking non-sampling characteristics into account as moderator variables, constitutes a useful complementary technique to synthesize research results. It usually employs meta-regression, based on least squares estimation of a model in which a specific effect measure observed in a series of studies is taken as the dependent variable. The set of explanatory variables frequently includes specific underlying causes for the phenomenon under consideration, and moderator variables representing, for instance, differences among research designs, time-periods, and locations covered in the original studies (see, e.g., Stanley and Jarell, 1989). In this paper we shall only deal with vote-counting and combining significance results. The meta-regression analysis is addressed in current work.

3.2 Vote-Counting

Simple vote-counting amounts to counting the number of studies (or results within studies) yielding positive, zero and negative effects. Such a procedure can be criticized on several grounds, including the fact that sample size is not taken into account, the fact that the method does not allow determining by how much the winner is winning and the fact that the statistical power is small (Bushman, 1994). We nevertheless present the results of such a procedure as a starting point for other types of analyses. It should also be pointed out, that vote-counting procedures are in principle meant for combining independent estimates. We have included multiple estimates from most studies, which makes it harder to justify independence. The impact of lacking independence on the accuracy of vote-counting is unknown.

One has to take into account the significance of the values at a level common to all studies. We have used a relatively high level of 10% based on a two-sided test. In the tables below we make a distinction between results on effect sizes according to a certain categorization. There is a multitude of variables that can be included, and hence a multitude of figures that can be produced. Therefore, a selection has been made, and for tables and figures we present just three dimensions.

First of all, we have made the distinction between exploratory, Leontief and econometric studies. Kalt's study is partly classified as econometric and partly as Leontief. Figure 1.A below, gives the number of zero (statistically insignificant), (statistically significant) negative and positive results per study. The striking features of the Figure are that the Leontief studies produce relatively many negative effects. Moreover, the share of insignificant results is high in all study types, although it should be taken into account that the Han and Braden study is quite dominant in the econometric approaches. There are many negative significant estimates compared to positive significant estimates. This supports the hypothesis of a negative impact of environmental stringency versus the Porter hypothesis, which claims that the impact is positive.

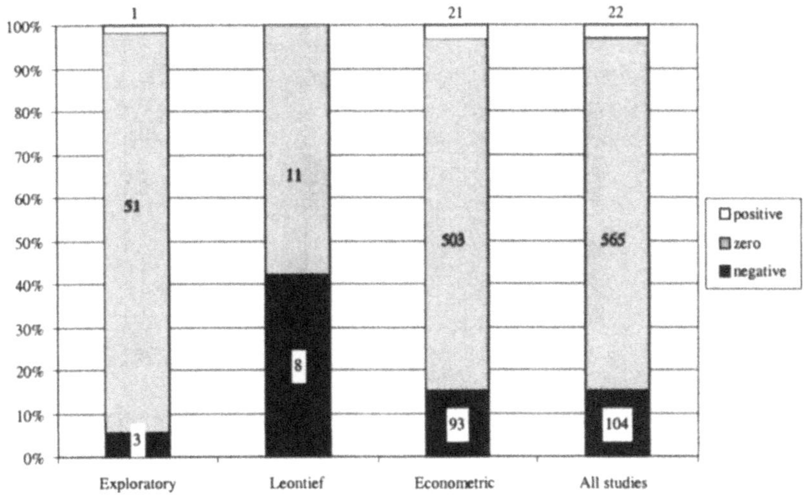

Figure 1a: Percentage and Number of Negative, Zero, and Positive Effect Sizes According to Study Approach.

A second distinction made follows from the way the effect sizes are calculated. One reason to study this is that one may for example argue that the picture of "all studies" in Figure 1a can be misleading because only the studies reporting regression coefficients are worth considering from a methodological perspective. As in Table 1, a distinction is made between regression coefficients, mean differences standardized by the control group, Glass' delta, standardized mean differences standardized by pooled standard deviation (Hedges' g), differences of proportions (Hedges' g-proportion), elasticities, differences from means and correlations. There are 691 observations in total, most of them regression coefficients (505) and elasticities (103). The results are summarized in Figure 1b. Results in terms of differences in means produce relatively many negative significant results again. Also with (pooled) standardized mean differences we obtain a relatively large share of negative significant results.

Finally, we have taken into account that stringency is measured in various ways. It can be measured by abatement costs, according to the Walter-Ugelow measure, the broad definition in Van Beers and Van den Bergh (VBVDB broad), the narrow definition of these authors (VBVDB narrow), the World Bank survey, the measure developed by Han and qualitative measures (Others). Figure 1c summarizes the results.

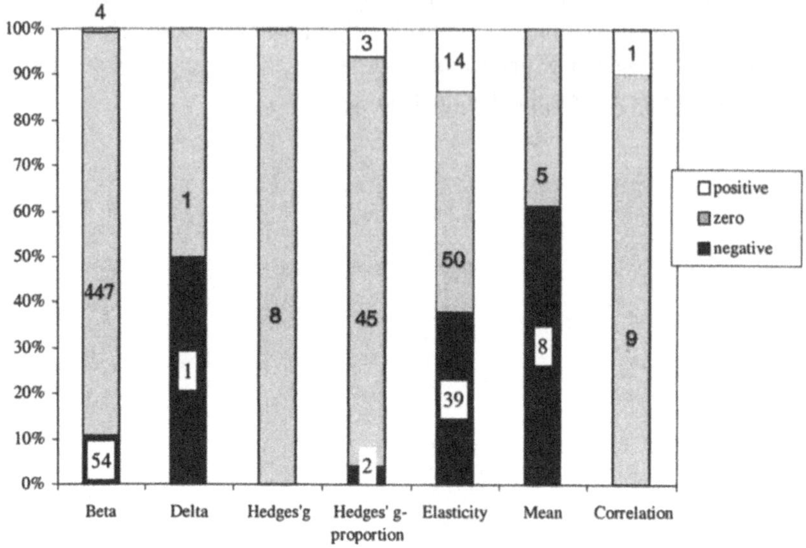

Figure 1b: Percentage and Number of Negative, Zero, and Positive Effect Sizes

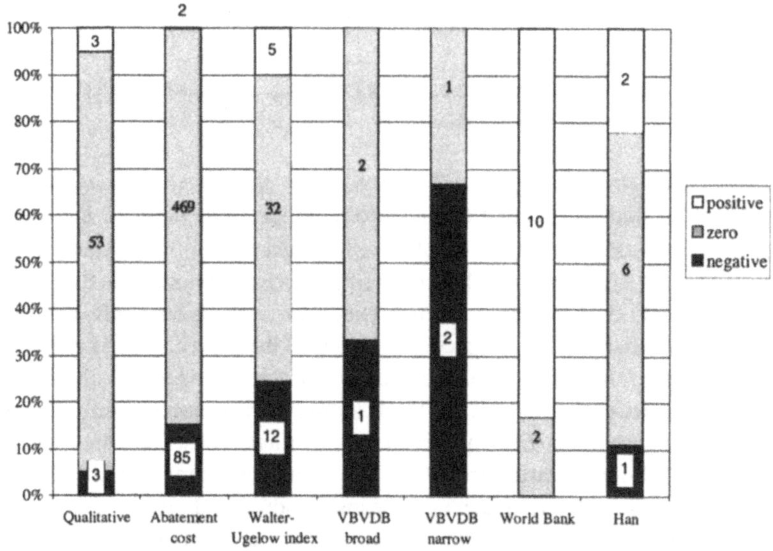

Figure 1c: Percentage and Number of Negative, Zero, and Positive Affect Sizes According to Type of Stringency.

The outcomes appear to differ drastically according to the stringency measure employed. Both measures employed by Van Beers and Van den Bergh yield relatively many negative and significant results (over 30%). Another remarkable feature is that with abatement costs as a measure relatively few statistically significant negative effects are found.

The Han and Braden (1996) study has produced several econometric estimates for which standard errors were not readily available. In the Appendix we constructed estimates of these standard errors. There we necessarily had to make assumptions that were on the safe side but were difficult to verify at the same time. Therefore, we have performed the vote-counting analysis also for a restricted data set. The restriction implied exclusion of all their estimates of elasticities, the time series estimates except that of the first year, which does not entail our approximation, and their panel estimates except those that are devoid of the time interaction term and the four panel estimates of the first period. Now there occurs an increase of the proportion of the statistically significant negative effects in the econometric studies from about 15% to 25%. Because many significantly negative elasticities disappear from the analysis, it turns out that the proportion of significantly negative elasticities decreases from about 35% to about 18%. Finally, studies with abatement costs now produce a relatively larger proportion of significantly negative results (from about 15% to about 38%). This is due to the fact that many of the insignificant outcomes from the Han and Braden study are not included anymore.

3.3 Combining Significance Level

All studies involved address more or less the same research question: is there a significant impact of environmental policy on trade performance? The particular context in which this question is answered differs per study, but since they have a common aim, it is worthwhile to see whether the information contained in each individual study can be combined to get a better insight. There are many long established statistical methods for combining significance levels. They all deal with the question of how probability values from independent studies can be combined. The hypothesis being tested can be expressed as follows:

$$H_0 : \vartheta_j = 0, \quad j = 1,2,...,k$$

where ϑ_j is the effect size in study j, and k is the number of studies. If the null hypothesis is rejected then at least one of the studies has a nonzero effect size. In the case at hand, the null hypothesis will be that none of the studies supports a significant negative impact of stringency of environmental regulation on trade performance. The alternative hypothesis is

$$H_1 : \vartheta_j \leq 0, \quad j = 1,2,...,k, \quad \vartheta_j < 0, \text{ for at least one } j.$$

Testing the null hypothesis requires the assumption that the population effects represented by ϑ_j-values all are in the same direction[12] (Cooper and Hedges, 1994: 220).

We consider four representative popular methods to combine significance levels.

- The minimum p-method, developed by Tippett (1931), rejects the null hypothesis that in all studies the effect size is zero if:

  $$\text{Min } (p_1, p_2, \ldots, p_k) < \alpha := 1 - (1 - \alpha^*)^{1/k}$$

  where α^* is the predetermined significance level for the combined significance test ($\alpha^* = 0.05$). The basis of this test is the fact that if the p-values are independent, then each of them is uniformly distributed.

- As a second method we consider the sum of z's method. It is based on the sum of the $z(p_i)$'s, where $z(p_i)$ is the z-value associated with p_i, divided by its standard deviation:

  $$\sum_{i=1}^{k} z(p_i) / \sqrt{k}$$

  So, each p_i-value is transformed to the normal score, which is then averaged.

- The third method is the sum of logs method.

  $$-2\sum^{k} \log(p_j)$$

  Under the null hypothesis this statistic has a χ^2-distribution with $2k$ degrees of freedom.

- Finally, we use the logit method. The expression

  $$-\sum^{k} \log\left(p_j / (1 - p_j)\right)[k\prod^{2}(5k + 2)/3(5k + 40)^{-1/2}$$

  is approximately t-distributed with $5k + 4$ degrees of freedom.

It can be seen from the upper rows of Table 2 that for all types of studies all four methods almost always reject the null hypothesis that the relationship between environmental stringency and trade performance is not significantly different from zero. Only the minimum p-method does not reject the null hypothesis for exploratory studies. The middle part of Table 2 shows that the null hypothesis is rejected by all methods if the results are reported in terms of regression coefficients, elasticities, and difference from the mean. It seems that the minimum p-method rejects the null hypothesis least often. This also holds for the case where

12 In cases where the effect size can take both signs this technique is not perfectly reliable.

we consider the different stringency measures (lower section of Table 2). In this case we also find that for studies with abatement costs as a stringency measure all methods reject the null hypothesis.

Table 2: Results of Combining *p*-Values According to Various Categorizations.

	Method of combining *p*-values			
	Minimum *p*		Sum of *z*	
Typology of study	Minimum *p*	Critical level	Value of test-statistic	Probability
According to approach				
Econometric	0.000001	0.000085	-12.818	0.000
Leontief	0.000003	0.002696	-8.043	0.000
Exploratory	0.006086	0.000932	-2.474	0.007
All studies	0.000001	0.000076	-14.151	0.000
According to effect size				
Beta	0.000022	0.000104	-9.208	0.000
Elasticity	0.000001	0.000498	-10.827	0.000
Correlation	0.071608	0.005116	0.805	0.790
Mean	0.000003	0.003938	-9.098	0.000
Hedges' *g*	0.081347	0.006391	-2.019	0.022
Hedges' *g*-proportion	0.006086	0.001165	-2.091	0.018
Delta	0.034233	0.025321	-2.164	0.015
According to type of stringency				
Walter-Ugelow	0.004937	0.001314	-2.866	0.002
Abatement cost	0.000001	0.000092	-15.352	0.000
Han	0.008941	0.005683	1.221	0.889
World Bank	0.367030	0.004265	10.524	1.000
Others	0.006086	0.000967	-2.881	0.002
VBVDB narrow	0.002093	0.016952	-3.124	0.001
VBVDB broad	0.031186	0.016952	-0.857	0.196

Table 2: continued.

Typology of study	Method of combining p-values			
	Sum of logs		Logit	
	Value of test-statistic	Probability	Value of test-statistic	Probability
According to approach				
Econometric	2325.800	0.000	15.543	0.000
Leontief	178.964	0.000	12.669	0.000
Exploratory	139.631	0.030	2.384	0.018
All studies	2644.395	0.000	17.110	0.000
According to effect size				
Beta	1434.926	0.000	9.947	0.000
Elasticity	885.956	0.000	16.288	0.000
Correlation	16.141	0.708	-1.553	0.126
Mean	165.386	0.000	15.526	0.000
Hedges' g	25.077	0.068	2.601	0.013
Hedges' g-proportion	105.707	0.096	2.213	0.028
Delta	11.203	0.024	4.339	0.001
According to type of stringency				
Walter-Ugelow	159.161	0.000	2.589	0.010
Abatement cost	2294.155	0.000	19.324	0.000
Han	18.018	0.454	-1.696	0.096
World Bank	2.254	1.000	-18.236	0.000
Others	136.430	0.025	2.985	0.003
VBVDB narrow	24.497	0.000	6.574	0.000
VBVDB broad	9.880	0.130	1.591	0.128

Table 2: continued.

Typology of study	Method of combining p-values: Sum of logs – Value of test-statistic	Sum of logs – Probability	Logit – Value of test-statistic	Logit – Probability
According to approach				
Econometric	2325.800	0.000	15.543	0.000
Leontief	178.964	0.000	12.669	0.000
Exploratory	139.631	0.030	2.384	0.018
All studies	2644.395	0.000	17.110	0.000
According to effect size				
Beta	1434.926	0.000	9.947	0.000
Elasticity	885.956	0.000	16.288	0.000
Correlation	16.141	0.708	-1.553	0.126
Mean	165.386	0.000	15.526	0.000
Hedges' g	25.077	0.068	2.601	0.013
Hedges' g-proportion	105.707	0.096	2.213	0.028
Delta	11.203	0.024	4.339	0.001
According to type of stringency				
Walter-Ugelow	159.161	0.000	2.589	0.010
Abatement cost	2294.155	0.000	19.324	0.000
Han	18.018	0.454	-1.696	0.096
World Bank	2.254	1.000	-18.236	0.000
Others	136.430	0.025	2.985	0.003
VBVDB narrow	24.497	0.000	6.574	0.000
VBVDB broad	9.880	0.130	1.591	0.128

As in the case of vote-counting we have performed the above analysis for the restricted data base. It turns out that the results are only slightly affected. There occur some changes, however. The null hypothesis that the relationship between environmental stringency and trade performance is not significantly different from zero, is not rejected anymore for studies using elasticities. But for abatement costs the conclusion we reached earlier, is reinforced: studies with abatement costs as the stringency measure are likely to produce significant results.

4 Conclusion

This paper is essentially a critical review of the empirical literature on environmental regulation and international trade flows, a literature that displays contradictory evidence.

It starts with the customary practice of outlining the salient features of each study and summarizing the results. The conclusion from this review, like similar review exercises, is that the empirical literature does not strongly support the hypothesis that the effect of environmental regulation on competitiveness is negative. The qualitative review identifies several controversial issues in the literature. Most prominent are the stringency measure, type of data, methods of investigation and sectoral disaggregation.

Simple vote-counting was employed as a first and crude way of comparing different estimates. It reveals that estimates vary over three factors: measure of stringency, study approach and effect size measure. However, a more formal comparison by means of meta-regression seems to be in order. Such analysis, while demanding in terms of data, is perhaps doable and would have made the analysis complete. But that is subject to current research. Subsequently, four types of meta-analytic techniques of combining significance levels are applied to the studies under discussion. The results suggest that there is almost always at least one study/estimate in the available literature that displays a statistically significant negative relationship between environmental stringency and international trade flows. Stated otherwise, the hypothesis of no effect is rejected in almost all tests. This still hold for what is called a restricted data base.

What are the insights one can gain for future primary research in this area? The environmental regulation-competitiveness linkage should be investigated with data on the industrial level rather than on the country level. This is to be preferred from a theoretical point of view, because environmental policy is usually industry related rather than general. But it also follows from our findings that the hypothesis of no effect of stringency of environmental policy on trade performance should be tested at a low level of aggregation: Rejection of the hypothesis at that level seems to warrant rejection at the higher level as well. In addition, the fact that environmental stringency differentials are likely to prevail more between North and South, inclusion of developing countries in the sample is desirable. Also, since most of the studies under scrutiny did not involve the European countries, it seems worthwhile to have a much closer look at this region, in relation to developing countries. Similarly, the diverse measures of stringency are one of the factors for the disparate evidence in the empirical literature. Therefore, any new empirical investigation of the relationship under discussion should come to grips with this problem before embarking on estimation by picking up one or the other controversial measures.

Appendix

The objective of this appendix is to introduce the so-called effect size used in the meta-analysis. The effect size in this study generally refers to the effect of stringent environmental policy on export performance. Meta-analysis offers several methods to obtain the effect size. Two of these methods are often used in the paper, and will therefore be described in some detail in this appendix. We also elaborate on the way the data from the study of Han and Braden are used.

Effect Size

Generally a distinction is made between a control group, with index *c*, consisting of n_c elements, and an experimental group, with index *e*, consisting of n_e elements. Two alternative effect size measures are Glass' Δ and Hedges' *g* defined as:

$$\Delta = \frac{M_e - M_c}{S_c} \text{ and } g = \frac{M_e - M_c}{S_p}.$$

where M_c and M_e are the means of the relevant variable in the control group and the experimental group, respectively, and S_c and S_p are the standard deviations in the control group and the pooled standard deviations respectively. The latter is defined by:

$$S_p = \sqrt{\frac{((n_c - 1)S_c)^2 + ((n_e - 1)S_e)^2}{n_c + n_e - 1}}.$$

where S_e is the standard deviation in the experimental group. In the meta-analytical literature it is argued that the pooled standard deviation is a better estimate of the population standard deviation, but choosing the standard deviation of the control group is a "very reasonable alternative" (Rosenthal, 1994: 232). We can use the fact that

$$\Delta\sqrt{\frac{n_c n_e}{n_c + n_e}} \text{ and } g\sqrt{\frac{n_c n_e}{n_c + n_e}}$$

have a *t*-distribution with n_c+n_e-2 degrees of freedom, under the commonly made assumption of a normal distribution of the original observations.

When the data are proportions, the effect size is defined by p_e-p_c, with a variance defined by

$$\frac{p_e(1-p_e)}{n_e}+\frac{p_c(1-p_c)}{n_c}$$

We can then make use of the normal distribution to derive confidence intervals and p-values.

Han and Braden (1996)

The study by Han and Braden provides the majority of the data for our analysis. First, for 19 product groups and for four years (1975, 1980, 1985 and 1989) they present elasticities giving the percentage change in net exports following a one percent increase in abatement expenditures AB. These elasticities are the result of a regression equation of the type:

$$X_{it} = \ldots + \beta_5 AB_{it} + \beta_6 AB_{it} t + \ldots$$

where t refers to time. The elasticity is:

$$\varepsilon_{it} = \frac{\partial X_{it}}{\partial AB_{it}} \frac{AB_{it}}{X_{it}}$$

We are interested in finding the standard error of the estimate of the elasticity. To that end an assumption has to be made, because the available data do not allow for a straightforward calculation. We have chosen to assume that in every year the ratio of abatement expenditures over net exports is the same, for all sectors. It is denoted by c. That implies that the standard error can be written as:

$$c\left(Var(\beta_5)+t^2 Var(\beta_6)+2t.rSE(\beta_5)SE(\beta_6)\right)^{1/2}$$

In this expression r is the correlation coefficient and the β's refer to the estimates. The time variable assumes the values 3, 8, 13 and 17. (1975 is 3 because the initial data are from 1973). The variances of the individual estimates are given in Han and Braden. Next we had to choose values for c and r. To be on the safe side we take c = 0.1 and r = 0.8. This procedure enables us to perform the usual tests on effect sizes.

References

Ashenfelter, O., C. Harmon, and H. Oosterbeek (1999), A Review of Estimates of the Schooling/Earnings Relationship, With Tests for Publication Bias, *Labour Economics* 6, 453-470.

Baumol, W. and W. Oates (1975), *The Theory of Environmental Policy*, New Jersey.

Becker, R. and V. Henderson (2000), Effects of Air Quality Regulations on Polluting Industries, *Journal of Political Economy* 108, 379-421.

Beers, C. van and J. van den Bergh (1997), An Empirical Multi-Country Analysis of the Impact of Environmental Regulations on Foreign Trade Flows, *Kyklos* 50, 29-46.

Beers, C. van and J. van den Bergh (1998), The Impact of Environmental Policy on Foreign Trade: An Empirical Multi-Country Analysis Based on Bilateral Trade Flows, *mimeo*, Free University Amsterdam.

Beers, C. van and J. van den Bergh (2000), The Impact of Environmental Policy on Foreign Trade: Tobey Revisited With a Bilateral Trade Flow, *mimeo*, Department of Economics, Delft University of Technology, Delft.

Bergh, J. van den, K. Button, P. Nijkamp, and G. Pepping (1997), *Meta-Analysis in Environmental Economics*, Dordrecht.

Bouman, M. (1998), *Environmental Costs and Capital Flight*, PhD thesis, Tinbergen Institute Research Series, University of Amsterdam, Amsterdam.

Bushman, B. (1994), Vote Counting Procedures in Meta-Analysis, in: Cooper, H. and L. Hedges (Eds.), *The Handbook of Research Synthesis*, New York, 193-214.

Button, K. and J. Kerr (1996), Effectiveness of Traffic Restraint Policies: A Simple Meta Regression Analysis, *International Journal of Transport Economics* 23, 213-225.

Button, K. and T. Weyman-Jones (1992), Ownership Structure, Institutional Organization and Measured X-Inefficiency, *American Economic Review* 82, *Papers and Proceedings*, 439-445.

Card, D. and A. Krueger (1995), Time-Series Minimum-Wage Studies: A Meta-Analysis, *American Economic Review* 85, 238-243.

Co, C., H. Folmer, T. Jeppesen, and J. List (1999), Impact of Environmental Regulation on Foreign Trade Flows: Methodological Problems and Possible Solutions, *working paper*, CentER Tilburg University, forthcoming in: L. Marsiliani, M. Rauscher, and C. Withagen, *The International Dimension of Environmental Policy*, Dordrecht.

Cooper, H. and L. Hedges (1994), *Handbook of Research Synthesis*, New York.

Dasgupta, S., R. Mody, S. Roy, and D. Wheeler (1995), Environmental Regulation and Development: A Cross-Country Empirical Analysis, *World Bank Policy Research Working Paper No 1448.*

Diakosavvas, D. (1994), The Impact of Environmental Policies on Agricultural Trade, *Journal of International Development* 6, 207-218.

Ekins, P. and S. Speck (1999), Competitiveness and Exemptions from Environmental Taxes in Europe, *Environmental and Resource Economics* 13, 369-396.

Greenstone, M. (1998), The Impacts of Environmental Regulations on Industrial Activity: Evidence from the 1970 and 1977 Clean Air Act Amendments and the Census of Manufactures, *Working Paper #408*, Princeton University.

Grossman, G. and A. Krueger (1993), Environmental Impacts of a North American Free Trade Agreement, in: Garber, P. (Ed.), *The U.S.-Mexican Free Trade Agreement*, Cambridge MA, 13-56.

Han (1996), *Environmental Regulation and International Competitiveness: Cross-Industry and Cross-Country Analyses*, PhD thesis, University of Illinois.

Han, K. and J. Braden (1996), Environment and Trade: New Evidence from U.S. Manufacturing, *mimeo*, University of Illinois.

Hedges, L. (1997), The Promise of Replication in Labour Economics, *Labour Economics* 4, 111-114.

Hedges, L. and I. Olkin (1980), Vote Counting Methods in Research Synthesis, *Psychological Bulletin* 88, 359-369.

Hedges, L. and I. Olkin (1985), *Statistical Methods for Meta-Analysis*, New York.

Helpman, E. and P. Krugman (1985), *Market Structure and Foreign Trade*, Cambridge MA.

Henderson, J. (1996), Effects of Air Quality Regulation, *American Economic Review* 86, 789-813.

Hettige, H., R. Lucas, and D. Wheeler (1992), The Toxic Intensity of Industrial Production: Global Patterns, Trends, and Trade Policy, *American Economic Review* 82, *Papers and Proceedings*, 478-481.

Jaffe, A., S. Peterson, P. Portney, and R. Stavins (1995), Environmental Regulation and the Competitiveness of US Manufacturing: What Does the Evidence Tell Us?, *Journal Economic Literature* 33, 132-163.

Jarrell, S. and T. Stanley (1990), A Meta-Analysis of the Union Non-Union Wage Gap, *Industrial and Labour Relations Review*, 54-67.

Jeppesen, T., H. Folmer, and M. Komen (2000), Environmental Policy, International Trade and Capital Movement: A Synopsis of the Macroeconomic Literature, in: Sterner, Th. (Ed.), *The Market and the Environment: The Effectiveness of Market-Based Policy Instruments for Environmental Reform*, Cheltenham, 83-113.

Jeppesen, T., J. List, and H. Folmer (2002), Environmental Regulations and New Plant Location Decisions: Evidence from a Meta-Analysis, *Journal of Regional Science 42, 19-49*

Kalt, J. (1988), The Impact of Domestic Environmental Regulatory Policies on U.S. International Competitiveness, in: Spence, M. and H. Hazard (Eds.), *International Competitiveness*, 221-262.

Leonard, H. (1988), *Pollution and the Struggle for the World Product*, Cambridge.

Light, R. and P. Smith (1971), Accumulating Evidence: Procedures for Resolving Contradictions Among Different Research Studies, *Harvard Educational Review* 41, 429-471.

List, J. and W. McHone (2000), *Effects of Environmental Regulations on Manufacturing Plant Births: Evidence from a Propensity Score Matching Estimator*, Paper presented at the European Science Foundation (EURESCO) meeting on the International Dimension of Environmental Policy, Kerkrade.

Low, P. and A. Yeats (1992), Do 'Dirty' Industries Migrate?, in: Low, P. (Ed.), *International Trade and the Environment,* World Bank, Washington DC.

Phillips, J. and E. Goss (1995), The Effects of State on Local Taxes on Economic Development: A Meta-Analysis, *Southern Economic Journal* 62, 320-333.

Porter, M. (1991), America's Green Strategy, *Scientific American*, 168.

Rauscher, M. (1997), *International Trade, Factor Movements, and the Environment,* Oxford.

Robison, H.D. (1988), Industrial Pollution Abatement: The Impact on Balance of Trade, *Canadian Journal of Economics* 1, 187-199.

Smith, V. (1989), Can We Measure the Economic Value of Environmental Amenities?, *Southern Economic Journal* 56, 865-878.

Smith, V. and Y. Kaoru (1990a), Signals or Noise: Explaining the Variation in Recreation Benefit Estimates, *American Journal of Agricultural Economics* 72, 419-433.

Smith, V. and Y. Kaoru (1990b), What Have We Learned Since Hotelling's Letter? A Meta-Analysis, *Economics Letters* 32, 267-272.

Smith, V. and L. Osborne (1996), Do Contingent Valuation Estimates Pass a "Scope" Test? A Meta-Analysis, *Journal of Environmental Economics and Management* 31, 287-301.

Sorsa, P. (1994), *Environmental Protection, North-South Trade, and the Uruguay Round,* Washington DC.

Stanley, H. and S. Jarrell (1989), Meta-Regression Analysis: A Quantitative Method of Literature Surveys, *Journal of Economic Surveys* 3, 161-170.

Thompson, G. (1998), International Competitiveness and Globalization: Framework for Analysis, Connections and Critiques, in: Barker, T. and J. Köhler (Eds.), *International Competitiveness and Environmental Policies,* Cheltenham, 13-31.

Tippett, L. (1931), *The Methods of Statistics*, London.

Tobey, J. (1993), The Impact of Domestic Environmental Policy on International Trade, in: Giersch, H. (Ed.), *Economic Progress and Environmental Concerns,* Berlin.

Ulph, A. (1997), International Trade and the Environment: A Survey of Recent Economic Analysis, in: Folmer, H. and T. Tietenberg (Eds.), *The International Yearbook of Environmental and Resource Economics 1997/1998*, Aldershot, 205-242.

United Nations (1982), *Changing Patterns of Trade in World Industries*, New York.

Walter, I. (1973), The Pollution Content of American Trade, *Western Economic Journal* 1, 61-70.

Walter, I. and J. Ugelow (1979), Environmental Policies in Developing Countries, *Ambio* 8, 102-109.

Xu, X. (1999), Do Stringent Environmental Regulations Reduce the International Competitiveness of Environmentally Sensitive Goods? A Global Perspective, *World Development* 27, 1215-1226.

Xu, X. (2000), International Trade and Environmental Regulations: Time Series Evidence and Cross Section Test, *Environmental and Resource Economics* 17, 233-257.

Trade, Technology, and Carbon Emissions: A CGE Analysis for West Germany

Heinz Welsch

Department of Economics, University of Oldenburg, 26111 Oldenburg, Germany, *welsch@uni-oldenburg.de*

Summary:
This paper examines the determinants of the substantial decline of West German production-related carbon intensity in the face of falling energy prices. A computable general equilibrium model is used to determine the simulated effects of observed changes of world energy prices and domestic energy policy on the sectoral patterns of carbon emissions, energy consumption, output, value added and other indicators of structural change. The structural changes not accounted for by energy prices and energy policy are attributed to changing patterns of productivity growth in Germany and the rest of the world (ROW) and changing patterns of ROW demand. Weights on these driving forces are selected by least squares. One key finding is that the contribution of ROW productivity and demand patterns to emission-relevant structural change unaccounted for by energy prices and energy policy is just under 30%. The remainder is split almost equally among patterns of domestic autonomous energy efficiency improvement and domestic labor efficiency patterns.

Keywords: Carbon emissions, structural change, technology, trade, computable general equilibrium.

1 Introduction

A well established literature has decomposed changes in economy-wide energy intensity into changes in energy intensity at the sectoral level and changes in the sectoral structure of the economy (see e.g. Boyd et al., 1987; Ang, 1994; Unander et al., 1999; de Bruyn, 2000, ch. 9). Based on such approaches, it has been found that in the second half of the 1980s, structural change has contributed one quarter to the observed change in energy intensity within West German industry (Unander et al., 1999). Structural change may have contributed even more to overall energy intensity, because the share of industry, which is more energy intensive than services, has been declining in West Germany, as in most other industrialized countries.

In addition to deindustrialization (defined as a decrease of the share of industry in domestic output), economic structural change has the following dimensions: (see, e.g., Saeger, 1997; Burda and Dluhosch, 2000):

- Globalization: the share of domestic output in total supply decreases.[1]
- Fragmentation: the share of intermediate inputs in domestic output increases.[2]

While deindustrialization and globalization tend to reduce the energy or carbon intensity of total supply[3], the impact of fragmentation is ambiguous, depending on the degree to which intermediate inputs are of domestic origin or imported from abroad.

Several candidates have been proposed as driving forces of these structural changes (see, e.g., Saeger, 1997; Berman, Bound, and Machin, 1998). They can be classified into changes in the structure of world trade and changes in the pattern of productivity improvement at home and abroad. For instance, deindustrialization in advanced economies, such as Germany, has been attributed to growing world demand for services (relative to commodity demand), strong productivity growth in the service sector of advanced economies, and strong productivity growth in the industrial sector of emerging (or developing) economies (Minford, Riley, and Nowell, 1997).

This paper provides a case study of the impact of trade and productivity (technology) on carbon emissions, taking into account that part of the impact may become effective via induced structural change. In addition to trade and technology, the role of energy policy as a contributor to structural change and the decrease of carbon intensity is examined. The analysis refers to West Germany in the period of 1985-1990. This period is particularly interesting because of the strong decline of carbon intensity in spite of falling energy prices: In 1985-1990 the production-related carbon emissions[4] in West Germany (excluding those from

---

1 Total supply is a composite of domestic output and imports. The counterpart to the decreasing share of domestic output in total supply is, of course, the falling share of domestic absorption in total demand.

2 The statistical phenomenon of fragmentation, as defined above, is the result of a disintegration of production processes along with outsourcing of activities (see Jones and Kierzkowski, 1990, and Burda and Dluhosch, 2000).

3 Deindustrialzation reduces not only the carbon and energy intensity of total supply, but also of domestic output; see Section 2.

4 The analysis focuses on production-related emissions in the sense that emissions from the government and household sector are not the subject of the paper. In addition, the study is restricted to West Germany because appropriate data on East Germany before the unification are not available; the changes in energy use and economic structure in the eastern part of Germany after the unification are difficult to analyse in purely

the government sector) decreased by 0.2% annually whereas the total private supply of goods increased by 4.2%. The carbon intensity of private goods supply thus decreased by approximately 4.4% per year. At the same time, the dollar prices of world energy decreased by 7.3% per year (coal), 5.5% (oil) and 2.3% (gas), see IEA (1995).

In contrast to energy policy (especially the expansion of nuclear power over the period considered), important aspects of trade and technology are difficult if not impossible to observe. Regarding trade in services, there exist at best rudimentary data, at least for the period considered.[5] As concerns productivity, what would seem to be needed is data on something like "autonomous labor productivity improvement", differentiated at least by the categories "industry" and "services". This is difficult to observe, as is the rate of "autonomous energy efficiency improvement".[6]

Given the unavailability of appropriate data on potentially important explanatory variables, the paper pursues the following methodological approach.[7]

(0) We provide various decompositions of production-related carbon emissions which reflect structural changes in the use of energy and in the overall economy.

(1) We employ a computable general equilibrium (CGE) model to determine how observed changes in energy prices and energy policy have influenced the structural change indicators from step (0).

(2) We use the CGE model to simulate the effect on the indicators from step (0) of arbitrary energy efficiency improvement rates and arbitrary variations of the above mentioned unobserved drivers of structural change.

(3) We employ regression analysis to determine the importance of these unobserved drivers in explaining that part of the indicators from step (0) that remain unexplained in step (1).

The paper is organized as follows. Section 2 describes the decomposition of production-related carbon emissions, resulting in the change indicators to be explained. Section 3 provides an outline of the CGE model employed and discusses the model's key exogenous drivers that may serve as explanatory variables for the change indicators. Section 4 considers the simulated contribu-

---

economic terms. Focusing on West Germany then implies restriction to the period 1985-1990 because separate data for the western part of unified Germany are not supplied by the statistical offices.

5 For the 1990s, WTO (2000) indicates that world export of commercial services grew stronger than merchandise export.

6 These concepts as well as the difficulty of observing them are discussed in Section 3.

7 A similar methodology has been used by Minford, Riley, and Nowell (1997) in a different context.

tions of observed energy policy and energy price changes. In Section 5 we analyze the simulated partial effects of arbitrary variations in the unobserved exogenous drivers. Section 6 uses regression analysis to determine the weights to be attached to these drivers. Section 7 concludes.

2 Indicators of Change in Energy Use and Economic Structure

Production-related carbon emissions and the associated energy consumption data are available for the 58 sectors of the German input-output table for selected years. Within the period of interest they are available for 1985 and 1990.[8] For the purpose of this paper the original data have been aggregated to eight non-government sectors[9], corresponding to the sectoral classification of the CGE model employed. Data on total supply, output, and value added (Statistisches Bundesamt, 1999) has also been aggregated to match these eight sectors.

To provide a first illustration of structural change in energy use, Table 1 displays the carbon intensity and the energy intensity of output in two broad sectoral aggregates, referred to as "industry" and "services".[10]

In 1985 the carbon intensity and the energy intensity, respectively, of industry output was 3.9 and 4.6 times that of service output. In 1990 the corresponding ratios were 3.2 and 4.0. Thus, the "efficiency gap" of industry relative to services became more narrow, with respect to both carbon and energy. The improvement in carbon efficiency was substantially stronger in industry (25.2%) than in services (7.5%). The difference in energy efficiency improvement was somewhat less pronounced (18.2% in industry vs. 6.6% in services). The differences between carbon efficiency improvements and energy efficiency improvements mainly reflect a substitution of electricity for fossil fuels.

8 I am grateful to Helmut Mayer (German Federal Statistical Office) for supplying unpublished data. These data are a result of the Statistical Office's comprehensive work on establishing a 'Satellite Account on Environment' and conform to the quality standards of official statistics (on sources and methods see Mayer and Stahmer, 1989).

9 The analysis is restricted to the non-government sectors because changes in the government sector are difficult to explain by economic considerations alone.

10 Note that these labels are suggestive, but depart somewhat from the usual terminology. "Industry" in this paper is meant to include agriculture, but to exclude the energy supply industries. "Services", as a broad concept, includes transportation and excludes non-market services.

Table 1: Carbon and Energy Intensity of Output [a)].

	Industry 1985	Services 1985	Industry 1990	Services 1990
Carbon intensity	93.62	24.00	70.03	22.18
Energy intensity	2175.29	472.24	1780.15	441.29

[a)] Kilotons of CO_2 per billion DM and terajoule per billion DM, respectively; output at 1991 prices.

Despite the narrowing of the industry's relative efficiency gap, the carbon and energy efficiency of output was at least 3 to 4 times higher in services than in industry. Since service output grew faster than industry output in the period considered (4.7% per year vs. 3.9% per year), it is evident that structural change played a major role in reducing overall carbon and energy intensity.

In a more systematic perspective, the carbon emissions in a particular sector i, C_i, can be decomposed in various ways, to reflect not only the structure of energy use and the sectoral structure of production, but also the share of value added in domestic output and the share of domestic output in total supply. Our decompositions are based on the following identities:

$$C_i = \frac{C_i}{F_i} \cdot \frac{F_i}{E_i} \cdot \frac{E_i}{Q_i} \cdot \frac{Q_i}{Q} \cdot Q, \tag{1}$$

$$C_i = \frac{C_i}{F_i} \cdot \frac{F_i}{E_i} \cdot \frac{E_i}{Q_i} \cdot \frac{Q_i}{X_i} \cdot \frac{X_i}{X} \cdot X, \tag{2}$$

$$C_i = \frac{C_i}{F_i} \cdot \frac{F_i}{E_i} \cdot \frac{E_i}{Q_i} \cdot \frac{Q_i}{Y_i} \cdot \frac{Y_i}{Y} \cdot Y, \tag{3}$$

where C_i = carbon emissions, F_i = emission-relevant energy use[11], E_i = total energy use, Q_i = gross output, Y_i = value added, X_i = total supply.

The first three ratios on the right hand side are the same in each equation. They refer to the structure of energy use in each particular sector. The ratio C_i / F_i basically reflects the composition of fossil energy input in terms of hard coal, soft coal, oil and gas, each having a different carbon content. The ratio F_i / E_i captures

11 Emission-relevant energy use is a subset of total energy use that excludes electricity, district heating, crude oil and lubricants. It corresponds quite closely to fossil energy use, hence the notation "*F*".

the composition of total energy use in terms of fossil and non-fossil (carbon-free) energy carriers. E_i / Q_i is the energy intensity of production.

The ratios Q_i / Q, X_i / X, and Y_i / Y in equations (1)-(3) give sector i's share in total domestic output, total supply, and value added, respectively. The ratio Q_i / X_i in equation (2) is an inverse measure of the degree of globalization in sector i (a high value of Q_i / X_i indicating a low degree of globalization). The ratio Q_i / Y_i in equation (3) measures fragmentation (a high value indicating a high share of intermediate inputs in sectoral output).

Our decomposition proceeds by computing the average annual percentage changes of the expressions showing up in equations (1)-(3). We deliberately avoid any attempt at allocating the residuals that arise when employing rates of change in a discrete-time framework. The residuals are rather small and do not affect our main conclusions.

Table 2 shows the decomposition of carbon emissions according to equation (1) for eight production sectors and a summary category.[12] It can be seen that carbon emissions (column C_i) decreased in all non-energy sectors except transport and services. Column C_i / F_i reveals a change of the mix of fossil energy carriers in favor of the less carbon intensive fuels (substitution of oil and gas for coal), especially (of course) in the industry sectors. According to column F_i / E_i there was a substitution of carbon-free energy (electricity) for fossil fuels in all sectors except the energy industry. The substitution of electricity for fuels was particularly pronounced in the equipment goods industry. As shown in column E_i / Q_i, the energy intensity of output decreased in all sectors, again with the exception of the energy industry. The equipment goods industry was especially successful in saving energy, whereas the transport sector showed the smallest improvement in energy efficiency. Overall, energy efficiency improvement (energy saving) was by far the most important contributor to the decrease of the carbon intensity of total output, amounting to 3.5% p.a. on average. Inter-fuel substitution and electrification were less important (just under 0.7% p.a.).

Finally, the column Q_i / Q illustrates the degree of deindustrialization that has taken place: All industry sectors except the equipment goods industry had declining output shares, whereas transportation and services increased their shares. The increasing share of transportation is consistent with the phenomenon of fragmentation, in that the outsourcing of activities leads to higher transport requirements.[13]

---

12 Note that the category "services" in Table 2 and Table 3 is more narrow than the broad category in Table 1 (cf. Footnote 5).

13 One may wonder to what extent the changes portrayed in Table 2 are already influenced by German unification. It is true that the unification implied a substantial boost to West German production in 1990. In fact, the growth rate of private output in 1990 was about twice the average rate in 1985-1989. It should be noted, however, that

Table 2: Average Annual Percentage Changes 1985-1990, I.

% / a	C_i	C_i / F_i	F_i / E_i	E_i / Q_i	Q_i / Q	Q
Agriculture	-1.77	-0.18	-0.46	-3.09	-1.97	/
Energy	0.22	-0.42	0.17	1.41	-4.86	/
Intermediate goods	-2.29	-0.71	-1.19	-2.87	-1.49	/
Equipment goods	-2.77	-0.72	-2.35	-5.51	1.72	/
Consumption goods	-0.63	-0.72	-0.85	-1.78	-1.24	/
Construction	-0.24	-0.23	-0.07	-3.45	-0.48	/
Transport	4.02	-0.02	-0.19	-1.37	1.52	/
Services	2.21	-0.09	-0.05	-2.23	0.55	/
Total	-0.18	-0.46	-0.23	-3.50	0.00	4.01

In Table 3 the decompositions according to equation (2) and equation (3) are displayed; the components common with equation (1) are not reiterated. The last entry of column Q_i / X_i shows that the share of domestic production in total supply decreased on average by 0.23% annually, indicating the degree of globalization over the period considered. This process was especially pronounced in industry, whereas transportation and services were not affected by it. Hence we have another indication of the phenomenon of deindustrialization.

Looking at the last entry of column Q_i / Y_i we find that the ratio of gross output to value added has been increasing on average. This is equivalent to an increase in the ratio of intermediate input to value added, i.e. fragmentation. On the sectoral level, fragmentation can be found in five out of our eight sectors. It has not taken place in agriculture, the intermediate goods sector and the service sector.

Looking at the sector shares in value added (Y_i / Y) we find an even stronger indication of deindustrialization than was evident in output shares: With respect to the share in value added, the equipment goods industry declined (whereas it increased with respect to output share). We here find a clear dichotomy between the transport and service sectors gaining shares and all other sectors losing.

the boost was by far most pronounced in the West German energy industry, whose growth rate in 1990 was 4.7%, while there had been an average annual decrease by 2.4% in 1985-1989. One can therefore conclude that without unification and the ensuing dismantling of East German energy industry (mainly for environmental reasons) the decline in West German energy intensity would have been even stronger than presented above. Notwithstanding such considerations, the challenge remains to explain the drop in energy and carbon intensity in 1985-1990 in the face of the energy price decline.

As illustrated above, the structural change in favor of the less energy intensive service sector may have contributed considerably to the observed decrease in overall carbon and energy intensity. In fact, the literature suggests that even within German industry, structural change may have contributed one quarter to the observed change in energy intensity (Unander et al., 1999).

Table 3: Average Annual Percentage Changes 1985-1990, II.

% / a	Q_i / X_i	X_i / X	X	Q_i / Y_i	Y_i / Y	Y
Agriculture	-0.22	-1.98	/	-0.69	-0.97	/
Energy	-0.31	-4.77	/	0.42	-4.94	/
Intermediate goods	-0.63	-1.10	/	-0.86	-0.32	/
Equipment goods	-0.77	2.25	/	2.31	-0.15	/
Consumption goods	-0.67	-0.81	/	1.66	-2.50	/
Construction	0.00	-0.70	/	1.62	-1.70	/
Transport	0.01	1.29	/	0.88	1.03	/
Services	0.03	0.30	/	-0.52	1.41	/
Total	-0.23	0.00	4.24	0.36	0.00	3.65

Our aim in this paper is not to provide another separation of the overall energy intensity change into aggregate measures of sectoral energy efficiency improvement and changing sectoral structure. Instead, we take the indicators of change in the structure of energy use and in the structure of the economy (globalization, deindustrialization, fragmentation) as representing an intermediate link in a causal chain from more fundamental drivers of structural change to changes in the release of carbon. In this view, the rates of change discussed in this section are taken as the variables to be explained. Overall, there are 70 indicators of change in energy use and economic structure which we seek to explain: The rates of change of the macro indicators C, F, E, Q, X, Y, and the rates of change of the ratios C_i / F_i, F_i / E_i, E_i / Q_i, Q_i / X_i, Q_i / Y_i, Q_i / Q, X_i / X and Y_i / Y.

3 Linking Structural Change Indicators to Driving Forces Utilizing a CGE Model

Evident drivers of change in energy use are improvements in energy technology, changes in world energy prices, and politically determined changes in the energy mix (e.g. expansion of nuclear power). The drivers of deindustrialization, globalization and fragmentation can be classified in a natural way into supply side (technology) and demand side variables. More specifically, the first set of variables comprises changes in the pattern of productivity improvement at home and abroad[14], whereas the second set refers to changes in the structure of world demand.

A key difficulty in any attempt at linking structural change to these potential drivers is that many of the latter are difficult to observe. As regards improvements in energy technology, the concept of "autonomous energy efficiency improvement" represents a purely notional parameter which is difficult to measure directly (see below). Similar measurement problems exist for other forms of productivity improvement, especially if they are to be measured for several sectors and on a world scale. Finally, with respect to the pattern of world demand, there exists only rudimentary data on trade in services, especially for the period considered.

For these reasons the methodology used in this paper is an indirect one. We use a CGE model to simulate the effects of observed changes in world energy prices and German energy policy on the indicators (rates of change) discussed in the preceding section. Furthermore, we simulate the (partial) effects of arbitrary changes in the unobserved "drivers" mentioned above. Finally, we use least squares to select the weights on the unobserved drivers that provide the best explanation for structural change not explained by the observable drivers.

The CGE model employed comprises West Germany vs. the exogenous rest of the world. It has 13 sectors, five of which refer to energy carriers: hard coal, soft coal, petroleum, gas and electricity. In this paper, the results relating to the energy sectors are displayed only for the summary category, referred to as energy industry. The other sectors are agriculture, intermediate goods, equipment goods, consumption goods, construction, transport, services, and government.

The model structure is particularly suitable for the current analysis because it incorporates exactly those exogenous inputs which constitute important potential drivers of change in energy use and economic structure in Germany:

- Domestic hard coal production and nuclear power generation,
- World market prices of hard coal, mineral oil, and natural gas,

---

14 This is broadly consistent with a comparative advantage view of structural change.

- Export prices of the rest of the world's industry and service sectors,
- Rest of the world's demand for import of goods and services,
- Energy-augmenting technological progress in Germany's industry and service sectors,
- Labour-augmenting technological progress in Germany's industry and service sectors.

The evolution of these exogenous inputs impacts on the German economy, as captured by the model, in various ways. The expansion of nuclear power over the period considered leads (cet. par.) to a reduction of fossil fuel input, whereas the reduction in world market prices of fossil fuels tends to enhance their utilization. Changes in the rest of the world's industry and service prices affect the structure of comparative advantage in Germany, as do changes in the sectoral pattern of German labor productivity. With respect to the level and structure of German production-related carbon emissions, a relatively strong increase in the rest of the world's industry productivity coupled with a relatively strong increase in the German service sector's labor productivity would produce structural change in favor of the less carbon-intensive German service sector. A high rate of energy-augmenting technological progress in German industry, relative to the service sectors, would explain the observation discussed in Section 2 that the carbon and energy intensity declined faster in industry than in services.

The remainder of this section gives a brief, non-technical description of the model and an outline of the following steps of the analysis. A more detailed and technical statement of the model formulation is given in Appendix A.

3.1 General Features

As mentioned above, the model has 13 sectors and comprises Germany (GER) vs. the rest of the world (ROW). ROW is captured in terms of exogenous export prices and import quantities for the 13 goods. Other exogenous parameters which determine the dynamics of the model are rates of labor-augmenting technical progress and energy-augmenting technical progress in the various German production sectors. Multiplying these efficiency rates by the quantities of labor or energy input gives input in "effective units".

The output of German hard coal mining is treated as exogenous. Import demand for hard coal is the residual from total demand (at exogenous ROW prices) less administered (and subsidized) domestic supply. The electricity industry is split up into various power generation technologies, one of which is nuclear power. The demand for fossil-fuel based electricity is the residual from total electricity demand and exogenous (administered) nuclear power supply (for details on the electricity sub-model see Welsch and Ochsen, 2001).

All aggregator functions (quantity and price aggregators) are of the Leontief or CES type. A nested production structure allows the substitution elasticities to differ among sub-aggregates. Consumer demand is modeled by means of the Linear Expenditure System.

The time treatment in the model is recursively dynamic, the capital stock being predetermined.[15] This implies intrinsic dynamics which need to be accommodated when using the model to identify the impacts of changes in the exogenous drivers (see below).

3.2 Market Clearance and Macro Closure

For each good there are two classes of markets: An international market and a domestic market. In addition, there is one labor markets (labor being mobile across sectors) and one capital market.

In the international market the world trade volume of each good is determined as being the sum across GER and ROW of import demands. In the domestic goods market the total quantity equals the sum of intermediate demands, consumption demand, government demand, investment demand, and export demand. Employment is determined as being the sum of labor demand across sectors.

In contrast to the demand side, which is captured by ordinary demand functions, supply is modeled in an inverse fashion, via supply prices. Because, in any model solution, the demand functions are evaluated at those supply prices, the overall quantities demanded are equilibrium quantities.

The capital market is treated differently. Capital market equilibrium requires that the value of macroeconomic net investment equals private savings less the budget deficit less the balance of current account. This condition is a way of stating the equality of income generated and income used. The way in which a model achieves this accounting identity is usually referred to as the macro closure. In the current model we use the interest rate as the closure variable.

3.3 Foreign Trade and Final Demand

As is common in computable general equilibrium models, foreign trade modeling follows the approach of Armington (1969), according to which imported and domestically produced goods of the same kind are treated as incomplete substitutes. Thus the aggregate amount of each good (Armington good) is divided among imports and domestic production. For exports, there is a similar structure:

---

15 The capital stocks in the base year reflect, in part, the high energy prices up to the year 1985. The model thus integrates the lasting effect of previous induced substitution of capital for energy.

The world trade volume of each good is allocated to exports from ROW and German exports. The incomplete substitutability between goods of different origin is captured by CES aggregator functions. Accordingly, the demand for imports and domestic production as well as the demand for German exports are determined by CES demand functions.

Consumption expenditures of the representative household are a fraction of available labor and capital income. The savings ratio is assumed to depend on the interest rate, with a constant elasticity. Total consumption expenditures are then allocated to consumption of the different goods, utilizing the Linear Expenditure System.

Government expenditure in nominal terms is determined as a constant fraction of nominal GDP in the previous period, and real government expenditure is obtained from this by division by the price of the sector non-market services.

Nominal macroeconomic investment is the sum of the various sectors' investment in value terms. The price of investment goods is sector specific, since each sector's capital good is characterized by its specific composition in terms of sectors of origin. A sector's real investment is the difference between the capital stock considered optimal for the next period and that part of the current capital stock that will survive until the next period. Finally, investment demand for a sector's goods is the sum of sectoral investment requirements, weighted by the (constant) coefficients of the capital composition matrix.

3.4 Factor Demand

Factor demand is derived from a five-stage nested production function for each sector, which allows for a flexible treatment of substitution possibilities.

At the top level, output is linked to an aggregate of energy, capital and labor (EKL) and to the various non-energy intermediate inputs via constant input coefficients. The EKL aggregate is further broken down into labor and an energy-capital aggregate. This choice of disaggregation reflects our interpretation of capital as a collection of facilities for using energy (for empirical evidence backing this kind of disaggregation see Kemfert and Welsch, 2000).

Next, energy-capital is separated into capital and energy. Energy, in turn, is an aggregate of fossil energy and electricity. Finally, fossil energy is composed of the four different fossil fuels distinguished in the model.

Factor demand is derived from profit maximization subject to the production structure just outlined. At the top level of the production process inputs are related to output via fixed coefficients. At the subsequent levels there are CES demand functions. The sectoral capital stock in operation in any period is the capital stock considered optimal in the previous period. Thus, capital is a quasi-fixed factor. Energy, being a variable factor, adjusts to the predetermined capital stock.

3.5 Prices and Taxes

Prices represent the supply side of the model. The exogenous driving forces of the price model are the export prices of the rest of the world.

Due to the assumption of perfect intersectoral mobility of labor there is a uniform wage rate. In agreement with the practice of wage formation in Germany (see Carruth and Schnabel, 1993), the current wage equals the wage of the previous period times the increase in labor productivity and in the price level, modified by the ratio between actual employment and "normal" employment. As mentioned above, the interest rate is the closure variable of the model.

The prices and taxes discussed so far are the "fundamentals" of the price model. All other prices are derivatives thereof. These derivatives are obtained from price aggregator functions dual to the quantity aggregators (production functions) discussed above. These price aggregator functions may be interpreted as marginal cost functions, or inverted supply functions. The assumption of constant returns to scale, implicit in the CES specification of the production functions, implies constant marginal costs, which are at the same time average costs. Hence there are no quantities among the arguments of the price functions. Only for the price of the energy-capital aggregate there is a dependence on the level of the aggregate and on the capital stock. This is due to capital being a fixed factor (in the short term), which implies that marginal and average costs differ and neither of them is constant.

3.6 Procedure of Subsequent Analysis

The remainder of the paper seeks to explain the average annual rates of change of C, F, E, Q, X, Y, C_i / F_i, F_i / E_i, E_i / Q_i, Q_i / X_i, Q_i / Y_i, Q_i / Q, X_i / X and Y_i / Y, jointly referred to as structural change indicators. The actual values of these rates of change are collected in a 70-vector *OBS*. The subsequent analysis proceeds in three steps.

Step 1 examines the simulated partial and combined effects of observed changes in domestic energy policy and world energy prices (Section 4). The simulated combined effects on the structural change indicators of energy policy and energy price changes are collected in a 70-vector SIM_0.

Step 2 examines the simulated effects of 1% p.a. changes of total factor productivity and import demand in ROW and 1% p.a. autonomous energy efficiency improvements and labor efficiency improvements in Germany

(Section 5). The simulated effects on the structural change indicators of these changes are collected in eight vectors of the dimension 70 x 1, as follows:

SIM_1: Effect of 1% p.a. rise in ROW industry prices[16],

SIM_2: Effect of 1% p.a. rise in ROW service prices,

SIM_3: Effect of 1% p.a. rise in ROW industry demand,

SIM_4: Effect of 1% p.a. rise in ROW industry demand,

SIM_5: Effect of 1% p.a. rise in autonomous energy efficiency in German industry,

SIM_6: Effect of 1% p.a. rise in autonomous energy efficiency in German services,

SIM_7: Effect of 1% p.a. rise in autonomous labor efficiency in German industry,

SIM_8: Effect of 1% p.a. rise in autonomous labor efficiency in German services.

Step 3 (Section 6) seeks to explain the divergence between the actual structural change vector (*OBS*) and the simulated effects of actual energy policy and energy prices (SIM_0) in terms of the drivers considered in step 2. This is achieved by applying ordinary least squares to the equation

$$OBS - SIM_0 = \sum_j \beta_j \cdot SIM_j + u, \tag{4}$$

where the summation runs from 1 to 8 and u is an error term. The estimates of the coefficients β_j are the annual percentage changes of the drivers from step 2 which provide the best explanation of *OBS*, after SIM_0 is taken into account.[17]

Before carrying out this work plan, it may be useful to clarify the notions of energy efficiency improvements and labor efficiency improvements involved in SIM_5 through SIM_8. It is important to note that these simulations relate to (unobservable) parameters of the production function, not to observed energy or labor productivity. A stylized representation of the production function (which omits sector indices) may clarify this:

$$Q = const \cdot f(a \cdot E, b \cdot L, K), \tag{5}$$

---

16 Changes in ROW prices are used to proxy changes in total factor productivity.

17 Minford, Riley, and Nowell (1997) use a similar combination of simulation and regression analysis (steps 2 and 3) in order to identify the contribution of productivity and other shocks to changes in 21 key macro-sectoral indicators in a two-region world model. The methodology of the present paper departs from that approach by adding step 1.

where E, L, K denote energy, labor and capital and *const* captures the constant input coefficient which links output Q to the energy-capital-labor composite. The parameters a and b are the efficiency factors of energy and labor. The energy and labor efficiency improvements to be determined in step 3 are the rates of change of a and b, not the rates of change of the observables Q/E and Q/L. Estimates of the former may well differ from observations of the latter, as can be illustrated by a back-of-the-envelope calculation: In a CES production framework the energy demand function[18] implies the following relationship between the rates of change of energy input, output, the output price (P_Q), the energy price (P_E), and the energy efficiency factor (denoting rates of change by 'hats'):

$$\hat{E} = \sigma \cdot \left(\hat{p}_Q - \hat{p}_E \right) + \left(\sigma - 1 \right) \cdot \hat{a} + \hat{Q}. \tag{6}$$

From the last row of Table 2 we have an observed overall rate of energy efficiency improvement $\hat{Q} - \hat{E} = 3.5\,\%$ p.a., where $\hat{Q} = 4.0\,\%$ p.a. and $\hat{E} = 0.5\,\%$ p.a. Letting $\hat{p}_Q - \hat{p}_E = -5\,\%$ p.a. capture the decline of energy prices, and fixing σ (the elasticity of substitution between energy and other inputs) at 0.3 implies $\hat{a} = 7.1\,\%$. Fixing σ at 0.4 would imply $\hat{a} = 9.2\,\%$. This illustrates that observed efficiency rates may be a bad guide to estimating the structural parameters we seek to identify.

4 Effects of National Energy Policy and World Energy Prices

Table 4 shows the simulated effects on the structural change indicators from Section 2 of observed changes in German energy policy and world energy prices.[19] The table is restricted to the indicators carbon emissions (C), emission relevant energy use (F), total energy use (E), gross output (Q), total supply (X) and value added (Y), added across all non-government sectors. Table B1 in Appendix B supplies similar information for the other indicators.

---

18 The generic form of the CES input demand function is stated in equation (38) of Appendix A.

19 Note that in order to determine these effects it is necessary to eliminate the intrinsic model dynamics mentioned in Section 3. This is achieved as follows: (a) The model is run with all exogenous variables frozen at their base year value. (b) The model is run with the exogenous variable(s) under consideration varied. (c) the vector of indicator variables resulting from simulation (b) is regressed on the corresponding vector from simulation (a). The residual from this regression represents the pure effect of the exogenous variable(s). This method of recovering the pure effects of changes in exogenous variables is used throughout all the simulations performed.

The first row reproduces the observed percentage annual changes discussed in Section 2. The second row shows the simulated effect of the administered reduction of German hard coal mining from 96.8 to 82.5 million tons per year. It had almost no effect on carbon emissions and emission-relevant energy use, because in the simulation domestic hard coal was replaced by other fossil fuels, especially imported coal.[20] It did, however, have a negative effect on total energy use, because hard coal mining itself is relatively energy intensive. The effect on total output, total goods supply and value added[21] was slightly positive. Thus, the signs of the impacts are as one would expect them.

Table 4: Effects of German Energy Policy and World Energy Prices on Selected Indicators.

% / a	*C*	*F*	*E*	*Q*	*X*	*Y*
Observed	-0.18	0.28	0.51	4.01	4.24	3.65
Domestic hard coal	0.00	0.01	-0.17	0.05	0.06	0.05
Nuclear power	-0.58	-0.50	-0.28	-0.01	0.00	0.05
Price of hard coal	1.81	1.62	1.01	0.20	0.24	0.35
Price of oil	0.53	0.58	0.45	-0.02	0.01	0.05
Price of gas	0.03	0.06	0.04	-0.01	-0.01	-0.01
Combined effect	1.73	1.71	1.02	0.18	0.27	0.36
Sum of partial effects	1.79	1.77	1.05	0.21	0.30	0.49

In a more detailed perspective (see Table B1), there is a strong simulated increase in the ratio of the energy sector's emission-relevant energy to total energy, due to imported coal (which is cheaper than domestic coal) and other fuels being used more intensively. On the other hand, domestic output of the energy industry decreases in relation to both, its total supply and value added, and the energy industry's share in the aggregate output, total supply and value added is also reduced. On balance, these countervailing effects imply a non-existing effect on aggregate emissions.

20 It must be noted that this simulation is counterfactual in that it reflects only the partial variation of domestic hard coal. In reality, the use of hard coal in electricity generation was partly replaced by nuclear power. In our simulation framework, the expansion of nuclear power is examined in a separate partial variation.

21 Note that value added is not exactly GDP because we consider only the non-government sectors.

The third row in Table 4 gives the simulated effects of the expansion of nuclear power generation from 115 TWh to 135 TWh. As expected, the effect on emissions and emission-relevant energy use is negative. We can conclude that the expansion of nuclear power reduced the annual rate of change of CO_2 emissions by almost 0.6 percentage points. It also reduced total energy use.[22] The rate of growth of value added was slightly increased (0.05 percentage points). In a detailed view (Table B1) we find a reduction of the ratio of emission-relevant energy to total energy in all sectors, but especially in the energy industry. The energy intensity of output increases in all sectors, except for the energy industry, where it decreases strongly (nuclear fuels are not classified as energy). The effect on the energy industry dominates that on the non-energy sectors; hence the aggregate energy intensity of output is reduced.

The next three rows give the simulated effects of changes in world energy prices (coal: -7.3% per year; oil: -5.5% per year; gas: -2.3% per year). They each imply an increase in emissions and energy use, and more or less pronounced increases in value added growth (except for the gas price). The fall of the coal price has a particularly strong effect on emissions; it also contributed substantially to the growth of output, total supply and, especially, value added. The effect of the oil price decrease on emissions is smaller; yet it almost offsets the effect of nuclear expansion.

These aggregate effects become more transparent from a more detailed perspective (see Appendix, Table B1). The decline in the hard coal price entails an increase in the share of emission-relevant energy in total energy, especially in the energy sector, and an increasing energy intensity in all sectors. The ratio of output to total supply decreases in the energy sector. The same holds for the decrease of the oil and gas price (imported fuels are substituted for domestic ones). The ratio of energy output to value added in the energy industry rises. As regards the oil price crash, it leads to increasing output shares of transportation and services, whereas the shares of industry decrease.

The row labeled "combined effect" shows the effects of varying the aforementioned drivers simultaneously. The simulated overall effect of national energy policy and the decline of world energy prices is a substantial boost to emissions and emission relevant energy and a somewhat smaller one to total energy use. There is also a sizeable effect on macroeconomic growth, especially value added. Indeed, the simulated growth rate of value added is 0.36 percentage points higher than it would have been otherwise.

Regarding the structural impacts, there is a strong increase in the ratio of emission-relevant energy to total energy and in the energy intensity of output. There is also a decline in the aggregate output-supply ratio (Q/X). Thus, according to these simulations, the development of world energy prices has contributed in a

22 Nuclear fuels are not counted as energy.

visible manner to globalization of the West German economy. It also contributed to deindustrialization (viz. the output shares in Table B1): The substitution of imported fuels for domestic ones triggered a relative decline of the domestic energy industry. This had negative demand spill-overs to other industry sectors, especially the equipment goods industry (being a supplier of intermediate inputs for the energy industry), whereas construction, transport and services benefited from this development.

If one compares the combined effect to the sum of the various partial effects, one finds that the former is somewhat lower than the latter, which is natural in view of general equilibrium adjustment effects. The deviation is highest with respect to value added; it amounts to more than a quarter of the sum of partial effects.

An important message from this exercise is that the simulated effects of energy policy and energy price changes are broadly consistent with observed patterns of deindustrialization and globalization. However, the observed increase of Q / Y (fragmentation) is not captured by the prediction. Quantitatively, there is a tremendous bias between the observed indicators and the prediction based on energy policy and energy prices.[23] With respect to emissions the bias between predicted and observed change amounts to about 1.9 percentage points. With respect to the macroeconomic indicators it is even stronger.

These deviations are, of course, not surprising, because neither improvements in energy or labor productivity at home nor changing demand and supply patterns abroad have been accounted for in these exercises. Their effect is the subject of the remainder of the paper.

5 Partial Effects of Productivity and Demand Changes

The following exogenous drivers are considered, each differentiated by "industry" (agriculture, intermediate goods, equipment goods, consumption goods, construction) and "services" (transport, other private services):

- Growth of total factor productivity in the rest of the world (ROW),
- Growth of import demand of ROW,
- Energy-augmenting technological progress in Germany (so-called autonomous energy efficiency improvement),

---

23 This statement holds a fortiori if the full vector of indicators is considered. In fact, there is a negative correlation between observed and predicted indicators (correlation coefficient: -0.6208).

- Labour-augmenting technological progress in Germany.

Each of these eight parameters was set at 1% per year. Since production in ROW is not modeled explicitly, factor productivity changes in ROW are captured by changes in output prices, where a price decrease corresponds to a productivity increase[24].

The simulated partial effects on our macro indicators are shown in Table 5. The effects on the other indicators are given in Table B2 in Appendix B. These effects can be interpreted as elasticities.

Table 5: Elasticity of Selected Indicators with Respect to Exogenous Drivers.

	C	*F*	*E*	*Q*	*X*	*Y*
ROW industry prices	.202	.201	.135	-.001	-.040	.044
ROW service prices	-.005	-.005	-.003	-.002	-.006	-.009
ROW industry demand	.074	.067	.039	.000	.050	.035
ROW service demand	-.002	-.003	-.001	-.003	-.004	-.020
Energy efficiency in industry	-.347	-.337	-.237	.047	.041	.125
Energy efficiency in service	-.087	-.082	-.060	.041	.032	.035
Labour efficiency in industry	.165	.153	.115	.184	.168	.121
Labour efficiency in service	.127	.118	.090	.169	.155	.055

It should be noted that the two first rows of Table 5 (and two first columns of Table B2) give the simulated effects of increases in ROW prices. An interesting result in Table 5 is the effect of ROW industry prices on German emissions. It suggests that a 1% increase in ROW's industry prices would have increased German production-related carbon emissions by 0.2%. Likewise, a corresponding drop in ROW industry prices would have reduced emissions by 0.2%; this is a considerable figure, especially if viewed in comparison with the effect of energy efficiency improvements in German industry. By contrast, a hypothetical 1% increase in ROW's service prices would have had a negative (but negligible) effect on German emissions. The reason for both such effects are the induced structural changes to the advantage/disadvantage of German industry (based on comparative advantage), relative to German services (cf. Table B2).

---

24 Aggregate German output is the numeraire in the model, i.e. the aggregate price of German output is fixed.

Similar reasoning applies to changes in demand: An increase in ROW demand for industrial goods boosts German emissions, whereas an increase in service demand reduces them.

The effects of domestic productivity improvements on emissions are straightforward: Energy efficiency improvements have a negative effect on emissions, whereas labor efficiency improvements raise emissions. Both of these effects are more pronounced if the productivity improvement takes place in industry than if taking place in services.

The effects of these productivity and demand shocks on emission-relevant energy use and energy use in general are qualitatively the same as those on emissions, but less pronounced.

Regarding domestic output, total supply, and value added, it is straightforward that they all are positively affected by domestic productivity improvements, both of energy and labor. They are also positively affected by rising demand for industry goods. Rising productivity in ROW's industry increases total supply in Germany (because of cheaper imports), but reduces value added (because of increased competition from abroad). The effects of ROW service productivity and service demand on output, total supply and value added are very small.

The impacts of ROW productivity and demand shocks on the sectoral structure are as expected (see Table B2). Especially, an increase in ROW industry productivity entails deindustrialization in Germany, whereas an increase in ROW service productivity would do the opposite.

6 Measuring the Size and Contribution of the Exogenous Drivers

Qualitatively, the effects of the drivers considered are as one would expect them to be. The aim of this section is to measure their size and their contribution to the observed development of the structural change indicators (*OBS*), taking into account the effect of energy prices and energy policy (SIM_0). As explained in Section 3, this is achieved by estimating equation (4).

The estimation results are displayed in Table 6. The coefficients give the estimated annual percentage growth rates of the exogenous drivers.

Regression 1 comprises the full list of explanatory variables. We find that all signs are as expected and that 6 out of 8 coefficients are statistically significant at the usual confidence levels. The coefficient on ROW service prices is dubious in terms of both its magnitude and t-statistic.[25] All other results appear to be plausible. Especially, we find that service demand has grown stronger than industry demand and that energy efficiency improvements were more pronounced in industry than in the service sector, whereas the opposite holds for labor efficiency improvements.

Table 6: Estimated Growth Rates of Exogenous Drivers (t-statistic in parenthesis).

	Regression 1	Regression 2	Regression 3
ROW industry prices	-1.64 (-1.87)	-1.61 (-1.85)	-1.69 (-1.91)
ROW service prices	13.12 (0.35)	/	50.95 (3.01)
ROW industry demand	4.01 (1.80)	3.91 (1.78)	4.65 (2.12)
ROW service demand	6.38 (1.13)	8.14 (3.23)	/
Energy efficiency in industry	6.97 (7.16)	6.99 (7.23)	6.94 (7.11)
Energy efficiency in service	5.00 (2.70)	5.11 (2.83)	4.97 (2.68)
Labour efficiency in industry	2.98 (3.77)	2.91 (3.85)	3.17 (4.09)
Labour efficiency in service	8.47 (4.93)	8.36 (4.97)	8.28 (4.83)
DW	1.93	1.95	1.87
R^2	.670	.668	.665
R^2 adjusted	.633	.636	.633

When we consider Regression 2, in which the regressor "ROW service prices" is omitted, we find that all remaining coefficients are significant at the 95% confidence level. Except for the coefficient on ROW service demand, all parameters are quite robust in comparison with Regression 1. The coefficient on ROW service demand is now larger in magnitude and highly significant.

The results for Regression 1 and Regression 2 suggest that it is hard to separate the effect of a change in ROW service prices (productivity) from that of a change in ROW service demand. If we omit ROW service demand instead of ROW service prices, an extremely unplausible estimate for the ROW service price is obtained (see Regression 3), whereas most other parameters are almost unaffected by this variation.

25 In their "North-South" framework, Minford, Riley, and Lowell (1997) find the "Southern" service productivity quite insignificant as an explanatory variable of structural changes in the world economy.

Overall, we find that the three regressions are in very good agreement with each other, except for the role attributed to ROW service prices and demand. With respect to the latter issue, Regression 2 seems to be the most plausible one, because it highlights the importance of demand growth for the growth of the German service sector, rather than stipulating an extreme divergence between the service sector productivity in Germany and abroad.

In assessing the overall quality of Regression 2, it should be noted that the dependent variable (the 70-vector $OBS\text{-}SIM_0$) exhibits quite a large variance (as it comprises mainly rates of change of ratio variables). The descriptive statistics of the dependent variable are as follows: minimum = -7.086, maximum = 3.970, mean = -0.7376, standard error = 2.1269. In spite of the rather erratic behavior of the dependent variable, the seven explanatory variables which were found significant explain almost 67% (R^2) of the variance of the dependent variable.[26]

The main findings based on Regression 2 are as follows.

- ROW industry prices have decreased by 1.6% per year, relative to the overall German output price level.
- Commodity demand of ROW has increased by only half the rate of service demand (3.9 vs. 8.1%).
- Energy efficiency in German industry has grown by almost 7%, while growing by 5% in the service sector.
- Labour productivity in German industry has grown by about 3%, while growing by more than 8% in the service sector.

In interpreting these results, it should be recalled that the efficiency rates estimated refer to structural parameters of the production technology, which may differ substantially from observed efficiency rates. As illustrated in Section 3, the autonomous energy efficiency improvements required to reconcile observed energy efficiency rates with falling energy prices may be in the range of 7 to 9%, if changes in the sectoral structure of the economy are disregarded. Therefore, our estimates (5 to 7% in the presence of structural change) should not be dismissed as implausible.

---

26 The goodness of fit varies substantially across the various components of the vector of dependent variables. For instance, while the fit is relatively poor for total energy input E (-0.51 ($OBS\text{-}SIM_0$) vs. -0.95 (predicted)), it is very good with respect to emission-relevant energy input F (-1.43 vs. -1.42). A similar divergence exists with respect to the accuracy with which sector-specific energy efficiency improvements are captured (for instance, consumption goods (-3.2 vs. -4.9) on the one hand, and agriculture (-4.6 vs. -4.4) on the other. Because the way in which the goodness of fit varies shows no evident pattern, we restrict ourselves to a purely statistical assessment of the quality of the regression, as provided in the main text.

The regression results allow an assessment of the relative importance of the various explanatory variables. Table 7 shows their percentage contribution to the overall explanatory power of the regressions. The following can be concluded:

- About 7% of the variation in our indicator variables not accounted for by energy policy and energy prices are due to the (relative) decline in ROW industry prices.
- About 18 to 20% are due to the pattern of ROW demand growth (based on Regressions 1 and 2).
- About 33 to 34% are due to the pattern of domestic energy efficiency improvement.
- About 38 to 39% are due to the pattern of domestic labor productivity improvement

If we disregard the energy efficiency issue we find that domestic (labor) productivity patterns (technology) are somewhat more important as drivers of structural change (about 40%) than trade-related patterns induced in ROW (about 28%).

Table 7: Percentage Contribution of Explanatory Variables.

	Regression 1	Regression 2	Regression 3
ROW industry prices	7.1	7.0	7.3
ROW service prices	3.2	/	12.6
ROW industry demand	6.8	6.6	7.8
ROW service demand	10.7	13.8	/
Energy efficiency in industry	23.7	24.0	23.5
Energy efficiency in service	9.9	10.2	9.8
Labour efficiency in industry	14.9	14.7	15.8
Labour efficiency in service	23.7	23.6	23.2

It should be noted that the results discussed in this section are based on the simulated effects of each of the driving forces varied separately. As to the validity of these results, a useful check consists in using the estimated growth rates as a joint simulation input and to check the resulting simulated indicator changes for consistency with the corresponding observations. Such an approach allows to examine whether the estimated growth rates may be biased because of neglected interactions.

Table 8 displays the coefficient of correlation between the vector of observed indicator changes and simulated changes arising from the estimated growth rates from Table 6 (and from observed changes in domestic energy policy and world energy prices). For all regressions considered the correlation is not statistically different from one. There is thus no indication that our results are infected by some sort of simultaneity bias. Moreover, comparing the various regressions suggests that Regression 2 yields the best agreement with the vector of observations.

Table 8: Coefficient of Correlation Between Observed and Simulated Changes (standard error in parenthesis).

Regression 1	Regression 2	Regression 3
0.9320	0.9342	0.8911
(0.1055)	(0.1052)	(0.1095)

7 Conclusions

This paper has examined the determinants of the decline of West German production-related carbon intensity in the second half of the 1980s. The decline is astonishing in view of the substantial drop of world energy prices over that period; it has triggered a substantial amount of research which aims at decomposing changes in aggregate carbon or energy intensity into a measure of energy intensity change at the sectoral level and a measure of change in the sectoral structure of the economy.

The present paper departs from this literature by attempting to identify what drives the structural changes that have contributed to the decrease of carbon intensity. The approach pursued is to use a computable general equilibrium model of the West German economy to determine the simulated effects of changes of world energy prices, domestic energy policy, productivity growth in Germany and abroad, and changing patterns of ROW demand. Weights on those driving forces that are unobservable are selected by least squares.

It is found that the expansion of nuclear power over the period considered is just sufficient to offset – in terms of its effect on carbon emissions – the drop of oil and gas prices, but not the additional effect of falling coal prices. Overall, the effects of energy prices and energy policy are qualitatively consistent with observed patterns of globalization and deindustrialization in the West German economy, but inconsistent with the phenomenon of fragmentation that is also present in the data.

The deviation, both qualitatively and quantitatively, between actual changes in energy use and economic structure and changes predicted by energy prices and energy policy are attributed to changing technological patterns in Germany and abroad, and changing patterns of ROW demand. It is found, for instance, that the effect on German production-related carbon emissions of a 1% increase in industry productivity abroad is more than half of the effect of a 1% autonomous energy efficiency increase in German industry. Overall, the contribution of ROW productivity and demand patterns to emission-relevant structural change not accounted for by energy prices and energy policy is just under 30%. The remainder is split almost equally among patterns of domestic autonomous energy efficiency improvement and domestic labor efficiency patterns.

These findings suggest that a substantial fraction of emission-relevant structural change in West Germany in 1985-1990 is trade-induced. Given that the speed of "globalization" has been increasing over the last decade, it can be expected that the role of trade as a determinant of carbon emissions has become even more important since then. Unfortunately, the structural breaks after the unification make it very difficult to examine these issues in a conventional economic framework. However, as more recent data become available that are less infected by these turbulences, it may be worthwhile reconsidering the link between technology, trade and carbon emissions in future research.

References

Ang, B.W. (1994), Decomposition of Industrial Energy Consumption: The Energy Intensity Approach, *Energy Economics* 16, 83-92.

Boyd, G., J.F. McDonald, M. Ross, and D.A. Hanson (1987), Separating the Changing Composition of U.S. Manufacturing Production from Energy Efficiency Improvements, *The Energy Journal* 8, 77-95.

Berman, E., J. Bound, and S. Machin (1998), Implications of Skill-Biased Technological Change: International Evidence, *Quarterly Journal of Economics* CXIII, 1245-1279.

Burda, M. and B. Dluhosch (2000), *Fragmentation, Globalization and Labour Markets*, CESifo Working Paper No. 352, Munich.

Burniaux, J.-M., J.P. Martin, G. Nicoletti, and J. Oliveira-Martins (1992), *GREEN – A Multi-Sector, Multi-Region General Equilibrium Model for Quantifying the Costs of Curbing CO_2 Emissions: A Technical Manual*, OECD Working Paper No. 116, Paris.

Carruth, A. and C. Schnabel (1993), The Determination of Contract Wages in West Germany, *Scandinavian Journal of Economics* 95, 297-310.

De Bruyn, S. (2000), *Economic Growth and the Environment*, Dordrecht.

EUROSTAT (1992), *National Accounts ESVG: Input-Output Tables 1985*, Luxembourg.

IEA (1995), *Energy Prices and Taxes*, Paris.

Jones, R.W. and H. Kierzkowski (1990), The Role of Services in Production and International Trade: A Theoretical Framework, in: Jones, R.W. and A.O. Krueger (Eds.), *The Political Economy of International Trade: Essays in Honor of Robert E. Baldwin*, Cambridge, Oxford, 31-48.

Kemfert, C. and H. Welsch (2000), Energy-Capital-Labour Substitution and the Economic Effects of CO_2 Abatement: Evidence for Germany, *Journal of Policy Modeling* 22, 641-660.

Mayer, H. and C. Stahmer (1989), *Energy Consumption and Sulphur Dioxide Emissions in the Federal Republic of Germany*, mimeo, Wiesbaden.

Minford, P., J. Riley, and E. Nowell (1997), Trade, Technology and Labour Markets in the World Economy, 1970-90: A Computable General Equilibrium Analysis, *Journal of Development Studies* 34, 1-34.

Saeger, S.S. (1997), Globalization and Deindustrialization: Myth and Reality in the OECD, *Weltwirtschaftliches Archiv* 133, 549-608.

Statistisches Bundesamt (1999), *Fachserie 18: Volkswirtschaftliche Gesamtrechnungen*, www.zr.statistik-bund.de/zeitreih.

Unander, F., S. Karbuz, L. Schipper, M. Khrushch, and M. Ting (1999), Manufacturing Energy Use in OECD Countries: Decomposition of Long-Term Trends, *Energy Policy* 27, 769-778.

Welsch, H. and C. Ochsen (2001), Dismantling of Nuclear Power in Germany: Sectoral and Macroeconomic Effects, *Energy Policy* 29, 279-289.

WTO (2000), *International Trade Statistics 2000*, Geneva.

Appendix A: Formal Statement of the Model

This Appendix provides a technical statement of the model described in Section 3 of the main text. In Sections A1-A4, demand functions and price functions are indicated simply by $f(.)$. The precise functional form follows from the functional specification of the aggregator functions and is discussed in Section A5. The country index GER is omitted unless necessary. Notation is introduced in order of appearance.

A1 Market Clearance and Macro Closure

World Trade Market

Notation: EX_i^W : world trade volume of good i; IM_i^{GER}; IM_i^{ROW} : import demand of Germany or rest of world.

$$EX_i^W = IM_i^{GER} + IM_i^{ROW} \qquad \textit{where } IM_i^{ROW} = \textit{exogenous}. \tag{1}$$

Domestic Goods Markets

Notation: X_i: total supply; XX_{ij}: intermediate demand; XC_i: consumption demand; XG_i: government demand; XEX_i: export demand; S: set of sectors.

$$X_i = \sum_{j \in S} XX_{ij} + XC_i + XG_i + XI_i + XEX_i \tag{2}$$

Labour Market

Notation: LAB: employment; L_j: sectoral labor demand.

$$LAB = \sum_{j \in S} L_j \tag{3}$$

Capital Market and Macro Closure

Notation: *IN*: investment; *DEPN*: depreciation; *YVN*: available income; *GN*: government expenditure; *TN* total tax revenue; *EXN* exports; *IMN*: imports; *s*: savings ratio, Symbols containing "*N*" refer to nominal variables.

$$IN - DEPN = s \cdot YVN - (GN - TN) - (EXN - IMN) \tag{4}$$

Price Normalization

Notation: *PC*: consumer price index.

$$PC \equiv 1 \qquad (numeraire) \tag{5}$$

A2 Foreign Trade and Final Demand

Imports

Notation: PIM_i: import price; IM_i: import volume; PX_i: price of total supply (Armington good); Q_i: domestic output; PQ_i: price of domestic output.

$$IMN = \sum_i PIM_i \cdot IM_i \tag{6}$$

$$IM_i = f(X_i, PX_i, PIM_i) \tag{7}$$

$$Q_i = f(X_i, PX_i, PQ_i) \tag{8}$$

Exports

Notation: PEX_i^W; PEX_i : export price of world, GER.

$$EXN = \sum_i PX_i \cdot XEX_i \tag{9}$$

$$XEX_i = f(EX_i^W, PEX_i^W, PEX_i) \tag{10}$$

Consumption

Notation: *CN*: consumption expenditures; *WAGE*: labor income; *NOS*: net operating surplus; *z*: real interest rate; a, b, γ_i: parameters

$$CN = (1-s) \cdot YVN = (1-s) \cdot (WAGE \; + NOS) \tag{11}$$

$$s = a \cdot z^b \tag{12}$$

$$XC_i = \overline{XC_i} + \gamma_i \frac{CN - \sum_j \cdot PX_j \cdot \overline{XC_j}}{PX_i} \tag{13}$$

Government Expenditure

Notation: YN: nominal GDP; PX_{NMSV}: price of non-market services; s_{GY}: parameter.

$$GN = s_{GY} \cdot YN_{-1} \tag{14}$$

$$XG = GN \,/\, PX_{NMSV} \tag{15}$$

Investment

Notation: I_j: sectoral investment; PI_j: sectoral purchase price of capital; K_j: capital stock at beginning of period; K_j^*: capital stock planned for next period; δ_j: depreciation rate; κ_{ij}: coefficient of capital composition matrix.

$$IN = \sum_{j \in S} PI_j \cdot I_j \tag{16}$$

$$I_j = K_j^* - (1-\delta_j) K_j \tag{17}$$

$$XI_i = \sum_{j \in S} \kappa_{ij} \cdot I_j \tag{18}$$

A3 Factor Demand

Notation: EKL_j: energy-capital-labor aggregate; L_j: labor; $PEKL_j$: price of energy-capital-labor aggregate; PL: wage rate; EK_j: energy-capital aggregate; PEK: price of energy-capital aggregate; K_j: capital stock; E_j: energy aggregate; $XXEC_{EL,j}$: electricity input; PE_j: price of energy aggregate; PEL_j: price of electricity; $XXEC_{F,j}$: fossil fuel aggregate; PF_j: price of fossil fuel aggregate; $XXEC_{F1,j}$... $XXEC_{F4,j}$: individual fossil fuel inputs; $PF1$,... $PF4$: price of individual fossil fuel inputs. Planned capital stock K^* is derived in Section A7.

$$XX_{ij} = f(Q_j) \tag{19}$$

$$EKL_j = f(Q_j) \tag{20}$$

$$L_j = f(EKL_j, PEKL_j, PL) \tag{21}$$

$$EK_j = f(EKL_j, PEKL_j, PEK_j) \tag{22}$$

$$K_j = K^*_{j,-1} \tag{23}$$

$$E_j = f(EK_j, K_j) \tag{24}$$

$$XXEC_{EL,j} = f(E_j, PE_j, PEL_j) \tag{25}$$

$$XXEC_{F,j} = f(E_j, PE_j, PF_j) \tag{26}$$

$$XXEC_{F1,j} = f(XXEC_{F,j}, PF_j, PF1_j) \tag{27a}$$

$$\vdots$$

$$XXEC_{F4,j} = f(XXEC_{F,j}, PF_j, PF4_j) \tag{27d}$$

A4 Prices and Taxes

Prices of Primary Inputs

Notation: Π: rate of nominal productivity growth.

$$PL = \Pi \cdot PL_{-1} \cdot \left(\frac{LAB}{\overline{LAB}} \right)^{\alpha} \tag{28}$$

$$z \leftarrow \text{ model closure} \tag{29}$$

Price Aggregates

$$PEX_i^W = f(PEX_i^{ROW}, PX_i) = PIM_i \tag{30}$$

$$PX_i = f(PQ_i, PIM_i) \tag{31}$$

$$PQ_i = f(PX_1, ..., PX_{13}, PEKL_i) \tag{32}$$

$$PEKL_i = f(PL, PEK_i) \tag{33}$$

$$PEK_i = f(EK_i, K_i, PE_i) \tag{34}$$

$$PE_i = f(PEL_i, PF_i) \tag{35}$$

$$PF_i = f(PF1_i, \ldots, PF4_i) \tag{36}$$

A5 Functional Forms

The quantity aggregator functions in the foreign trade model are specified as CES functions. The production functions are of a nested Leontief/CES form.

Cost minimization on the basis of these quantity aggregators yields Hicksian demand functions. Prices are obtained from the corresponding price aggregators.

Consider the generic CES quantity aggregator function (production function)

$$X = \left[\sum_{i=1}^{n} d_i (g_i X_i)^{\frac{\sigma-1}{\sigma}} \right]^{\frac{\sigma}{\sigma-1}} \tag{37}$$

where

X_i = input of good i in quantity units,

$g_i X_i$ = input of good i in effective units,

σ = elasticity of substitution,

d_i = distribution parameter.

If $g_i \geq 1$ increases in time, this is referred to as factor-augmenting technical progress. If, conversely, $g_i \equiv 1$; the technology is stationary.

If all inputs are variable, the demand functions corresponding to (37) are

$$X_i = d_i^{\sigma} \left(\frac{PX}{PX_i} \right)^{\sigma} g_i^{\sigma-1} X, \tag{38}$$

and the price aggregator is of the form

$$PX = \left[\sum_i d_i^\sigma \left(\frac{PX_i}{g_i} \right)^{1-\sigma} \right]^{\frac{1}{1-\sigma}} \tag{39}$$

These formulas are valid for the foreign trade model and the production model, except for the top level and the EK level.

At the top level, demand is obtained from fixed input-output coefficients, and the price aggregator is the weighted average of input prices.

At the EK level, there are only two inputs to the quantity aggregator (37), one of which (capital) is fixed. The demand for the other factor (energy) is then obtained by rearranging the equation (37).

A6 Parameterization

The model is calibrated against input-output tables (EUROSTAT, 1992) for 1985. The key exogenous drivers (i.e., world energy prices, nuclear capacity, renewable energy, domestic hard coal mining) 1985-1990 have been set according to their historical development. The elasticity assumptions used are given in Table A1. The default value for the Armington elasticities of the traditional tradable goods is 3.0. Lower values (typically 0.1 to 0.3) are applied in the case of services (see Burniaux et al., 1992, for a review of estimates of Armington elasticities).

Table A1: Elasticity Parameters.

Armington elasticities:	
Export of GER vs. export of ROW	0.1-3.0
Domestic products vs. Imports	0.1-3.0
Substitution elasticities in production:	
Energy/capital vs. Labour	0.6
Energy vs. Capital	0.3-0.7
Electricity vs. Fuel	0.6
Fuel vs. Fuel	0.8
Reaction elasticities:	
Savings ratio w.r.t. interest rate	0.0
Wage rate w.r.t. employment	0.5

Appendix B: Additional Tables

Table B1: Effects of German Energy Policy and World Energy Prices.

1 = Observed
2 = Domestic Hard Coal
3 = Nuclear Power
4 = Price of Hard Coal
5 = Price of Oil
6 = Price of Gas
7 = Combined Effect

		1	2	3	4	5	6	7
Ci / Fi	Agriculture	-1.180	0	0	0.029	0.020	-0.007	0.034
	Energy	-0.421	0.006	-0.022	0.013	-0.014	-0.014	-0.024
	Intermediates	-0.713	-0.004	-0.006	0.408	-0.040	-0.058	0.289
	Equipment	-0.715	-0.006	-0.010	0.161	0.166	-0.064	0.231
	Consumption	-0.722	-0.002	-0.007	0.156	0.129	-0.063	0.192
	Construction	-0.230	-0.003	-0.003	0.023	-0.005	-0.006	0.016
	Transport	-0.020	-0.001	-0.001	0.001	0.002	-0.002	0.003
	Services	-0.093	-0.002	-0.005	0.003	0.125	-0.030	0.093
Fi / Ei	Agriculture	-0.460	-0.002	-0.044	-0.083	0.466	0	0.358
	Energy	0.174	0.228	-0.822	1.158	-0.240	0.008	0.372
	Intermediates	-1.181	0	-0.017	0.178	0.053	0.011	0.210
	Equipment	-2.348	0.003	-0.048	0.012	0.361	0.058	0.377
	Consumption	-0.847	0	-0.042	0.010	0.290	0.044	0.307
	Construction	-0.072	0.003	-0.003	-0.011	0.075	0.004	0.058
	Transport	-0.193	-0.002	-0.018	-0.050	0.220	-0.002	0.158
	Services	-0.054	-0.001	-0.060	-0.142	0.618	0.019	0.463
Ei / Qi	Agriculture	-3.090	0.018	0.017	0.057	1.325	0.011	1.493
	Energy	1.410	0.037	-0.524	0.615	-0.034	0.015	0.108
	Intermediates	-2.867	-0.007	0.065	1.901	0.221	0.079	2.111
	Equipment	-5.510	-0.009	0.125	0.616	0.743	0.133	1.576
	Consumption	-1.780	-0.003	0.099	0.561	0.672	0.132	1.410
	Construction	-3.450	-0.017	0.050	0.216	1.490	0.003	1.835
	Transport	-1.367	-0.015	0.056	0.131	1.559	0.001	1.727
	Services	-2.230	-0.052	0.194	0.492	1.962	0.101	2.549

Table B1: continued.

Qi /Xi	Agriculture	-0.220	0.005	-0.014	0.143	0.193	-0.003	0.190
	Energy	-0.310	-0.331	0.068	-0.387	-0.311	-0.020	-0.936
	Intermediates	-0.630	0.022	-0.032	0.028	0.013	0.004	0.045
	Equipment	-0.770	0.024	-0.046	-0.060	-0.042	-0.003	-0.110
	Consumption	-0.670	0.025	-0.040	-0.042	-0.046	-0.005	-0.093
	Construction	0	0	0	-0.001	0	0	-0.001
	Transport	0.010	0.005	-0.006	-0.011	-0.004	-0.001	-0.012
	Services	0.030	0.002	-0.003	-0.006	-0.004	0	-0.009
Qi / Yi	Agriculture	-0.690	-0.059	0.076	0.480	0.421	0.013	0.632
	Energy	0.420	-0.179	0.039	1.311	0.500	0.020	1.351
	Intermediates	-0.860	-0.014	0.072	0.108	0.118	-0.002	0.294
	Equipment	2.310	0	0.019	0.072	0.098	0.002	0.233
	Consumption	1.660	0.018	0.018	0.068	-0.029	-0.010	0.112
	Construction	1.620	-0.012	-0.167	-0.510	0.076	0.028	0.262
	Transport	0.880	0.056	-0.009	0.054	-0.111	-0.016	0.059
	Services	-0.520	0.085	-0.110	-0.248	-0.337	-0.036	-0.448
Qi / Q	Agriculture	-1.970	0.006	-0.027	0.046	0.266	0.002	0.151
	Energy	-4.860	-0.418	0.037	0.023	0.287	0.019	-0.041
	Intermediates	-1.490	0.056	-0.079	-0.041	-0.041	0.011	-0.063
	Equipment	1.720	0.077	-0.072	-0.179	-0.275	-0.022	-0.453
	Consumption	-1.240	0.023	-0.048	-0.160	-0.015	0.002	-0.172
	Construction	-0.480	0.015	0.132	0.310	-0.271	-0.021	0.055
	Transport	1.520	0.007	-0.001	-0.055	0.068	0.004	0.046
	Services	0.550	0.009	0.057	0.123	0.148	0.009	0.339

Table B1: continued.

Xi / X	Agriculture	-1.980	-0.008	-0.023	-0.134	0.036	0.003	-0.131
	Energy	-4.770	-0.097	-0.040	0.375	0.561	0.036	0.805
	Intermediates	-1.100	0.023	-0.058	-0.106	-0.092	0.003	-0.199
	Equipment	2.250	0.043	-0.036	-0.156	-0.270	-0.023	-0.434
	Consumption	-0.810	-0.013	-0.019	-0.154	-0.008	0.003	-0.170
	Construction	-0.700	0.005	0.123	0.275	-0.308	-0.025	-0.036
	Transport	1.290	-0.008	-0.005	-0.081	0.035	0.003	-0.035
	Services	0.300	-0.002	0.049	0.092	0.115	0.006	0.255
Yi / Y	Agriculture	-0.970	0.063	-0.176	-0.589	-0.232	-0.016	-0.661
	Energy	-4.940	-0.241	-0.073	-1.442	-0.289	-0.007	-1.570
	Intermediates	-0.320	0.068	-0.222	-0.302	-0.236	0.007	-0.536
	Equipment	-0.150	0.075	-0.162	-0.405	-0.450	-0.030	-0.864
	Consumption	-2.500	0.003	-0.136	-0.381	-0.063	0.007	-0.462
	Construction	-1.700	0.024	0.229	0.666	-0.425	-0.055	-0.385
	Transport	1.030	-0.051	-0.062	-0.263	0.103	0.015	-0.192
	Services	1.410	-0.079	0.096	0.217	0.408	0.040	0.608

Table B2: Elasticity of Indicators with Respect to Exogenous Drivers.

1 = ROW Industry Prices
2 = ROW Service Prices
3 = ROW Industry Demand
4 = ROW Service Demand
5 = Energy Efficiency in Industry
6 = Energy Efficiency in Service
7 = Labour Efficiency in Industry
8 = Labour Efficiency in Service

		1	2	3	4	5	6	7	8
Ci / Fi	Agriculture	0.003	0	0.001	0	-0.003	-0.001	0	0.001
	Energy	0	0	0.003	0.001	-0.009	-0.001	0.004	0.005
	Intermediates	0.004	0	0.004	-0.002	-0.014	-0.005	0.007	0.006
	Equipment	0.006	-0.001	0.004	-0.002	-0.017	-0.008	0.007	0.007
	Consumption	0.003	0	0.004	-0.001	-0.016	-0.004	0.005	0.008
	Construction	0	-0.001	0.001	-0.003	-0.002	-0.003	0.001	0
	Transport	0.001	0	-0.001	0.001	0	-0.001	0	0
	Services	0.004	0	0.003	0	-0.009	-0.002	0.004	0.004
Fi / Ei	Agriculture	0.036	-0.002	0.010	-0.002	-0.004	-0.003	-0.005	0.001
	Energy	0.071	-0.003	0.069	0.011	-0.148	-0.060	0.144	0.070
	Intermediates	0.017	0	0.004	0	0.002	-0.002	-0.003	-0.001
	Equipment	0.048	0.001	0.013	0.001	0.010	-0.002	-0.013	-0.005
	Consumption	0.035	-0.001	0.006	-0.001	0.007	-0.003	-0.015	-0.009
	Construction	0.006	-0.001	0.002	0.003	0.001	0	0.001	0.002
	Transport	0.016	-0.001	0.004	-0.002	-0.004	-0.003	-0.003	-0.001
	Services	0.049	-0.001	0.013	-0.001	0	-0.009	-0.010	-0.002
Ei / Qi	Agriculture	0.263	-0.004	-0.030	-0.027	-0.574	0.014	-0.069	0.063
	Energy	0.060	-0.001	0.041	0.011	-0.064	-0.026	0.055	0.031
	Intermediates	0.136	-0.002	0.015	-0.019	-0.587	0.006	-0.209	-0.017
	Equipment	0.130	-0.002	-0.005	-0.011	-0.505	0.020	-0.312	-0.060
	Consumption	0.113	-0.001	0.003	-0.015	-0.537	0.017	-0.251	-0.029
	Construction	0.178	0	0.033	-0.011	-0.623	0.011	-0.254	-0.029
	Transport	0.116	0	0.017	0.008	0.046	-0.646	0.128	-0.313
	Services	-0.040	-0.012	0.021	-0.042	0.106	-0.370	0.416	-0.199

Table B2: continued.

Qi / Xi	Agriculture	0.080	0.002	-0.100	-0.009	0.009	0	-0.018	-0.053
	Energy	-0.043	0.002	-0.014	0.003	0.037	0.017	-0.025	-0.012
	Intermediates	0.113	0	-0.099	-0.006	0.028	0.014	0.032	0.020
	Equipment	0.110	0.002	-0.083	0.003	-0.019	0.012	0.064	0.018
	Consumption	0.095	0.001	-0.082	-0.002	-0.014	0.013	0.041	0.010
	Construction	0	0	0	0	0	0.001	0.001	0
	Transport	-0.020	0.021	-0.006	-0.001	-0.003	0.006	-0.022	0.028
	Services	-0.005	0.008	-0.002	0.001	-0.001	0.001	-0.009	0.005
Qi / Yi	Agriculture	-0.759	0.010	0.095	0.068	-0.020	-0.022	0.153	-0.233
	Energy	-0.413	-0.020	-0.105	-0.115	-0.145	-0.116	0.201	0.092
	Intermediates	-0.218	0.005	-0.097	0.026	0.021	0.016	0.606	-0.133
	Equipment	-0.050	0.008	-0.036	0.041	0.031	0.013	0.711	-0.085
	Consumption	-0.166	0.008	-0.037	0.034	0.006	0.020	0.567	-0.117
	Construction	0.216	0.018	0.151	0.118	-0.176	0.085	0.641	-0.120
	Transport	-0.006	-0.030	-0.047	-0.230	0.035	0.020	-0.221	0.617
	Services	0.072	0.017	-0.055	0.052	-0.050	0.042	-0.522	0.297
Qi / Q	Agriculture	0.137	0.001	-0.065	-0.008	-0.007	-0.002	-0.069	-0.111
	Energy	0.039	-0.001	0.008	0.002	-0.149	-0.079	-0.109	-0.110
	Intermediates	0.190	-0.002	0.051	-0.025	0.056	0.017	0.031	-0.010
	Equipment	0.017	-0.001	0.038	-0.021	-0.071	0.021	0.128	-0.037
	Consumption	0.097	-0.001	0.002	0	-0.036	0.004	-0.009	-0.055
	Construction	-0.412	-0.013	-0.211	-0.102	0.079	-0.103	0.057	-0.031
	Transport	-0.062	0.030	-0.011	0.205	-0.006	0.019	-0.139	0.078
	Services	-0.044	0.002	0.009	0.020	0.054	0.017	-0.049	0.078

Table B2: continued.

Xi / X	Agriculture	0.096	0.002	-0.014	0.001	-0.010	0.006	-0.034	-0.044
	Energy	0.120	0.001	-0.028	0	-0.180	-0.086	-0.067	-0.086
	Intermediates	0.115	0.002	0.100	-0.019	0.033	0.011	0.016	-0.016
	Equipment	-0.055	0.001	0.070	-0.024	-0.046	0.018	0.081	-0.042
	Consumption	0.040	0.002	0.034	0.002	-0.015	0	-0.034	-0.052
	Construction	-0.375	-0.009	-0.260	-0.101	0.085	-0.095	0.073	-0.017
	Transport	-0.004	0.013	-0.054	0.207	0.003	0.021	-0.099	0.064
	Services	0	-0.002	-0.039	0.020	0.062	0.025	-0.024	0.088
Yi / Y	Agriculture	0.850	-0.002	-0.194	-0.060	-0.064	0.025	-0.158	0.236
	Energy	0.407	0.026	0.077	0.133	-0.081	0.044	-0.247	-0.089
	Intermediates	0.363	0.001	0.113	-0.035	-0.043	0.007	-0.511	0.236
	Equipment	0.022	-0.001	0.038	-0.047	-0.179	0.013	-0.520	0.162
	Consumption	0.218	0	0.005	-0.017	-0.118	-0.010	-0.512	0.176
	Construction	-0.675	-0.024	-0.396	-0.204	0.176	-0.182	-0.521	0.203
	Transport	-0.101	0.067	0.002	0.451	-0.118	0.004	0.145	-0.425
	Services	-0.162	-0.009	0.030	-0.015	0.026	-0.018	0.536	-0.105

Environmental Policies in Open Economies and Leakage Problems

Michael Rauscher

Department of Economics, University of Rostock, Ulmenstr. 69, 18057 Rostock, Germany, *michael.rauscher@wiwi.uni-rostock.de*

Summary:
Pollution leakage is an important issue when countries address international environmental problems unilaterally. Leakage in this context is the increase in foreign emissions after a reduction in domestic emissions which results from international market interdependencies. In this paper, I consider two markets, the first one for a polluting input (e.g. energy) and the second one for internationally mobile capital. Leakage is decomposed into its two components and the magnitudes of the effects are determined for a calibrated version of the model. Towards the end of the paper, normative questions related to the optimality of environmental policies are addressed.

Keywords: International trade and the environment, foreign direct investment, pollution leakage.

1 Introduction

Environmentalists and sector-specific interest groups often demand that environmental policies in open economies be accompanied by trade policies that eliminate the cost advantages that producers enjoy in less regulated countries. Various arguments are used to support this idea. Industry interest groups claim that firms producing in low-regulation countries are given an unfair advantage in international competition and that, therefore, some kind of countervailing duty is needed to re-establish fair competition. Environmentalists often argue along similar lines when they say that foreign environmental standards are too lax and that domestic policies should be used to save foreign countries, particularly developing ones, from becoming pollution havens. To most trade economists (and probably to many environmental economists, too) these arguments are hardly convincing. A country's environmental policy should be determined by its endowment with environmental resources. This depends on physical and geographical conditions but also on the willingness of its people to pay for environmental quality. Since countries differ in these respects, environmental policies should differ as well, and it does not make sense to introduce policy measures that level the playing-field. There is, however, still another argument in favor of trade intervention. This is the leakage problem. Due to the interdependence of international commodity and factor markets, tighter environmental policies in one country may lead to higher emissions in other countries. In the case of transfrontier pollution spill-overs, the regulating country suffers from the increase in emissions as well and, thus, benefits less from stricter environmental policies. The effectiveness of unilateral environmental policies is undermined. The optimal solution to this problem requires international cooperation. In a second-best world, however, it turns out that trade interventions can be justified to cope with the leakage issue This paper is an attempt to re-assess the leakage problem, to give some rough quantitative estimates of its magnitude, and to discuss trade interventions.

There are three channels through which leakage problems are generated. The first one is trade in final goods. If domestic environmental regulation becomes stricter, then the domestic production of environmentally intensive goods becomes relatively more expensive and the comparative advantage in this production is shifted abroad. Emissions in the rest of the world tend to increase and this creates leakage problems. The second channel is factor movements. If a country tightens its environmental standards, this reduces the productivity of the other factors of production. If they are internationally mobile, they tend to emigrate and move to countries with less stringent environmental policies. If they are employed there, additional demand for the use of environmental resources in production is generated and emissions increase. In the recent debate, this relocation of factors has been termed 'environmental capital flight'. The third channel is trade in primary goods or factors of production. This is particularly important in the case of energy resources. If a country introduces a tax on greenhouse gases, then its energy demand tends to be reduced. This leads to reduction in world energy prices and to an increase in demand in the rest of the world. Thus, greenhouse gas emissions are increased there and the domestic policy is (at least partially) eroded. In the present paper, I deal with the second and third channels only. Trade in final goods is not considered. Of course, this is a limitation of the model, albeit probably a less severe one than one might think at a first glance. Factor movements and trade in final goods are substitutes in the model setting I use and, thus, not much would be added by complicating the model by the introduction of trade in final commodities.[1]

Probably, the first paper that established optimal tariffs as an instrument to deal with transfrontier pollution problems is that by Markusen (1975). For a generalization see Rauscher (1997, ch. 5). Leakage in the context of international factor movements has been considered by Rauscher (1997, ch. 3). A partial-equilibrium

[1] See Markusen (1983) for situations in which trade and factor movements are not substitutes but complements.

model of carbon leakage through energy markets is discussed by Hoel (1994). With the notable exception of this paper, the theoretical literature on international environmental problems has, however, largely neglected trade in polluting intermediate products such as fossil fuels. In contrast, this type of trade plays an important role in applied general-equilibrium models. These models usually address positive rather than normative questions: what is the impact of changes in environmental policies on factor allocation, international trade and output? Examples are Oliveira-Martins, Burniaux, Martin (1993), Felder, Rutherford (1993), Manne, Oliveira-Martins (1994), OECD (1995), Burniaux, Oliveira-Martins (2000), Kuik, Verbruggen (2001). It turns out that the results are highly sensitive to the parameters used in the models, e.g. energy demand and supply elasticities and the degree of future integration of China with its huge coal reserves into the world economy. Carbon leakage figures derived from these models range from merely 3 to 40% of original emission reductions. It should be noted that several of these models do not consider the possibility of international factor movements, that may accelerate carbon-leakage. International factor movements are taken into account by Ulph (1994) in a non-competitive partial-equilibrium model. He calibrates it for the fertilizer industry and finds even higher leakage figures of more than 60%. The overall impression from this literature is that the empirical results are still a rather mixed bag and do not provide a reliable guideline for estimating the practical relevance of pollution leakage.

The present paper uses a simple model to re-consider the leakage problem. I abstract from trade in final goods and consider only factor movements and trade in primary commodities. Realistic parameter values will be inserted to calibrate the model and to do some sensitivity analyses. The paper is organized as follows. The next section will be used to present a basic model with factor mobility and energy trade. Section 3 looks at the leakage problem, i.e. the effect of a reduction in domestic emissions on foreign emissions. Section 4 is concerned with the optimal response of the government to the leakage problem in terms of environmental policy. In Section 5, I will address first-best environmental policies, i.e. the government can use the whole range of policy instruments including trade interventions. Section 6, presents some considerations concerning non-competitive market structure. The final part of the paper summarizes the results.

2 The Model

Consider a world consisting of two countries, the home country and the rest of the world. All variables and parameters of the home country are represented by lower-case letters and the variables and parameters of the rest of the world by the corresponding upper-case letters. There are two variable factors of production, capital and energy, k and e, which are employed in a production process described by a well-behaved neoclassical production function:

$$q = f(k,e), \tag{1}$$

$$Q = F(K,E) \tag{1'}$$

There may be other factors of production such as land and labor but they are taken as given here. For the calibration of the model, we assume a CES (constant elasticity of substitution) technology such that

$$f(k,e) = \left(ak^{\rho} + be^{\rho} + l\right)^{1/\rho}, \tag{2}$$

$$F(K,E) = \left(aK^{\rho} + bE^{\rho} + L\right)^{1/\rho}, \tag{2'}$$

where l and L represent the impact of factors of production not considered explicitly here. The elasticity of substitution is $1/(\rho\text{-}1)$. Moreover, $\rho<1$ and $a+b<1$.

The calibration of the model is based on some stylized facts. In most countries, the GDP shares of capital income and expenditure for primary energy are roughly 0.3 and 0.1, respectively. Moreover, the price elasticity of energy demand is less than unity. Perfect competition implies that marginal productivities equal factor prices. For the case of a CES production function, we have

$$\frac{f_k k}{f} = \frac{ak^{\rho}}{ak^{\rho} + be^{\rho} + c} = 0.3, \tag{3a}$$

$$\frac{f_e e}{f} = \frac{be^{\rho}}{ak^{\rho} + be^{\rho} + c} = 0.1, \tag{3b}$$

$$\eta = \frac{f_e}{ef_{ee}} = \frac{1}{(\rho-1)\left(1-\dfrac{be^{\rho}}{ak^{\rho}+be^{\rho}+c}\right)} = \frac{1}{0.9(\rho-1)}. \tag{3c}$$

Arguments of the production function have been omitted for convenience and subscripts denote partial derivatives. From (3b), we have that ρ should be negative, i.e. the production function should be inelastic. For ρ = -1, the price elasticity is -0.56; for ρ=-3, it is -0.28. Probably, the latter value is more realistic.

There are international capital movements. The countries are endowed with capital stocks k^0 and K^0, respectively. Capital is fully employed:

$$k^0 + K^0 = k + K. \tag{4}$$

Under the assumption of perfect capital mobility and perfect competition, the allocation of capital is determined by

$$f_k + \sigma = F_K + \Sigma, \tag{5}$$

where the subscripts denote partial derivatives and σ and Σ are subsidies on the use of capital. If they are negative, they denote source taxes.

Energy is produced in the home and the foreign countries. Both countries use a part of their GDP to provide energy resources. These costs are $g(e^S)$ and $G(E^S)$ where, e^S and E^S denote domestic and foreign energy supply, respectively. We have

$$g'\left(e^s\right) = P + \theta, \tag{6}$$

$$G'\left(E^s\right) = P + \Theta, \tag{6'}$$

where the world market price is P and θ and Θ are import taxes (or export subsidies if the country is an energy exporter). Inversion of these functions gives the energy supply functions, $e^S = s(P+\theta)$ and $E^S = S(P+\Theta)$.

The world energy market is in equilibrium if

$$s(P+\theta) + S(P+\Theta) = e + E. \tag{7}$$

Energy use is taxed at rates t and T, such that energy demand is determined by

$$f_e = P + \theta + t, \tag{8}$$

$$F_E = P + \Theta + T. \tag{8'}$$

Energy use creates global environmental problems such as the greenhouse effect. Environmental externalities are independent of the location of the source of emissions and environmental damages are evaluated with damage functions $d(e+E)$ and $D(e+E)$ that have positive and non-decreasing partial derivatives. National welfare is

$$w = f(k,e) - g\big(s(P+\theta)\big) + \big(s(P+\theta) - e\big)P + (k^0 - k)(f_k + \sigma) - d(e+E), \tag{9}$$

$$W = F(K,E) - G\big(S(P+\Theta)\big) + \big(S(P+\Theta) - E\big)P + (K^0 - K)(F_K + \Sigma) - D(e+E). \tag{9'}$$

Its components are output, energy supply costs, income from energy exports, rent income from capital invested abroad, and environmental damage. Of course, some welfare components can be negative if the country under consideration is an importer of energy or of capital services. This concludes the exposition of the model.

3 The Leakage Problem

Assume the home country wishes to use a stricter environmental policy, e.g. by raising the energy tax. This has the following adverse effects on the domestic environment. On the one hand, an increase in domestic energy prices reduces domestic energy demand. If the home country is large, this has a significant effect on global energy demand and world market prices decline. Thus, energy demand is going to rise in the rest of the world. On the other hand, domestic capital productivity is reduced when energy prices rise and capital will move abroad. This raises the productivity of the other factors of production in the rest of the world, and energy demand will rise there.

The effects of a change in domestic energy use on foreign energy use is determined by total differentiation of:

$$f_k+\sigma=F_K+\Sigma, \tag{5}$$

$$s(P+\theta)+S(P+\Theta)=e+E, \tag{7}$$

$$F_E=P+\Theta+T. \tag{8'}$$

Assume that foreign energy taxes and import duties are given. Then, the comparative statics are

$$\begin{pmatrix} F_{KK}+f_{kk} & F_{KE} & 0 \\ F_{KE} & F_{EE} & -1 \\ 0 & 1 & -s'-S' \end{pmatrix}\begin{pmatrix} dK \\ dE \\ dP \end{pmatrix}=\begin{pmatrix} f_{ke} \\ 0 \\ -1 \end{pmatrix}de \tag{10}$$

and the leakage effect is

$$\frac{dE}{de}=\frac{(s'+S')f_{ke}F_{KE}-f_{kk}-F_{kk}}{-(s'+S')\left((f_{kk}+F_{KK})F_{EE}-F_{KE}^2\right)+f_{kk}+F_{KK}}<0. \tag{11}$$

This can be rewritten if we express $(s'+S')$ in terms of elasticities. This is done under the assumption that there are no trade interventions, i.e. $\theta=\Theta=0$:

$$\begin{aligned} s'+S' &= \frac{(s'+S')P}{e+E}\ \frac{e+E}{E}\ \frac{P+\Theta+T}{P+\Theta}\ \frac{F_{EE}E}{P+T}\ \frac{1}{F_{EE}} \\ &= \frac{\varepsilon^W}{\eta^F}\ \frac{e+E}{E}\ \frac{P+\Theta+T}{P+\Theta}\ \frac{1}{F_{EE}} \equiv \frac{\xi}{F_{EE}} \end{aligned}$$

where ε^W and η^F are the price elasticities of world energy supply and foreign energy demand, respectively. The parameter ζ is introduced for notational convenience. This can now be used in equation (11).

$$\frac{dE}{de} = \frac{\xi\ f_{ke}F_{KE} - (f_{kk} + F_{KK})F_{EE}}{-\ \xi\left((f_{kk} + F_{KK})F_{EE} - F_{KE}^2\right) + (f_{kk} + F_{KK})F_{EE}}. \tag{12}$$

To obtain some quantitative results, assume that both countries differ only in size:

$$\frac{k}{K} = \frac{e}{E} = \frac{f(k,e)}{F(K,E)} = \beta.$$

This implies that the marginal productivities in the two countries are identical and for the second derivatives with respect to the variables i and j we have $f_{ij}=\beta F_{ij}$. Thus, the leakage effect is

$$\frac{dE}{de} = \frac{\beta\xi\ F_{KE}^2 - (1+\beta)F_{KK}F_{EE}}{-\ \xi\left((1+\beta)F_{KK}F_{EE} - F_{KE}^2\right) + (1+\beta)F_{KK}F_{EE}},$$

which in the case of a CES function turns out to be

$$\frac{dE}{de} = -\frac{\beta\xi - (1+\beta)\left(1 - \frac{F}{KF_K}\right)\left(1 - \frac{F}{EF_E}\right)}{(1+\beta)(\xi - 1)\left(1 - \frac{F}{KF_K}\right)\left(1 - \frac{F}{EF_E}\right) - \xi}. \tag{12'}$$

Using assumptions (3a) and (3b), we arrive at

$$\frac{dE}{de} = \frac{21(1+\beta) - \beta\xi}{21(1+\beta)(\xi - 1) - \xi}. \tag{12''}$$

This can be calibrated easily with realistic values of the elasticities, the relative country sizes and the levels of energy taxation abroad. It is assumed that ε^W/η^F takes values between -0.5 and -2,[2] and that foreign energy tax rates are 0, $(P+\Theta)$, and $2(P+\Theta)$. Moreover, we look at cases were countries accounting for 20 or 50% of world energy demand reduce their energy input, i.e. β equals 0.25 or 1. See Table 1. There is quite some variety in the leakage results. The most realistic parameter combinations produce leakage effects in the range of 15 to 35%. Thus, these simple back-of-the-envelope calculations reproduce the leakage effects that have been obtained from computable general-equilibrium models such as Oliveira-Martins, Burniaux, and Martin (1993); Felder and Rutherford (1993);

2 This is a realistic parameter change. See, e.g., Huntington (1992) for an overview.

Manne and Oliveira-Martins (1994); and OECD (1995). See Burniaux and Oliveira-Martins (2000) for an overview and comparison. The disadvantage of my one-sector model, of course, is the simplistic approach to calibration It neglects the complexity of inter-sectoral interdependencies of the economy and the existence of barriers to trade and factor movements. The advantage is that the effects of parameter changes are very obvious and intuitive.

Table 1: Leakage in a Competitive-Markets Framework.

ε^W/η^F	T	$\frac{dE}{de}$ $(\beta = 0.25)$	$\frac{dE}{de}$ $(\beta = 1)$
-0.5	0	-0.628	-0.518
-1.0	0	-0.459	-0.355
-2.0	0	-0.301	-0.223
-0.5	P	-0.459	-0.355
-1.0	P	-0.301	-0.223
-2.0	P	-0.180	-0.135
-0.5	$2P$	-0.363	-0.273
-1.0	$2P$	-0.225	-0.167
-2.0	$2P$	-0.130	-0.101

In order to get a better intuition for the results and to interpret them economically, I will separate the two sources of leakage. Scenario A will consider energy-market effects only whereas scenario B will be restricted to the effects of international capital movements.

Scenario A: Immobile Capital. Capital mobility is restricted and leakage effects are due to the decline in world energy prices only. Then we have:

$$\frac{dE}{de} = -\frac{1}{1-(s'+S')F_{EE}}. \tag{11A}$$

Two things are remarkable here. First, the leakage effect is bound from below by -1. Thus it is not possible that a reduction in domestic emissions by one unit is accompanied by a more-than-one-unit increase in foreign emissions. Moreover, leakage is the more significant the less elastic energy supply is. The explanation for this is straightforward. With inelastic supply, a reduction in domestic energy use has a strong price effect. This in turn implies that foreign energy demand increases substantially. As has been done in the general case, equation (11A) can be rewritten in terms of elasticities if we assume that there are no trade taxes in the initial situation. Then, we have

$$\frac{dE}{de} = - \frac{1}{1 - \frac{\varepsilon^W}{\eta^F} \frac{P+T}{P} \Big/ \frac{E}{e+E}} \tag{12A}$$

or

$$\frac{dE}{de} = \frac{1}{\xi - 1}. \tag{12A'}$$

Besides the impact of the price elasticity of energy supply, additional effects can be identified.

- Leakage is the more important, the larger the price elasticity of foreign demand energy. A large price elasticity of demand implies that the energy-price effects of tighter domestic environmental policies cause substantial demand increases abroad.
- If the rest of the world uses high energy taxes, the leakage effect is small. For in the case of high energy taxes, the relative price change compared to the initial situation is small.
- If the rest of the world is large compared to the home country, the leakage effect is substantial. This is due to the fact that only a small number of countries use stricter environmental standards whereas a large number of countries reacts to lower energy prices. An emission reduction by one unit undertaken by Luxembourg has a larger leakage effect than the same reduction undertaken by, say, the USA.

In order to assess the magnitude of leakage, I use the same parameter values as before. The results are contained in Table 2 and it is seen that the leakage effects differ only slightly from the ones derived from the original model. Differences between the two scenarios are less than 2.5 percentage points. Energy-market interdependencies appear to be the most important source of leakage. However, since the effects of energy-market interdependencies and capital movements are linked in a non-trivial way in this model (see equation 11), this conjecture requires a bit of additional investigation. Therefore, we will now consider the leakage effects of international capital movements.

Table 2: Energy Markets and Leakage in the Competitive Model with Immobile Capital.

ε^W/η^F	T	$\frac{dE}{de}$ $(\beta=0.25)$	$\frac{dE}{de}$ $(\beta=1)$
-0.5	0	-0.614	-0.500
-1.0	0	-0.444	-0.333
-2.0	0	-0.286	-0.200
-0.5	P	-0.444	-0.333
-1.0	P	-0.286	-0.200
-2.0	P	-0.167	-0.111
-0.5	$2P$	-0.348	-0.250
-1.0	$2P$	-0.211	-0.143
-2.0	$2P$	-0.118	-0.077

Scenario B: No Trade in Energy. All leakage effects are due to environmental capital flight. The leakage effect in this case is:

$$\frac{dE}{de} = -\frac{f_{ke}F_{KE}}{\left(f_{kk}+F_{KK}\right)F_{EE}-F_{KE}^2}. \tag{11B}$$

To obtain quantitative results, assume again that both countries differ only in size. The size of the home country is β times that of the rest of the world. In the case of a CES production function, the leakage effect turns out to be

$$\frac{dE}{de} = \frac{-\beta}{(1+\beta)\left(1-\frac{F}{KF_K}\right)\left(1-\frac{F}{EF_E}\right)-1}. \tag{12B}$$

Using the assumptions (3a) and (3b), we get

$$\frac{dE}{de} = \frac{-\beta}{21(1+\beta)-1}. \tag{12B'}$$

Thus, the leakage effect must be less than 1/21. If β equals 0.25 or 1, i.e. if the countries restricting their environmental policies own one fifth or half or of the world capital stock, leakage effects are 1/101 or 1/41, respectively. These rather small values indicate that the major part of the leakage problem is indeed due to energy-market interdependencies but not to environmental capital flight. This view is supported by the results contained by Kuik, Verbruggen (2001). They also find – albeit in a more general computable general-equilibrium framework – that the major part of leakage is due to energy-market effects.

Result 1:

In the case of perfectly competitive markets, leakage effects in this simple model are to be expected in the range of 15 to 35%. The major part (some 90%) of this is explained by the effects of energy taxes on world energy prices. The impact of environmental capital flight is almost negligible.

One should note that compared to many multi-sectoral computable general-equilibrium models, these figures are rather high. See Burniaux, Oliveira-Martins (2000) for an overview. Probably the underlying intuition is that general-equilibrium market interdependencies dampen the leakage effects.

Finally the question arises as to whether leakage effects are stronger if there are two sources of leakage than if there is only one source. The appendix shows that this is not generally the case and that the counter-intuitive result is possible for unrealistic parameter constellations only.

4 Optimal Environmental Policy: The Second Best

In the process of determining its environmental policy, a country should take leakage effects into account. In a first step, assume that only one policy instrument is used. This is an emission tax or a tradable-permits scheme. Given that strict environmental policies induce leakage, i.e. an increase in emissions abroad, which causes a negative spill-over to the domestic environment, the domestic government is driven to utilize less-strict environmental policies than if leakage did not matter. To verify this, I determine the optimal environmental policy by deriving the welfare function, equation (9), with respect to domestic emissions. This yields

$$\begin{aligned} t+\theta &= f_e - P \\ &= d' + \frac{dE}{de} d' + \left(k^0 - k\right)\left(f_{kk}\frac{dK}{de} - f_{ke}\right) - \sigma\frac{dK}{de} - \left(e^s - e\right)\frac{dP}{de} + s'\theta\frac{dP}{de} \end{aligned} \tag{13}$$

with

$$\frac{dK}{de} = \frac{-(s'+S')f_{ke}F_{EE} + F_{KE}}{-(s'+S')\left(\left(f_{kk}+F_{KK}\right)F_{EE} - F_{KE}^2\right) + f_{kk} + F_{KK}} < 0 \tag{14}$$

and

$$\frac{dP}{de} = \frac{f_{ke}F_{KE} - \left(\left(f_{kk} + F_{KK}\right)F_{EE} - F_{KE}^2\right)}{-(s' + S')\left(\left(f_{kk} + F_{KK}\right)F_{EE} - F_{KE}^2\right) + f_{kk} + F_{KK}} \tag{15}$$

from equation (10) and dE/de as in equation (11). Laxer environmental standards at home imply that less capital is located abroad (equation 14). The effect on energy prices is ambiguous. There is at least the theoretical possibility that an increase in domestic energy use reduces world energy prices. This, however, requires a very large value of f_{ke}. The underlying mechanism is that laxer environmental policies at home lead to capital movements from abroad. With less capital, the foreign country will demand less energy. Although this theoretically possible, it is very unlikely. Thus, under realistic circumstances will laxer environmental tend to raise world energy prices via the increase in domestic energy demand.

Equation (13) states that the difference between marginal energy productivity and the world market price of energy consists of six components.

- d'. This is the marginal domestic environmental damage. In the simple closed-economy or first-best small open economy model, this would be the only component of an emission tax. Then a Pigouvian tax, that just internalizes the domestic environmental damage, would be optimal.
- $\frac{dE}{de}d'$. This is the leakage effect. It tends to reduce domestic environmental taxes. High environmental taxes induce additional pollution from abroad. This undermines the environmental benefit of tight environmental policies. Thus, domestic environmental policies should be relaxed compared to the case of autarky.
- $\left(k^o - k\right)\left(f_{kk}\frac{dK}{de} - f_{ke}\right)$. This is the rate-of-return effect or factor-terms-of-trade effect. Its sign depends on whether the country is an exporter or importer of the mobile factor. If the country is capital-poor, it benefits from a low factor price and should, therefore, impose a high emissions tax. A high emission tax reduces emissions, this has a negative impact on capital productivity and, therefore, reduces the rate of return. If the country is capital-rich, it benefits from a low energy tax energy, which raises demand for the mobile factor of production and, thus, induces a higher rental rate in the international capital market.
- $-\sigma\frac{dK}{de}$. This is a fiscal externality. If σ is negative, there is a source tax on mobile capital. The government is interested in attracting the tax base and, thus, charges low emission taxes. In the case of a positive subsidy, there is an incentive to impose a high emission tax rate. See Wilson (1997), for instance.

- $-\left(e^s - e\right)\frac{dP}{de}$. This is the energy-market terms-of-trade effect. High emission taxes reduce the demand for energy and thus, under realistic parameter constellations, raise the energy price. This is good if the country is an exporter of energy and bad if it is an importer. Energy-importing countries tend to have laxer environmental policies than energy-exporting countries.

- $s'\theta\frac{dP}{de}$. Finally, this is a second fiscal externality. As lax environmental policies raise the energy price, they also increase domestic supply. This reduces the tariff revenue if the country is an importer of energy and θ is a positive tariff. If the country is an exporter of energy and θ is a positive export subsidy, the subsidy payments are increased. Thus, both, importers and exporters of energy have an incentive to use stricter environmental standards if $\theta > 0$.

What are the effects of such environmental policies if all countries employ them? In order to concentrate on the leakage effect, assume that the countries under consideration are identical and that there are no taxes on capital. Then the rate-of-return, fiscal-externality and terms-of-trade effects vanish. And we have

$$t \;=\; f_e(k,e) - P \;=\; \left(1+\frac{dE}{de}\right)d'(2e) \tag{16a}$$

whereas the optimal tax rate (indicated by an asterisk) is

$$t^* \;=\; f_e(k,e^*) - P^* \;=\; 2d'(2e^*), \tag{16b}$$

where e^* is the corresponding emission level and P^* is the world market energy price prevailing in such a situation. In general, this is different from the price level of the non-cooperative case. It is seen that leakage aggravates the transfrontier pollution problem. In the cooperative case, countries internalize twice their own share of the pollution externality (n times their own share in the n countries case). Acting non-cooperatively and taking the leakage problem into account, they internalize even less than their own shares.

5 Optimal Environmental and Trade Policies: The First-Best Non-Cooperative Policy

Given that there are six effects and only one policy instrument, it is obvious that this policy is not first-best. It would be better to use additional policy instruments such as tariffs and capital taxes or subsidies. In order to keep the algebra as simple as possible, I will not determine the optimal policy by taking derivatives of the welfare function with respect to taxes and tariffs but, instead, determine optimal levels of domestic energy use, domestic energy supply, and the capital stock. The corresponding tax rates follow from equations (8), (6), and (5), respectively.

The restrictions the domestic welfare maximization problem is subject to are the energy-market equilibrium, equation (7), and marginal-productivity condition determining foreign energy demand, equation (8'). They are slightly rewritten here:

$$e^s + S(P+\Theta) = e + E\,, \tag{17}$$

$$F_E\left(K^0 + k^0 - k, E\right) = P + \Theta + T\,. \tag{18}$$

Foreign taxes are taken as given. The endogenous variables are the world market price of energy, P, and the foreign energy consumption, E. Comparative statics are determined by

$$\begin{pmatrix} S' & -1 \\ -1 & F_{EE} \end{pmatrix}\begin{pmatrix} dP \\ dE \end{pmatrix} = \begin{pmatrix} 0 & 1 & -1 \\ F_{KE} & 0 & 0 \end{pmatrix}\begin{pmatrix} dk \\ de \\ de^S \end{pmatrix}. \tag{19}$$

This implies that

$$\frac{dP}{dk} = -\frac{F_{KE}}{1-S'F_{EE}} < 0, \tag{20a}$$

$$\frac{dE}{dk} = -\frac{S'F_{KE}}{1-S'F_{EE}} < 0, \tag{20b}$$

$$\frac{dP}{de} = -\frac{dP}{de^s} = -\frac{F_{EE}}{1-S'F_{EE}} > 0, \tag{20c}$$

$$\frac{dE}{de} = -\frac{dE}{de^s} = -\frac{1}{1 - S' F_{EE}} < 0. \tag{20d}$$

An increase in domestic capital can only be achieved at the expense of the foreign capital stock. Thus, foreign energy demand is reduced. For given energy demand and supply in the home country, this results in a decline in the energy price. An increase in domestic energy demand (supply), of course, increases (reduces) the price of energy. This causes a leakage effect that has the expected sign. The impacts of changes in domestic energy demand and supply exhibit opposite signs but the same magnitudes, which is also intuitive.

The leakage effect derived here differs from that derived previously. Compare equation (11). It, however, closely resembles the leakage effect derived for the scenario of factor immobility, equation (11A). In contrast to the previous sections of the paper, the domestic government can now control the world market prices of energy and the capital allocation directly and, therefore, can take these variables as given when choosing its environmental policy. Rewriting equation (20d) in terms of elasticities yields:

$$\frac{dE}{de} = -\frac{1}{1 - \frac{\varepsilon^F}{\eta^F} \frac{P+\Theta+T}{P+\Theta} \frac{S}{E}} \tag{21}$$

where ε^F and η^F are the foreign price elasticities of energy supply and demand, respectively. The leakage effect is the stronger

- the less price-elastic foreign energy supply,
- the more price-elastic foreign energy demand,
- the lower foreign energy taxes,
- the larger foreign energy demand in relation to foreign energy supply.

The first three effects have been discussed earlier. The fourth one differs from the one derived in Section 3. If foreign energy demand is substantial compared to supply, a larger share of foreign energy consumption is imported, and thus directly affected by the price changes induced by domestic environmental policies.

The optimal environmental and trade policies can now be determined by maximizing the slightly modified welfare function,

$$w = f(k,e) - g(s^e) + (s^e - e)P + (k^0 - k)(F_K + \Sigma) - d(e+E)$$

with respect to energy demand, energy supply, and capital, taking account of the impacts on foreign energy consumption and the world market price. The first-order conditions are

$$f_e - P - d' + \left(s^e - e\right)\frac{dP}{de} + \left(k^0 - k\right)F_{KE}\frac{dE}{de} - d'\frac{dE}{de} = 0, \tag{22a}$$

$$P - g' + \left(s^e - e\right)\frac{dP}{de^s} + \left(k^0 - k\right)F_{KE}\frac{dE}{de^s} - d'\frac{dE}{de^s} = 0, \tag{22b}$$

$$f_k - F_K - \Sigma + \left(s^e - e\right)\frac{dP}{dk} + \left(k^0 - k\right)\left(F_{KE}\frac{dE}{dk} - F_{KK}\right) - d'\frac{dE}{dk} = 0. \tag{22c}$$

Using the private sector's first-order conditions, equations (5), (6), and (8), and, moreover, noting that $dE/de = -dE/de^s$, the optimal policy instruments turn out to be

$$t = d', \tag{23a}$$

$$\theta = d'\frac{dE}{de} - \left(s^e - e\right)\frac{dP}{de} - \left(k^0 - k\right)F_{KE}\frac{dE}{de} \tag{23b}$$

$$\sigma = -d'\frac{dE}{dk} + \left(s^e - e\right)\frac{dP}{dk} + \left(k^0 - k\right)\left(F_{KE}\frac{dE}{dk} - F_{KK}\right). \tag{23c}$$

Equation (23a) states that the optimal environmental tax rate is a Pigouvian tax rate, which just internalizes domestic environmental damage. The leakage and terms-of-trade effects are taken into account by trade interventions in the energy and factor markets. In (23b) and (23c), it is seen that each of the policy instruments consists of three components: an impact on foreign emissions and two terms-of-trade effects, related to the energy price and to the rental rate of capital. Under the assumption that the two countries are sufficiently similar, the terms-of-trade effects vanish.

From (23b), we then get

$$\theta = \frac{dE}{de}t. \tag{23b'}$$

Since, dE/de is negative, this means that energy imports should be subsidized and energy exports should be taxed. This confirms Hoel's (1994) result, derived from a partial-equilibrium energy market model. The underlying rationale is that these instruments induce additional demand for energy on the world market. This increases the price and reduces emissions elsewhere. It is not surprising that this border tax adjustment becomes the larger the more substantial the leakage effect is. For the calibration of the model used in Section 3, the subsidy or tax can be in the range of 15 to 35% of the emission tax rate.

Using (20b) and (20c), we obtain from (23c)

$$\sigma = -S'F_{KE}\frac{dE}{de}t\,.$$

Using elasticities, assuming $E=S$, and exploiting the properties of the CES function, one can rewrite the latter such that

$$\sigma = -\varepsilon^F(1-\rho)\frac{F_E E}{F}\frac{dE}{de}\frac{t}{P+\Theta+T}F_K\,. \qquad \text{(23c')}$$

From equation (23c'), we have that domestic use of capital should be subsidized. This reduces environmental capital flight and, therefore, leakage. The effect depends on the leakage rate, but it is rather small. Note that the terms by which F_K is multiplied are all dimensionless. The price elasticity of foreign energy supply, ε^F, is likely to be less than one. (1-ρ) is likely to be in the interval (2;5). The production elasticity of energy is unlikely to exceed 0.1. The next term is the leakage effect, which is probably less than 0.4 in absolute value. Finally, there is a ratio of the domestic tax rate over the foreign energy price, which in most real cases is also less than one. Using realistic values for these parameters, one arrives at the conclusion that the optimal subsidy rate amounts to 1 or 2% of the capital rental rate. This is an almost negligible magnitude.

The following proposition summarizes the results of this section.

Result 2:

Optimal trade interventions to cope with leakage include import subsidies or export taxes on energy and subsidies on the domestic use of capital. The trade interventions in the energy market are of substantial magnitude whereas the rate of subsidization of capital is negligible.

6 Non-Competitive Markets and the Indivisibilities of Investment

All the results of the preceding section were based on the assumption of competitive markets. This section will introduce an imperfection here and this will change the results concerning the capital-flight component of leakage dramatically. Besides non-competitive market structure, indivisibilities will be important: a certain minimum size of an investment is required; otherwise no investment will be undertaken at all. This implies, that the capital stock is not a continuous variable as in the preceding section. Moreover, it makes some sense to consider this in a partial-equilibrium setting. Models of this type are considered by Markusen, Morey, and Olewiler (1995); Rauscher (1995) and Hoel (1997). Markusen, Morey, and Olewiler (1995) consider a situation where there is one polluting firm which decides upon the location of its plants. High firm-specific fixed costs prevent the entry of additional firms. There are two kinds of fixed costs, that of being in the market and that of setting up a plant. The variable costs include pollution abatement and trade costs. The firm may build a plant in the home country, in the foreign country, in both of them or in neither of them, and the decision is influenced by the environmental policies in the two countries. This model turns out to be rather complex even in the case of only one single firm. Markusen, Morey, and Olewiler (1995), therefore, use a numerical example to derive some results. Rauscher (1995) and Hoel (1997) try to avoid this complexity by neglecting trade costs. The model can now be analyzed by simple diagrammatic methods. In what follows, a two-country version of my original *n*-country model (Rauscher, 1995) will be presented here.

There are two identical countries, the home country and the foreign country, and a single firm which supplies the world market with a final good. The fixed cost is so high that there can be only one firm in the market. Since trade costs are negligible, the firm opens at most a single plant. As the firm is small in the energy market, the world market price of energy, P, is exogenously given and, in particular, is not affected by the location decision. Thus, with otherwise identical countries, the firm locates in the country with lower emission tax rate. If, however, both countries charge extremely high energy taxes, the firm decides not to invest at all since the operating profit does not cover the fixed cost.

The carbon-leakage effect is straightforward. If the home country is the host, it can tighten its environmental policy as long as its emission tax rate is smaller than the foreign tax rate. If the domestic emission tax rate exceeds the foreign tax rate, the firm invests abroad. Thus the leakage effect is

$$\frac{dE}{de} = \{0,-1\} \tag{24}$$

In real-world situations, where location decisions are multidimensional issues, matters may be even worse and leakage effects may be much stronger. As an example consider a case of a plant in an industrialized country. The owner enjoys the benefits of being located in a developed country with good infrastructure and low political risks but pays high emission taxes. If environmental policy is tightened even further, she may be willing to relocate to a developing country that does not offer these benefits but has much lower taxes. Thus, emissions may be increased in the new host country by more than they have been reduced in the initial host country.

Optimal environmental policies in this situation are simple and disastrous. The governments are assumed to maximize social welfare, which is the sum of consumer surplus, tax revenue, and the utility derived from environmental quality. The firm's profits are not part of the objective function since the owners of the firm are assumed to be residents of a third country. According to the environmental-damage function, pollution is independent of the geographical localization of the polluting firm. This is a characteristic of a global environmental problem. Moreover, due to the absence of trade costs, the consumer surplus is also independent of the location of the firm. Thus, the tax revenue is the only target variable of environmental policy. Only the host country can enjoy a tax revenue, the other country collects no taxes at all. Thus, by marginally undercutting the other country's emission tax, a country can appropriate the whole tax revenue at a marginal opportunity cost. This induces a race to the bottom in environmental taxes.

As in Section 4, the environmental taxes discussed here are no first-best policies. Besides the objective of internalizing the environmental externality, there is an additional policy objective that is taken care of by the emission tax: the appropriation of the tax revenue. Thus, one may conjecture that it would be better to avail of more than one instrument. Subsidies might be better instruments to attract the foreign firm than low emission taxes. Let σ be a lump-sum subsidy paid by the domestic government to the foreign firm in case the investment is undertaken at home. Assume that the foreign government offers the potential investor a combination of subsidies and environmental regulation such that a minimum profit level Π^0 is guaranteed. The home country can host the firm if it offers a profit $\Pi^0+\varepsilon$. Thus, the objective of the domestic government is to maximize its net benefit,

$$te-\sigma-d(e), \tag{25}$$

subject to the condition that the investor's profit is $\Pi^0+\varepsilon$. Let $f(k,e)$ be the output of the firm (where the production function now contains an increasing-returns segment corresponding to the fixed set-up costs of the plant). Let $R(.)$ denote the revenue function. The firm's profits are

$$\pi = R\left(f\left(k,e\right)\right)+\sigma-te-rk\,. \tag{26}$$

Then the objective of the policy maker is to maximize the Lagrangean function

$$L = te-\sigma-d(e)+\lambda\left[R\left(f(k,e)\right)+\sigma-te-rk-\Pi^0-\varepsilon\right] \tag{27}$$

with respect to t and σ, where λ is the Lagrangean multiplier. Maximization with respect to σ yields $\lambda=1$. Taking this into account, maximization with respect to the tax rate yields

$$\left(t-d'\right)\frac{de}{dt}+\pi_e\frac{de}{dt}+\pi_k\frac{dk}{dt} = 0.$$

The second and third terms on the right-hand side contain the investor's first-order conditions and, therefore, vanish. This implies:

$$t = d'.$$

The emission tax rate equals the marginal environmental damage. The underlying reason for this result is the presence of a second policy instrument. Directly subsidizing foreign direct investment is the first-best approach to attracting foreign capital. The optimal emission tax only internalizes the environmental externality.

This confirms the results that have been derived earlier in this paper. If the appropriate policy instruments are available, the optimal emission tax is the Pigouvian tax. If the set of policy instruments is, however, limited, the effects of environmental tax competition tend to be disastrous. The downward race continues until the bottom, i.e. zero regulation is reached. Fortunately, this is a only a benchmark scenario derived in a very stylized partial-equilibrium model setting. In the general equilibrium and with market structures closer to perfect competition, conclusions are much less pessimistic.

7 Summary and Conclusions

The paper has addressed environmental leakage problems in a model of energy trade and international capital mobility. The following conclusions can be drawn. Leakage problems tend to be more substantial in the non-competitive partial-equilibrium model framework than under perfect competition and in general equilibrium. The most important channel through which leakage effects are generated appears to be the market for primary goods, in particular energy. Capital relocation has less significant effects. Since the model framework used to establish this result is one in which factor movements and trade are substitutes, one may (with caution) conclude that the results carry over to models with trade in final commodities. Changes in the pattern of specialization are likely to be much less important sources of leakage than changes in supply and demand in markets for primary goods. The other main result is that the existence of leakage aggravates the free-rider problem in a world with transfrontier pollution. Emission taxes are less than Pigouvian taxes if no additional policy instruments are available. If these instruments are used, the optimal emission tax rates are the Pigouvian tax rates as in situations without leakage. Leakage effects are taken into account by export taxes or import subsidies on energy and by the subsidization of the domestic use of mobile capital. The tax rates are substantial for energy and small for capital.

It is an interesting question as to what extent these results carry over to more realistic general equilibrium models. What are the optimal trade interventions if the complexity of interactions between markets for many factors, intermediate inputs, and final commodities are considered. It is a challenge for users of computable general-equilibrium models to provide more detailed and more accurate estimates of optimal policies than those derived from this stylized model.

Appendix

In Section 3, we consider two simplified versions of the competitive model in order to isolate the effects of energy price changes and environmental capital flight. This appendix establishes conditions under which the combined effects yield stronger leakage results than the isolated single effects.

In a first step, compare the general case and scenario A, where there is no capital mobility. Assume that the hypothesis is true. Then

$$\frac{-1}{-(s'+S')F_{EE}+1} > \frac{(s'+S')f_{ke}F_{KE}-f_{kk}-F_{kk}}{-(s'+S')\left(\left(f_{kk}+F_{KK}\right)F_{EE}-F_{KE}^2\right)+f_{kk}+F_{KK}}.$$

This can be rearranged such that

$$F_{KE} < f_{ke}(1-(s'+S')F_{EE}).$$

Of course, this is not generally true. In order to get an intuition what one should expect for the real world, I use the assumptions on production functions and price elasticities made before. Then we have

$$\beta < 1-\zeta \quad \text{where} \quad \zeta = -(1+\beta)\left|\frac{\varepsilon^W}{\eta^F}\right|\frac{P+T}{P}.$$

It follows that the counter-intuitive result is possible only if $\beta>1$ and $\left|\frac{\varepsilon^W}{\eta^F}\right|\frac{P+T}{P} < 1.$

Since energy taxes are quite high in many countries and the price elasticity of energy supply is unlikely to be much larger than the price elasticity of demand, this condition is unlikely to be met in reality.

In case B, we have

$$-\frac{f_{ke}F_{KE}}{\left(f_{kk}+F_{KK}\right)F_{EE}-F_{KE}^{2}} > \frac{(s'+S')f_{ke}F_{KE}-f_{kk}-F_{kk}}{-(s'+S')\left(\left(f_{kk}+F_{KK}\right)F_{EE}-F_{KE}^{2}\right)+f_{kk}+F_{KK}}.$$

Rearranging terms yields:

$$\frac{f_{ke}F_{KE}}{\left(f_{kk}+F_{KK}\right)F_{EE}-F_{KE}^{2}} < 1.$$

This is not generally true as well, but for the practically relevant parameter constellations the left-hand side of this inequality is less than 0.05 and, thus, this condition is likely to be met.

References

Burniaux, J.M. and J. Oliveira-Martins (2000), *Carbon Emission Leakages: A General Equilibrium View,* OECD Department of Economics Working Papers No. 242, Paris.

Felder, S. and T.F. Rutherford (1993), Unilateral CO_2 Reductions and Carbon Leakage: The Consequences of International Trade in Oil and Basic Materials, *Journal of Environmental Economics and Management* 25, 162-176.

Hoel, M. (1994), Efficient Climate Policy in the Presence of Free Riders, *Journal of Environmental Economics and Management* 27, 259-274.

Hoel, M. (1997), Environmental Policy with Endogenous Plant Locations, *Scandinavian Journal of Economics* 99, 241-259.

Huntington, H.G. (1992), Inferred Demand and Supply Elasticities from a Comparison of World Oil Models, in: Sterner, T. (Ed.), *International Energy Economics*, Chapman & Hall, London, 239-261.

Kuik, O.H. (2001), *The Kyoto Regime, Changing Patterns of International Trade, and Carbon Leakage*, unpublished manuscript, Verbruggen.

Manne, A.S. and J. Oliveira Martins (1994), *Comparison of Model Structure and Policy Scenarios: GREEN and 12RT*, OECD Department of Economics Working Papers No. 146, Paris.

Markusen, J.R. (1975), International Externalities and Optimal Tax Structures, *Journal of International Economics* 5, 15-29.

Markusen, J.R. (1983), Factor Movements and Commodity Trade as Complements, *Journal of International Economics* 14, 341-356.

Markusen, J.R., E.R. Morey, and N. Olewiler (1995), Noncooperative Equilibria in Regional Environmental Policies When Plant Locations Are Endogenous, *Journal of Public Economics* 56, 55-77.

OECD (1995), *Global Warming: Economic Dimensions and Policy Reponses*, Paris.

Oliveira-Martins, J., J.-M. Burniaux, and J.P. Martin (1992), Trade and the Effectiveness of Unilateral CO_2 Abatement Policies: Evidence from GREEN, *OECD Economic Studies* 19, 123-140.

Rauscher, M. (1995), Environmental Regulation and the Location of Polluting Industries, *International Tax and Public Finance* 2, 229-244.

Rauscher, M. (1997), *International Trade, Factor Movements, and the Environment*, Oxford.

Ulph, A. (1994), Environmental Policy, Plant Location and Government Protection, in: Carraro, C. (Ed.), *Trade, Innovation, Environment*, Dordrecht, 123-163.

Ulph, A. (1997), Environmental Policy and International Trade: A Survey of Recent Economic Analysis, in: Folmer, H. and T. Tietenberg (Eds.), *International Handbook of Environmental and Resource Economics 1987/88*, Cheltenham, 205-242.

Wilson, J.D. (1997), Capital Mobility and Environmental Standards: Is There a Theoretical Basis for a Race to the Bottom?, in: Bhagwati, J. and R.E. Hudec (Eds.), *Fair Trade and Harmonization: Prerequisites for Free Trade? Volume 1: Economic Analysis*, Cambridge, 393-427.

Pollution Charges and Incentives

Franz Wirl

Chair Industry, Energy and Environment, University of Vienna, Brünnerstr. 72, 1210 Vienna, Austria, *franz.wirl@univie.ac.at*

Summary:
Most investigations of the consequences of environmental policies and Pigouvian taxes (pollution charges, in particular carbon taxes) using CGE models reveal an only modest impact on economic activity; some report even positive economic side effects. Yet the slowdown in economic growth and in particular of productivity, see Conrad and Wastl (1995), suggest a stronger and negative influence in particular if one accounts for the fortunately very low energy prices that accompanied the ambitious goals of environmental policies in the late 1980s and the 1990s. This paper attempts to explain this apparent difference. In particular it will be shown that environmental policies, here investigated for pollution charges, call for a modification of incentives such that the power of optimal incentives is reduced to which workers will respond with less effort. This friction, which may add up to significant numbers for the economy at large, is neglected in both the theoretical literature and in the application of CGE-models. Accounting for this friction and the empirical evidence so far suggests to take the optimistic findings concerning the consequences of environmental policies on economic growth at least cum grano salis.

Keywords: Private information, principal-agent model, pollution charges, reduced incentives.

1 Introduction

The economic consequences of environmental policies can be ranked in the following way:

(1) Environmental policies, in particular standards and taxes, need not harm the economy but may be actually beneficial. This creation of a win-win situation has been espoused most prominently by Michael Porter, e.g. in Porter and van der Linde (1995ab). Lovins and Hawken (1999) argue that improving resource efficiency can be profitable; Lovins (1985) promised gigabucks of profits for energy conservation programs that were carried out predominately

in the United States throughout the 1980s and 1990s. This optimism is obviously supported by Greens, but more surprisingly, also by other politicians such as e.g. the former US Vice-president Al Gore (1993), and even by some entrepreneurs, e.g. Schmidheiny (1992). This optimistic point of view is to some extent even shared by some economic institutions – the WIFO in Austria (Köppl et al., 1995), and the DIW in Germany that promise positive economic consequences at least on unemployment in case of a 'green' tax reform. This relates to the claim that environmental taxes, in particular energy taxes, allow for a second dividend. That is, environmental taxes do not only reduce emissions, but also improve the efficiency of the tax system by lowering the tax burden. In short, 'how to make lots of money and save the planet too'.

(2) However, contrary to the above expressed enthusiasm, a number of theoretical papers, starting with the seminal contribution of Bovenberg and DeMoij (1994), prove that environmental policies – even in its most efficient and market conform way of a pollution (or 'energy', 'carbon') tax – cannot raise output in an optimal second best setting. That is, the so-called double dividend hypothesis is, unfortunately, infeasible.

(3) This theoretical result derived in Bovenberg and DeMoij (1994) complies with applied general equilibrium models, yet these tend to report that the consequences of past environmental policies are rather modest, for example Wirl (1989) for Austria, Wirl and Hoffmann (1993) for the FRG. A similar result holds in general equilibrium models quantifying the consequences of the much debated carbon taxes, for example, Conrad and Schröder (1988) and Welsch and Hoster (1994) calculate that the impact of a German CO_2 tax would be small on economic activity.

(4) Yet a number of purely empirical papers report a larger decline in economic growth and in particular in economic activity than the general equilibrium models suggest. Early examples for the United States are Gollob-Roberts (1983) and Färe et al. (1986) for power plants and Gray (1987) for the aggregate economy; these studies find that environmental regulation contributed significantly to the slow-down of productivity growth. Conrad and Wastl (1995), of which the major results are re-produced in Figure 1, consider the more recent German experience. This figure highlights the decline in total factor productivity for most sectors during 1986-1991, the period characterized by a much tougher environmental regulation. This observation contradicts strongly the claim of win-win, second dividends and alike and cautions the optimism associated with the results of general equilibrium models.

The purpose of this paper is to reconcile the divergence between general equilibrium models' results and observed declines in growth and/or productivity. The general equilibrium analysis reduces the analysis to a rather technical reoptimization of the demand and input mix. Yet this assumption of a rather frictionless re-organization of production processes ignores a crucial aspect: Controlling environmental inputs (or respectively, negative outputs like emissions, sewage, waste, etc.) complicates production management and thereby weakens incentives. This friction due to a weakening of incentives associated with environmental policies is, of course, ignored in the general equilibrium models and in related approaches but apparently shows up in the empirical data. The weakening of incentives is due to the fact that the additional task of environmental concern requires less 'powerful' incentives. Section 2 introduces a principal-agent model assuming that environmental regulation is specified in its most economic form as a pollution charge. Section 3 considers different scenarios: no pollution charge, a pollution charge when pollution cannot be monitored at the level of each worker (but only for the entire firm), finally, both tasks, production and pollution, can be monitored by the management, the principal. Section 4 compares and discusses the results.

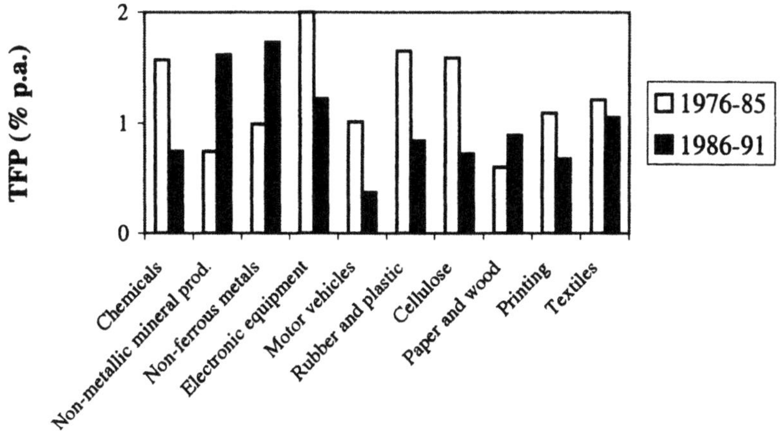

Figure 1: Improvements of Total Factor Productivity (TFP) in Germany Based on Tables in Conrad and Wastl (1995).

2 The Model

The environmental policy consists of a linear pollution tax, τ, per unit pollution p. This in turn requires that the managers of firms and production units, the principals in our principal-agent framework, must account for an additional and so far ignored output of their production unit or entire firm. The managers, interested in maximizing the firm's expected profit, entrust production to a team or group of workers. They offer the workers corresponding incentives to mitigate the agency relation resulting from private information that the agents possess about their individual productivity (hidden information) and efforts (hidden action) that cause disutility of labor.

Environmental concerns, in particular pollution charges, add a new dimension to the production process. It is assumed that pollution is observable at the level of the firm, because otherwise the instrument of a pollution tax is not implementable at all. However, two different assumptions are made with respect to the observability of pollution at the level of each worker:

(1) Pollution released by each individual worker cannot be monitored or is not monitored due to too high monitoring costs. This is a realistic assumption, at least in short and medium term. Indeed many firms do not measure pollution, the level of sewage, waste and alike, produced by a particular worker or at a particular location. However, this information is, of course, observable at the level of the firm, since corresponding charges have to be paid. In this case, the agent has to carry out two tasks – producing and taking care of pollution – of which only one, production, is observable and can thus be a part of the incentive schemes. As a consequence, the environmental task must be indirectly addressed, as it was the case in a different context in Holmström and Milgrom (1991).

(2) Pollution by each worker can be monitored (at zero variable costs). This set-up transgresses the multi-tasking framework of Holmström and Milgrom (1991) because each task, producing and polluting, is observable and can be rewarded.

Given this straightforward set-up, it is quite surprising that such managerial aspects and related frictions due to environmental policies are neglected in the corresponding literature and also in the policy analysis. Indeed, investigations of aspects of monitoring are restricted to regulatory issues, e.g. see the survey of Jaffe, Peterson, and Portnoy (1995).

2.1 Agents – 'Workers'

The workers exercise the effort e that causes disutility $D(e, t)$, which is increasing and convex in effort, $D_e \geq 0$ and $D_{ee} > 0$, and depends on the workers efficiency (or as it turns out, productivity). The individual productivity is the worker's private information, where t denotes the productivity type of a worker. More productive workers, types with a larger t, face less disutility, $D_t < 0$, which holds also at the margin, $D_{et} < 0$. All following results apply also if t refers either directly to productivity or different circumstances that otherwise identical workers face. Hence, the assumption of 't' as a disutility parameter is not crucial. Indeed, we refer to t as a productivity parameter, whether it is given by a worker's individual traits or by circumstances. Effort e can be devoted to production, q, and a, called abatement effort, that affects individual pollution either by taking care, prevention or literal abatement.

$$e = q + a. \tag{1}$$

For simplicity assume that production effort is measured in terms of output,

$$x = q. \tag{2}$$

The amount of pollution generated by a worker depends on output and abatement effort

$$p = P(x, a). \tag{3}$$

This function P is increasing and convex in output, $P_x \geq 0$, $P_{xx} \geq 0$, and declining and convex (to reflect the law of diminishing returns) in abatement efforts: $P_a < 0$, $P_{aa} > 0$. This assumption implies that prescribing output x and pollution p allows to infer the individual efforts q and a, i.e. invert (2) and (3), because the determinant of the corresponding Jacobian is $P_x > 0$.

This invertibility allows to solve (2) and (3) for efforts, $q = x$ and $a = A(x, p)$ depending on the outputs (x, p). Hence, $e = x + A(x, p)$ determines the worker's disutility associated with both tasks, which is denoted by

$$\delta = \begin{cases} \delta(x, p, t) = D(x + A(x, p), t) & \text{for } a > 0. \\ \delta(x, t) = D(x, t) & \text{for } a = 0. \end{cases} \tag{4}$$

For the following, only the specification of δ is relevant and all underlying descriptions that result in the same δ are equivalent. This is the reason, why attaching the private information parameter either to disutility, ability or circumstances, is irrelevant.

In order to differentiate between these two assumptions about the information available to the principal, I use D in the case that pollution is not observable, and δ if pollution is observable. This reduction of δ to D applies to the benchmark – no pollution taxes – and to multi-tasking, when p is not monitored at the level of the agent. The reason is that linking the wage w to output, the agent, sets $a = 0$, since individual abatement is costly for the agent but not rewarded. The assumptions about the agent's implicit disutility δ, some of them already mentioned, are summarized below.

Assumptions: The (implied) disutility δ has the following properties:

$$\delta_x > 0,\ \delta_{xx} > 0,\ \delta_t < 0,\ \delta_{xt} < 0,\ \delta_{xtt} \geq 0,\ \delta_{xxt} \leq 0$$

$$\delta_p < 0,\ \delta_{pp} > 0,\ \delta_{pt} < 0,\ \delta_{ptt} \leq 0,\ \delta_{ppt} \leq 0,\ \delta_{xp} < 0,\ \delta_{xpt} > 0.$$

2.2 The Principal – the Manager or Owner of a Firm

The manager or owner of the firm, thus agency frictions between owners and mangers are assumed away in this analysis, maximizes the expected profit, revenues minus the expenses for pollution charges, and the wages w paid to the workers:

$$\pi = E_t\left[x - \tau p - w\right] = \int_{\underline{t}}^{\bar{t}} \left[x(t) - \tau p(t) - w(t)\right] f(t)dt. \tag{5}$$

This objective (5) assumes implicitly that the firm is a price taker in output markets; without loss in generality we can set this price equal to 1. The expectation is taken with respect to the workers' private information, their productivity type t. As is common in this kind of literature, the principal has a prior distribution over the agents' types $F(t)$ with support over $[\underline{t}, \bar{t}]$ and density $f(t)$, otherwise the manager is unable to form expectations, such that the hazard rate,

$$h(t) := f(t)/\left(1 - F(t)\right), \tag{6}$$

is monotonically increasing. The assumption of a distribution function stipulates that the number of workers is normalized to one and aggregate output corresponds to the mean output.

2.3 The First Best and the CGE Approach

In the absence of environmental concerns, the first-best solution results from:

$$\max_x \; x - D(x, t), \tag{7}$$

which is characterized by the following first-order condition:

$$1 - D_x = 0. \tag{8}$$

Substituting the resulting production into the emission function P determines the corresponding pollution. Aggregating over the types using the density f determines then aggregate output and pollution.

Introducing pollution charges, the corresponding first-best solution is obtained from

$$\max_{x, p} \; x - \delta(x, p, t) - \tau p \tag{9}$$

with the first-order conditions:

$$\delta_x = 1, \tag{10}$$

$$\delta_p = \tau, \tag{11}$$

which are sufficient given the convexity of D. Indeed, this first-best outcome (10) and (11) falls probably already below the CGE outcome. The reason is that (10) and (11) imply a reduction of efforts (the proof is a simplified version of the one in Section 3.3), while the CGE models leave the effort at least of those employed (and assuming anyway full employment) unchanged and just distribute the tasks (optimally) between production and abatement.

3 Optimal Contracts

The design of the optimal contracts must account for the agent's strategic reaction to the incentive, because the 'first-best' is not implementable. The incentive schemes depend on the information available to the agent. I sketch the solution of this problem for the benchmark, no pollution charge, or $\tau = 0$ (for details see e.g. Fudenberg and Tirole, 1992), but skip the derivation of the first-order conditions in the other two cases.

3.1 Status Quo, No Pollution Charge

The optimal incentive (in its direct form) consists of production and wage schedules

$$\left\{\left(x(t),\ w(t)\right) \mathit{for}\ t \in \left[\underline{t}, \bar{t}\right]\right\}.$$

Confronted with this offer each individual worker pretends that type $\hat{t}$ that maximizes the personal benefit:

$$U(\hat{t},t) = w(\hat{t}) - D\left(x(\hat{t}),\ t\right),$$

since $e = q = x$. It is easy to see that the first-best solution is not implementable, because the workers will not carry out their 'assigned' tasks. The reason is that demanding the first-best output according to (8) yet offering a wage that just compensates for disutility, $w = D(x, t)$, will induce the workers to pretend a less efficient type, $\hat{t} < t$, probably the least efficient type, $\hat{t} = \underline{t}$, because this allows them to accrue rents, while truth telling results in zero net benefits to the agent.

Invoking the revelation principle, the manager maximizes the expected aggregate profit π subject to the incentive compatibility constraint

$$u(t) := U(t, t) \geq U(\hat{t}, t) \text{ for all } \hat{t} \neq t,$$

and the individual rationality constraint:

$$u(t) \geq u_0.$$

As usual, the reservation wage u_0 accounting for the agents' outside options is exogenously given, independent of the types and normalized to zero, $u_0 = 0$, however, we will comment on institutionally set reservation wages in the final section. The revelation principle states that the restriction to contracts that induce the agents to reveal their true type does not lower profits. This in turn allows to eliminate wages from the objective, $w(t) = u(t) + \delta(x(t), t)$. Applying the envelope theorem to $u(t)$, the following control problem results for solving the benchmark, wage contracts absent pollution charges:

$$\max_{\{x(t)\}} \int_{\underline{t}}^{\bar{t}} \left[x(t) - D\big(x(t), t\big) - u(t) \right] f(t)\, dt, \tag{12}$$

$$\frac{du}{dt} = -D_t, \; u(\underline{t}) = u_0 = 0. \tag{13}$$

Defining the corresponding Hamiltonian, $H = [x(t) - D(x, t) - u]\, f - \lambda D_t$, the following first-order optimality conditions result according to Feichtinger and Hartl (1986):

$$H_x = (1 - D_x) f - \lambda D_{tx} = 0, \tag{14}$$

$$\dot{\lambda} = f, \; \lambda(\bar{t}) = 0. \tag{15}$$

Integrating the costate variable differential equation (15) accounting for the transversality condition, $\lambda(\bar{t}) = 0$, yields $\lambda(t) = F(t) - 1$. Substituting this solution into the maximum principle (14) and introducing the already in (6) defined hazard rate, $h = f/(1 - F)$, yields the following characterization of the optimal prescription of production

$$1 - D_x(x, t) = -D_{xt}(x, t) / h(t). \tag{16}$$

This 'relaxed program' (compare Fudenberg and Tirole, 1992), is optimal if it is monotonically increasing in the efficiency parameter, which is ensured due to the assumptions:

$$\frac{dx}{dt} = \frac{\dfrac{D_{xtt} h - D_{xt} \dot{h}}{h^2} - D_{xt}}{D_{xx} - D_{xxt} / h} > 0. \tag{17}$$

The so far missing wage follows from integrating (13) using the normalized boundary condition $u_0 = 0$ and the definition of $u(t)$:

$$w(t) = \underbrace{\int_{\underline{t}}^{t} -D_v\big(x(v),v\big)dv}_{u(t)} + D\big(x(t),t\big). \tag{18}$$

Eliminating the private information parameter t from the so far calculated schedules $\{(x(t), w(t))$ for $t \in [\underline{t}, \overline{t}\,]\}$ allows to express the wage as a (non-linear) function of the output, $w = W(x)$; an example will be shown in Section 4.

3.2 Optimal Contracts when Individual Pollution Is Not Monitored

If the agent's pollution cannot be monitored, the manager faces a multi-task problem similar to the one analyzed in Holmström and Milgrom (1991). While Holmström and Milgrom (1991) restrict the analysis to linear contracts, general and consequently nonlinear contracts are derived in the following. A consequence of multi-tasking when the task is not observable is that rewarding only the observable output produces the familiar result – you get what you pay for – that is not in the interest of the principal anymore. In particular, the incentive of the previous section that rewards production efforts optimally is not efficient anymore because its leads to too much pollution for which the firm must pay now. Since it is impossible to link rewards to the unobservable pollution, rewarding on the basis of aggregate pollution will not help in sufficiently large production units either. All the managers can do is to design incentives that are optimal accounting for this spill-over from production on pollution. That is, the managers have to solve the following problem:

$$\max_{\{x(t)\}} \int_{\underline{t}}^{\overline{t}} \Big[x(t) - \tau P\big(x(t),0\big) - D\big(x(t),t\big) - u(t)\Big] f(t)dt, \tag{19}$$

$$\frac{du}{dt} = -D_t,\ u(\underline{t}) = u_0 = 0 \cdot \tag{20}$$

The necessary optimality condition for an optimal 'prescription' of output can be expressed in the following way:

$$1 - D_x = -D_{xt} / h + \tau P_x. \tag{21}$$

Hence, the output assigned to each type is reduced accounting for the associated externality. The reason is that a pollution charge lowers the effective price of output, $1 - \tau P_x$, which must equal the marginal costs, $D_x - D_{xt} / h$, which in turn consist of the agent's costs of incremental effort, D_x, and the agency costs, $-D_{xt}/h$. As a consequence, pollution charges flatten the optimal incentive to which the agents respond with less efforts.

3.3 Optimal Contracts with Observable Individual Pollution

When pollution is observable, the multi-tasking argument does not apply anymore, since abatement can be directly rewarded. More precisely, the principal can prescribe both tasks – production quotas and pollution allowances as functions of the agents' types.

$$\max_{\{x(t),p(t)\}} \int_{\underline{t}}^{\bar{t}} \left[x(t) - \tau p(t) - \delta\big(x(t), p(t), t\big) - u(t) \right] f(t)dt, \tag{22}$$

$$\frac{du}{dt} = -\delta_t, \quad u(\underline{t}) = u_0 = 0. \tag{23}$$

Nevertheless, still a reduction in efforts result. That is, the reduction in output exceeds the intermediate friction due to necessary abatement. It is this additional managerial friction that is not accounted for in traditional approaches such as in the CGE models.

In this case, the first-order conditions can be reduced to the following pair of implicit relations:

$$1 - \delta_x(x, p, t) = -\, \delta_{xt}(x, p, t) \,/\, h(t), \tag{24}$$

$$-\,\tau - \delta_p(x, p, t) = -\, \delta_{pt}(x, p, t) \,/\, h(t). \tag{25}$$

Solving this set of non-linear equations is in general impossible, unless for particular simple specifications (see next section). The second condition for the optimal level of pollution may be dominated by the boundary solution, $a = 0$, thus $p = P(x, 0)$.

The reduction in efforts after introducing a pollution charge is economically clear, since this amounts to a reduction in the price of output from 1 to $1 - \tau P_x$. Less obvious is, how far this holds once monitoring is possible. Indeed, an argument similar to Porter's would not exclude the possibility that monitoring allows for an improvement in efforts following his win-win 'logic'. In order to trace the impact of pollution charges on workers' efforts we have to compute

$$\frac{\partial e}{\partial \tau} = \frac{\partial x}{\partial \tau} + \frac{\partial A}{\partial x}\frac{\partial x}{\partial \tau} + \frac{\partial A}{\partial p}\frac{\partial p}{\partial \tau} = \frac{\left(P_a - P_x\right)x_\tau + p_\tau}{P_a} < 0. \tag{26}$$

This derivative is obviously negative, if the derivatives of production and pollution with respect to the pollution charge have the 'proper', expected signs,

$x_\tau < 0$ and $p_\tau < 0$, since $A_x = -P_x / P_a > 0$ and $A_p = 1 / P_a < 0$. Application of the implicit function theorem, see e.g. Rudin (1974), to the first-order conditions (24) and (25), gives

$$\begin{pmatrix} x_\tau \\ p_\tau \end{pmatrix} = -\begin{pmatrix} -\delta_{xx} + \delta_{xxt}/h & -\delta_{xp} + \delta_{xpt}/h \\ -\delta_{xp} + \delta_{pxt}/h & -\delta_{pp} + \delta_{ppt}/h \end{pmatrix}^{-1} \begin{pmatrix} 0 \\ -1 \end{pmatrix} = \frac{1}{\det} \begin{pmatrix} \delta_{xp} - \delta_{xpt}/h \\ \delta_{xxt}/h - \delta_{xx} \end{pmatrix} \tag{27}$$

This proves that the partial derivatives x_τ and p_τ have indeed the proper negative signs due to the assumptions about the (implied) disutility $\delta(x, p, t)$ if in addition the determinant (denoted det), which is necessary to calculate the inverse matrix,

$$\det = \{\underbrace{[\delta_{pp}\delta_{xx} - \delta_{xp}{}^2]}_{+} - \underbrace{[\overset{+}{\delta_{pp}}\overset{-}{\delta_{xxt}} + \overset{+}{\delta_{xx}}\overset{-}{\delta_{ppt}}]\overset{+}{h}}_{+} + \underbrace{\overset{-}{\delta_{ppt}}\overset{-}{\delta_{xxt}}}_{+} + \underbrace{\overset{+}{\delta_{xpt}}[2\overset{-}{\delta_{xp}} - \overset{+}{\delta_{xpt}}]}_{-}\} / h^2 \tag{28}$$

is positive. This determinant is positive, if the first three terms, which are definitely positive for the assumptions made (the first due to the convexity of δ), outweigh the last and negative term. Therefore, given this additional technical proviso that det > 0, which is hard to interpret given that it depends on mixed third order derivatives and that is satisfied in the following numerical examples (and other tried specifications), pollution charges reduce workers' efforts even if pollution is individually monitored.[1]

Economically and intuitively, one expects also that

(1) additional monitoring leads to an increase in efforts compared with the multi-tasking agency set-up when pollution is not observable at an individual level,

(2) this increase in efforts serves both 'outputs', 'goods' production x is increased and pollution p is reduced.

The reason for the first point is that the additional instrument will be only invoked if it reduces shirking. Indeed, a simple example will show that abatement is not always mandated for all agents so that in this case the results of the previous section apply ($a = 0$). The second point follows, because once abatement is optimal, the marginal production costs accounting for the external costs decline, $\delta_x < \tau P_x(x, 0)$, so that equating to the price yields a higher output. An arithmetic derivation is omitted due to the convincing economic logic and because establishing these properties theoretically is tedious and depends again on third order derivatives.

An interesting implication of this increase in efforts due to individual monitoring is that the expenditures for monitoring equipment are partially financed by the reduction in shirking. In other words, accounting for this feedback, investing in

1 Porter's thesis cannot be entirely ruled out, if one introduces pollution as an additional signalling device.

such equipment can turn out to be profitable, when a standard cost-benefit comparison suggests the opposite.

4 Comparison and Examples

A numerical example

$$D(e, t) = \tfrac{1}{2}\gamma e^2 / t,$$

$$P(x, a) = x / (1 + \alpha a) \;\Rightarrow\; P(x, 0) = x \text{ and } A(x, p) = (x / p - 1) / \alpha,$$

$$f(t) = 1 / (\bar{t} - \underline{t}), \text{ for } t \in [\underline{t}, \bar{t}], \text{ otherwise } f(t) = 0, \text{ thus } (1 / h(t)) = (\bar{t} - t),$$

is introduced to highlight the theoretical properties and the differences between the different outcomes. The model is normalized so that all redundant parameters are eliminated. Disutility D is simply quadratic in effort e and reciprocal to the type t such that higher types have less disutility. Pollution is linear in output and reduced by abatement efforts as assumed (i.e. satisfying the law of diminishing returns). This assumption is, of course, optimistic concerning the impact of environmental policies and pollution charges in particular since pollution is most likely convex in output. The crucial function of disutility implied by the tasks is given by

$$\delta = \tfrac{1}{2}\gamma\,[x + (x / p - 1) / \alpha]^2 / t.$$

Clearly, δ satisfies all the assumptions. The final assumption is that the types are uniformly distributed, which leads to a particularly simple, since linear, inverse hazard rate $1/h$. The purpose of this specification is to allow for plausible examples and for analytical solutions. An interesting consequence of this specification is that it allows for a pooling of the agents such that the less efficient ones, those that produce anyway little of output, and thus pollution and abatement, are exempted from abatement efforts.

The scenario of multi-tasking leads right away to the 'optimal' production that is quadratic in the types:

$$x = t^2(1 - \tau) / (\gamma\bar{t}).$$

The special case, $\tau = 0$, corresponds to the benchmark, no pollution charge. The relative reduction in output equals in this simple specification the tax rate. The reason for this is that the flattened incentive – see Figure 2 – induces the workers to reduce efforts, i.e. to shirk. This response of workers providing less effort due to necessarily weakened incentives, which may add up to a significant share, is overlooked in the CGE and related approaches.

Allowing for abatement yields for the optimal production task:

$$x(t) = \frac{\gamma p \bar{t}(1+\alpha p) + \alpha^2 p^2 t^2}{\gamma \bar{t}(1+\alpha p)^2},$$

which depends only indirectly – via the necessary abatement to reach pollution p – on the tax τ due to the supposed linearity of P in x. Substitution of this production target into the condition for 'optimal' pollution (25) yields a cubic relation in p that allows for an analytical solution, which, however, is suppressed because this cumbersome expression adds little further economic insight.

Figure 3 compares the efforts associated with the different scenarios. A stunning observation, although already established on theoretical grounds, is in this example that the slack introduced by pollution charges exceeds the actual abatement efforts by far. Reducing the efficiency of abatement efforts, setting $\alpha = 1$, implies that the rather inefficient types ($t < 7$, thus roughly $^2/_5$ of the labor force) should not worry about pollution at all and should thus produce as in the case of no monitoring. On the positive side monitoring allows to mitigate this negative impact of pollution charges substantially, i.e. to increase the efforts of the workers. In this example, the induced increase in effort is substantial, which underlines the claim that installing monitoring equipment can reclaim a significant portion of its costs by increasing efforts.

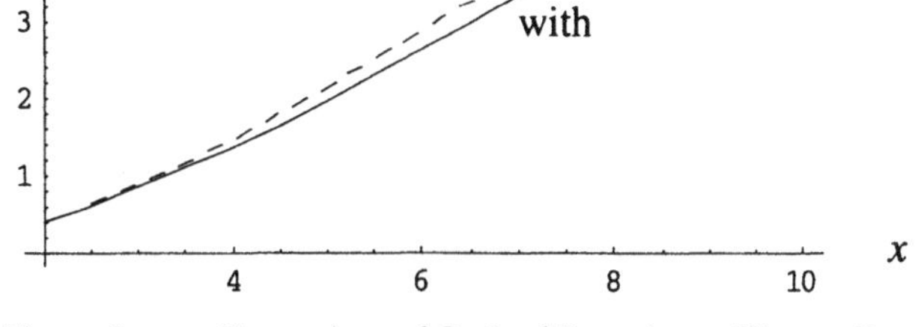

Figure 2: Comparison of Optimal Incentives – Wage w Depending on Output x – with and without Pollution Charges, when Individual Abatement Cannot Be Monitored, $\underline{t} = 5$, $\bar{t} = 10$, $\gamma = 1$.

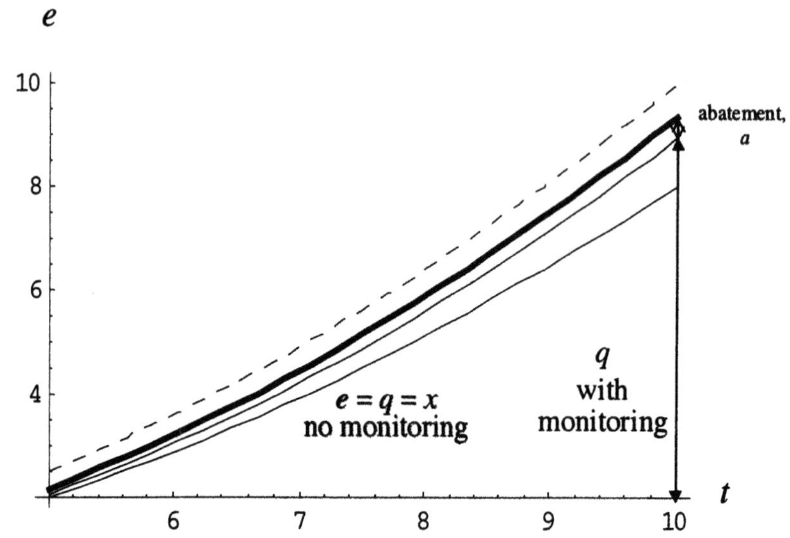

Figure 3: Comparison of the Efforts, e, for $\underline{t} = 5$, $\bar{t} = 10$, $\gamma = 1$, $\alpha = 10$, Dashing = No Pollution Charge.

5 Final Remarks

An interesting feature of the model is that it provides a partial explanation of the meager economic record – low growth and high unemployment rates, in particular in Europe – in the late 1980s and early 1990s despite the dramatic energy price cut in 1986. That is, the simultaneous initiation of ambitious environmental policies aggravated by the managerial friction discussed in this paper explains the apparent 'asymmetry' to energy price changes: an increase in energy prices is more harmful than a decline is beneficial.

The fact that the European unions' demand for wages did not account for the additional economic burden implicit in environmental regulations may provide a partial explanation of the much worse unemployment record in continental Europe. This follows from this model, because fixing the agent's reservation payoff u_0 (by minimum wages and unemployment benefits) yet introducing pollution charges reduces the 'efficiency' of 'marginal' workers to which the firms react by dismissing them; compare Wirl (2000), who shows in addition that inducing unemployment can be optimal for the unions even when accounting for the lot of the unemployed.

This paper used asymmetric information to explain a difference between traditional reasoning and stylized facts. Unfortunately, the aspects of private information are neglected in many environmental policy issues. This oversight may be substantial, because private information often calls for a complete reversal of policy recommendations as shown by Lewis (1996). A topical example in this direction is the neglect of possible strategic manipulations associated with 'flexible' mechanisms to reduce GHG emissions, like joint implementation and clean development measures (compare Hagem, 1996 and Wirl et al., 1998). The consequences of this oversight are observable in the case of conservation programs that did not deliver the promised 'gigabucks' (Lovins, 1985), but only 'stranded costs' (see Wirl, 2000). Another aspect that is in my opinion insufficiently treated in environmental economics concerns positive issues since the bulk of the literature seeks optimal interventions for benevolent, omnipotent, and in most cases omniscient dictators. This is, of course, naive, and a recent exception is e.g. Fredriksson (2001).

References

Bovenberg, L.A. and R.A. de Mooij (1994), Environmental Levies and Distortionary Taxation, *American Economic Review* 84, 1085-1089.

Conrad, K. and M. Schröder (1988), Effekte einer Emissionssteuer auf sektorale Produktionsstrukturen am Beispiel Baden-Würtemberg, in: Siebert, H. (Ed.), *Umweltschutz für Luft und Wasser*, Heidelberg, 1-32.

Conrad, K. and D. Wastl (1995), The Impact of Environmental Regulation on Productivity in German Industries, *Empirical Economics* 20, 615-633.

Färe, R., S. Grosskopf, and C. Pasurka (1986), Effects on Relative Efficiency in Electric Power Generation Due to Environmental Controls, *Resources and Energy* 8, 167-184.

Fredriksson, P.G. (2001), How Pollution Taxes May Increase Pollution and Reduce Net Revenues, *Public Choice* 107, 65-85.

Fudenberg, D. and J. Tirole (1992), *Game Theory*, 2nd printing, Cambridge (Mass.).

Gollob, F.M. and M.J. Roberts (1983), Environmental Regulations and Productivity Growth: The Case of Fossil Fueled Electric Power Plants, *Journal of Political Economy* 91, 654-674.

Gore, A. (1993), *Earth in the Balance: Ecology and the Human Spirit*.

Gray, W.B. (1987), The Cost of Regulation: OSHA, EPA, and the Productivity Slowdown, *American Economic Review* 77, 998-1006.

Hagem, C. (1996), Joint Implementation Under Asymmetric Information and Strategic Behaviour, *Environmental and Resource Economic* 8 (4), 431-447.

Holmström, B. and P. Milgrom (1991), Multitask Principal-Agent Analysis: Incentives, Contracts, Asset Ownership and Job Design, *Journal of Law, Economics and Organization* 7 (Special Issue), 24-52.

Jaffe, A.B., S.R. Peterson, and P.R. Portnoy (1995), Environmental Regulation and the Competitiveness of U.S. Manufacturing: What Does the Evidence Tell Us?, *Journal of Economic Literature* XXXIII, 132-163.

Köppl, A., K. Kratena, C. Pichl, F. Schebeck, S. Schleicher, and M. Wüger (1995), *Makroökonomische und sektorale Auswirkungen einer umweltorientierten Energiebesteuerung in Österreich*, Wien.

Lewis, T.R. (1996), Protecting the Environment When Costs and Benefit Are Privately Known, *Rand Journal of Economics* 27, 819-847.

Lovins, A.B., L.H. Lovins, and P. Hawken (1999), A Road for Natural Capitalism, *Harvard Business Review* (May-June), 145-158.

Lovins, A.B. (1985), Saving Gigabucks With Negawatts, *Public Utilities Fortnightly* 115 (6), 19-26.

Porter, M.E. and C. van der Linde (1995a), Green and Competitive: Ending the Stalemate, *Harvard Business Review* (Sept.-Oct.), 120-134.

Porter, M.E. and C. van der Linde (1995b), *Journal of Economic Perspectives.*

Rudin, W. (1976), *Principles of Mathematical Analysis*, 3rd edition, New York.

Schmidheiny, S. (with the Business Council for Sustainable Development) (1992), *Changing Course: A Global Business Perspective on Development and the Environment.*

Welsch, H. and F. Coster (1994), A General Equilibrium Analysis of European Carbon/Energy Taxation: Model Structure and Preliminary Results, presented at the Ausschuss für Umwelt- und Ressourcenökonomie in Ladenburg near Mannheim, 29. April.

Wirl, F. (1989), Impact of Environmental Regulation on Economic Activity – Austria, *EMPIRICA* 16, 209-233.

Wirl, F. (2000), Lessons from the Utility Conservation Programmes in the United States, *Energy Journal* 21 (1), 101-122.

Wirl, F., C. Huber, and I.O. Walker (1998), Joint Implementation: Strategic Reactions and Possible Remedies, *Environmental and Resource Economics* 12, 203-224.

Wirl, F. and H.J. Hoffmann (1993), Environmental Regulation in the F.R.G.: Its Economic Costs and the Sectoral Impact, *Journal of Environmental Management* 37, 103-115.

An Economic Assessment of the Kyoto Protocol Using a Global Model Based on the Marginal Abatement Costs of 12 Regions

ZhongXiang Zhang

Research Program, East-West Center, 1601 East-West Road, Honolulu, HI 96848-1601, USA, *ZhangZ@EastWestCenter.org*

Summary:
The Kyoto Protocol incorporates emissions trading, joint implementation, and the clean development mechanism to help Annex 1 countries to meet their Kyoto targets at a lower overall cost. Using a global model based on the marginal abatement costs of 12 countries and regions, this paper estimates the contributions of the three Kyoto flexibility mechanisms to meet the total greenhouse gas emissions reductions required of Annex 1 countries under the three trading scenarios respectively. Our results clearly demonstrate that the fewer the restrictions on the use of flexibility mechanisms the gains from their use are greater. The gains are unevenly distributed, however, with Annex 1 countries that have the highest autarkic marginal abatement costs tending to benefit the most. Our results also indicate that restrictions on the use of flexibility mechanisms not only reduce potential of the Annex 1 countries' efficiency gains, but also are not beneficial to developing countries because they restrict the total financial flows to developing countries under the clean development mechanism.

Keywords: Marginal abatement costs, emissions trading, clean development mechanism, joint implementation, Kyoto Protocol, greenhouse gases.

1 Introduction

The Kyoto Protocol to the United Nations Framework Convention on Climate Change (UNFCCC), agreed in December 1997, Kyoto, is the first international environmental agreement that sets legally binding emissions targets for a basket of six greenhouse gases and timetables for Annex 1 countries (i.e., the OECD countries and countries with economies in transition). Together, Annex 1 countries must reduce their emissions of six greenhouse gases by 5.2% below 1990 levels over the commitment period 2008-2012 (UNFCCC, 1997). Reflecting the underlying principle of the UNFCCC, which states "policies and measures to deal with climate change should be cost-effective so as to ensure global benefits at the lowest possible cost", the Kyoto Protocol incorporates a variety of provisions for flexibility mechanisms through which the costs of abating emissions can be lowered. Article 6 authorizes the transfer or acquisition of "emission reduction units" from joint implementation (JI) projects among Annex 1 Parties. Article 12 establishes the so-called "clean development mechanism" (CDM). Through the mechanism, Annex 1 countries will be able to obtain the certified emissions reductions from clean development projects jointly implemented with non-Annex 1 countries (i.e., developing countries), and use them to count towards meeting their commitments under the Kyoto Protocol. In addition to the two project-based mechanisms, the Kyoto Protocol accepts the concept of emissions trading in principle, under which one Annex B (an annex to the Kyoto Protocol that lists the quantified emission limitation or reduction commitment per Party) country or its sub-national entities would be allowed to purchase the rights to emit greenhouse gases (GHG) from other Annex B countries or their regulated entities that are able to cut GHG emissions below their assigned amounts or their targets. Currently, international climate change negotiations are finalizing the rules, procedures and institutions needed to make these flexibility mechanisms fully operational.

To what extent Annex 1 countries' compliance costs can be lowered depends on the extent to which the flexibility mechanisms will be allowed to contribute to meet their Kyoto targets. Under the Kyoto Protocol, each of the Articles defining the three flexibility mechanisms carries wording that the use of the mechanism must be supplemental to domestic actions. Whether the supplementarity clauses will be translated into a concrete ceiling, and if so, how should a concrete ceiling on the use of the three flexible mechanisms be defined remain to be determined. At the June 1999 Sessions of the Subsidiary Bodies of the UNFCCC, the EU (European Union, 1999) has tabled a proposal for concrete ceilings on the use of flexibility mechanisms. The EU proposal calls for the limits on both buying countries and selling countries. For a buying country, the maximum purchase for GHG emission reduction units via all three flexible mechanisms cannot exceed the higher of the following two alternatives:

(1) 5% of {(its base year emissions multiplied by 5 + its assigned amount)/2};

(2) 50% of the difference between its annual actual emissions in any year between 1994 and 2002, multiplied by 5, and its assigned amount.

The EU proposal is based on quantities already agreed upon or emissions that will be observed before the proposed restriction becomes applicable (Ellerman and Wing, 2000). The difference between the two alternatives is that the first is based mainly on the Kyoto Protocol's quantified emission limitation or reduction commitments, whereas the second takes the actual emission reduction efforts of buying countries as its basis. One reason behind the two alternatives is that industrialized countries whose emissions are already very high on a per capita basis should take the lead in reducing their own emissions so that developing countries are encouraged to follow suit and take on emissions commits at a later date. Another reason has been to urge Annex 1 countries to stimulate technical innovation domestically by raising marginal abatement costs of buying countries, although it is unclear to what extent a stimulus of increased technical innovation in buying countries would remain. Motivated by alleviating the concern about hot air, the EU proposal also sets the rule for a selling country. Similar to the first alternative for a buying country, the EU proposal specifies that the maximum allowed sale for GHG emission reduction units via all three flexible mechanisms cannot exceed the amount calculated by: 5% of {(its base year emissions multiplied by 5 + its assigned amount)/2}, referred to hereafter as Alternative 1. This proposed restriction on transfers provides an indirect way of implementing supplementarity since the higher market price as a result of the restriction on the amount of hot air for sale restricts the acquisitions from what would otherwise have occurred.

Under the EU proposal, "however, the ceiling on net acquisitions and on net transfers can be increased to the extent that an Annex B Party achieves emissions reductions larger than the relevant ceiling in the commitment period through domestic action undertaken after 1993, if demonstrated by the Party in a verifiable manner and subject to the expert review process to be developed under Article 8 of the Kyoto Protocol." (European Union, 1999). This is the so-called however clause. It allows an importing (exporting) country to purchase (sell) more than the amount defined by the above alternatives if verifiable domestic abatement by the importing (exporting) country can be demonstrated. Thus, the however clause effectively raises the importing ceiling and allows an importing country to purchase emissions reductions from abroad up to 50% of the emission reduction requirement, provided that the country can verify a similar volume of domestic abatement undertaken after 1993.

Given the emissions reductions required of Annex 1 countries, part of the needed reductions will take place domestically, whereas the rest will come from emissions trading and JI with other Annex 1 countries, and acquisitions of the certified CDM credits from non-Annex 1 countries. Using a global model based on marginal abatement costs of 12 countries and regions, this paper aims to examine the implications of the differing extent of use of these flexibility mechanisms for both Annex 1 countries and non-Annex 1 countries as well as for the market price of permits under the no limits scenario, under the EU ceilings with the however clause scenario, and under the no hot air scenario. Section 2 describes the 12-region's marginal abatement cost-based model used in determining the division of abatement actions at home and abroad. Section 3 discusses the data used. Section 4 presents the results. Finally, Section 5 draws up the main conclusions.

In writing the paper, we have taken a balanced approach combining the calculations with policy analysis in order to put the numbers into perspective and thus facilitate the understanding of what the numbers say. Wherever necessary we have compared our estimates with those from a variety of economic modeling studies. Although not aimed to provide a completely rigorous explanation for the differences between these estimates, such a comparison will indicate the range of such estimates and identify the sources of the differences, thus providing the broad perspective on the issues examined.

2 The Global Model Based on Marginal Abatement Costs of 12 Countries and Regions

Given the emissions reductions required of Annex 1 countries, part of the needed reductions will take place domestically, whereas the rest will come from emissions trading and JI with other Annex 1 countries, and acquisitions of the certified CDM credits from non-Annex 1 countries. The relative differential between the marginal cost of domestic abatement and the international price of emissions permits would apportion the total emissions reductions into domestic reductions and the demand for GHG offsets. Thus, in order to divide up abatement actions at home and abroad, we need to determine the marginal abatement costs for all the countries. Because it is a daunting task to estimate the marginal abatement cost for each individual country, we do so at a regional level. We divide the world into twelve regions. The acronyms for the twelve regions are given in Table 1. The first six regions are Annex 1 regions, whereas the other six are non-Annex 1 regions whose emissions are unconstrained under the Kyoto Protocol. In defining these regions, we have attempted to employ the minimal level of disaggregation necessary for the purpose of this study.

Table 1: Definitions of Countries and Regions.

Annex 1 countries and regions	Non-Annex 1 countries and regions
1. US: United States	7. EEX: Energy Exporting Countries
2. JP: Japan	8. CN: China
3. EU: European Union	9. IN: India
4. OOE: Other OECD Countries	10. DAE: Dynamic Asian Economies
5. EE: Eastern Europe	11. BR: Brazil
6. FSU: Former Soviet Union	12. ROW: Rest of the World

Now let us describe the model for determining the division of abatement actions at home and abroad. To start with, we assume that the marginal abatement cost function for region i is of the quadratic form:

$$MCA_i = a_i Q_i^2 + b_i Q_i$$

where MCA_i is the marginal cost of domestic abatement for region i, Q_i is the amount of greenhouse gas abatement undertaken domestically in million tons of carbon, and a_i and b_i are coefficients. By integration, the total cost of domestic abatement, TCD_i, is:

$$TCD_i = \int_0^{Q_i} (a_i q_i^2 + b_i q_i) dq_i = \tfrac{1}{3} \times a_i Q_i^3 + \tfrac{1}{2} \times b_i Q_i^2 \text{ for every } i. \quad (1)$$

Denoting the total emissions reductions required of the Kyoto-constrained Annex 1 region i by D_i and the amount of hot air that is allocated to region i by H_i, the total autarkic abatement cost, that is, the total abatement cost in the absence of emissions trading (i.e., in the no trading case), TCA_i, is calculated as

$$TCA_i = \int_0^{D_i - H_i} (a_i q_i^2 + b_i q_i) dq_i = \tfrac{1}{3} \times a_i (D_i - H_i)^3 + \tfrac{1}{2} \times b_i (D_i - H_i)^2 \quad (2)$$

for $i = US, JP, EU, OOE, EE$.

Emissions trading helps a region with high autarkic marginal cost to lower its compliance cost by avoiding the undertaking of more costly domestic actions. In this case, the region undertakes domestic abatement Q_i ($Q_i < D_i$) at the marginal cost

$$a_i Q_i^2 + b_i Q_i = MCA_i \text{ for } i = US, JP, EU, OOE, EE, \quad (3)$$

and meets the remaining demand (D_i-Q_i-H_i) via purchasing the "right to emit" at the international price p. So, the total remaining demand of all purchasing regions, TD, is:

$$TD = \sum_i (D_i - Q_i - H_i) \quad i = US, JP, EU, OOE, EE. \quad (4)$$

Measured as percentages of the total abatement costs in the absence of emissions trading, the reductions in abatement costs for purchasing region i is:

$$G_i = \frac{TCA_i - (TCD_i + p(D_i - Q_i - H_i))}{TCA_i} \quad i = US, JP, EU, OOE, EE \quad (5)$$

where G_i is the gains from emissions trading for purchasing region i. On the other hand, for the Kyoto-unconstrained region with lower marginal cost, emissions trading provides an incentive for it to undertake otherwise no domestic abatement actions and sell the permits generated to those higher cost regions at the international price p

$$a_i Q_i^2 + b_i Q_i = p \quad \text{for } i = FSU, EEX, CN, IN, DAE, BR, ROW. \quad (6)$$

Thus, the total amount of emissions permits available for sale, *TS*, are:

$$TS = \sum_i Q_i \quad i = FSU, EEX, CN, IN, DAE, BR, ROW, \tag{7}$$

of which, the value of the total non-Annex 1 countries' sales for permits to the Annex 1 regions, that is, the value of the CDM market, *VCDM*, is:

$$VCDM = \sum_i pQ_i \quad i = EEX, CN, IN, DAE, BR, ROW. \tag{8}$$

Subtracting their own abatement costs, the net value of the CDM market, or the net gain of non-Annex 1 regions, *NCDM*, is derived as

$$NCDM = \sum_i (pQ_i - TCD_i) \quad i = EEX, CN, IN, DAE, BR, ROW. \tag{9}$$

At the equilibrium, the total amount of demand for emissions permits are equal to the total supply so that we have

$$\sum_i H_i + TS = TD \quad i = US, JP, EU, OOE, EE, FSU. \tag{10}$$

This completes the summarized description of the general model. When it is used to examine the following cases, some specific settings are involved.

- Case A. No Limits Scenario

Even if no limits are imposed on the use of flexibility mechanisms, it is still in the interest of a purchasing country to abate its own emissions up to the point where the marginal cost of doing so is equal to the prevailing price of permits. Thus, equation (3) merges with equation (6) so that we have

$$a_i Q_i^2 + b_i Q_i = MCA_i = p \quad \text{for every } i. \tag{11}$$

This implies that the marginal cost of domestic abatement for each region is the same and that there is no distinction between the international price and domestic prices.

- Case B. Supplementarity Restrictions Scenarios

When a uniform formula defining the restrictions on the use of flexibility mechanisms is applied to all the regions, some regions might be allowed to purchase more than needed. Put another way, it is more costly for the regions to purchase part of the allowed acquisitions than to abate them domestically. To remove the unnecessary part, we set equation (3) into

$$a_i Q_i^2 + b_i Q_i = MCA_i \geq p \quad \text{for } i = US, JP, EU, OOE, EE. \tag{12}$$

As a result, a region is going to purchase permits only if it becomes more costly to undertake emissions abatement on its own. By contrast, for a region that is allowed to meet part of its commitments via the purchase of emissions permits, it must rely on domestic abatement capabilities. The lesser extent it is allowed to purchase permits abroad, the higher the domestic prices, and hence the larger the distinction between the international price and domestic prices.

- Case C. No Hot Air Scenario

Under no hot air scenario, trading in hot air is not allowed. Thus, equation (10) becomes

$$TS = TD. \tag{13}$$

3 Data

To run the above model, we first need the aggregate magnitude of emissions reductions required of each Annex 1 region and the size of hot air in 2010. The former represents the amount of the mandated reductions from projected business-as-usual (BAU) emissions levels, whereas the latter represents those assigned amounts under the Kyoto Protocol that exceed anticipated emissions requirements even in the absence of any limitation. The two types of data are derived from the individual national communications (to the UNFCCC) of the following 35 Annex 1 countries with emissions targets: Australia, Austria, Belgium, Bulgaria, Canada, Czech Republic, Denmark, Estonia, Finland, France, Germany, Greece, Hungary, Iceland, Ireland, Italy, Japan, Latvia, Lithuania, Luxembourg, the Netherlands, New Zealand, Norway, Poland, Portugal, Romania, Russian Federation, Slovakia, Slovenia, Spain, Sweden, Switzerland, Ukraine, the United Kingdom, and the United States. This involves three steps. The first step is to determine GHG emissions for each Annex 1 country in the base year. The second step is to determine the Kyoto target for each Annex 1 country in 2010. The third step is to estimate baseline GHG emissions for each Annex 1 country in 2010. By adding up the amount of the mandated reductions from projected baseline emissions levels for each Annex 1 country, the aggregate magnitude of emissions reductions required of Annex 1 countries in 2010 is estimated to be 620.6 million tons of carbon (MtC) equivalent, as given in Table 2. Similarly, the size of hot air in 2010 is calculated to be 105.0 MtC.[1] See Zhang (1999) for detailed discussion on procedures and results.

---

1 Note that in some regions there is the co-existence of hot air and the required emissions reductions within the same region. This is simply because of the sums across countries within each of these regions. For an individual Annex 1 country, it is either required to reduce its emissions to meet the Kyoto targets or not required if it has hot air.

Table 2: Annex 1 Regions' Emissions Reductions Required and the Size of Hot Air in 2010.

Annex 1 regions	Emissions reductions required in 2010 (MtC)	The size of hot air in 2010 (MtC)
United States	423.9	0
Japan	71.2	0
European Union	40.6	12.7
Other OECD Countries	57.3	0
Eastern Europe	27.6	10.9
Former Soviet Union	-	81.4
Annex 1 Total	620.6	105.0

Source: Zhang (1999).

The second set of data are the maximum allowed acquisitions and the maximum allowed transfers in 2010. Applying the two EU ceiling alternatives to each Annex 1 country, and assuming that countries would wish to use the higher allowed acquisitions in 2010, we have calculated the maximum allowed acquisitions in 2010 for those Annex 1 countries whose emissions targets in 2010 are below their projected BAU emissions (as given in Table 3) without considering the however clause. When the however clause is factored in, the above ceilings on the allowed acquisitions are relaxed up to 50% of the difference between projected baseline emissions and the Kyoto targets in 2010. Following the same procedure as one in calculating the maximum allowed acquisitions, we have calculated the maximum allowed transfers in 2010 for those Annex 1 countries whose emissions targets in 2010 are above their projected BAU emissions. See Zhang (1999) for detailed discussion on procedures and results.

Table 3: Allowed Acquisitions and Transfers of GHG Emission Reduction Units Under the Two EU Ceiling Alternatives.

Annex 1 regions	Maximum allowed acquisitions in 2010 without considering the however clause (MtC)	Maximum allowed transfers in 2010 (MtC)
United States	136.8	-
Japan	31.6	-
European Union	56.4	10.9
Other OECD Countries	23.3	-
Eastern Europe	14.1	4.4
Former Soviet Union	-	54.9
Annex 1 Total	261.9	70.2

Source: Zhang (1999).

The third set of data are coefficients of the marginal abatement cost function for each of the above twelve regions. They are taken from Ellerman and Decaux

(1998) who have estimated the marginal abatement cost (at 1985 US$ per ton of carbon) functions for the above twelve regions running the Emissions Prediction and Policy Analysis (EPPA) model under progressively stringent carbon constraints for the year 2010 and then econometrically estimating the EPPA runs for the amount of abated carbon emissions and the corresponding marginal abatement costs. The results are given in Table 4. It can be seen that the assumed cost functions fit very well to the results from the EPPA runs because R^2 is very close to one.

Table 4: Coefficients of the Marginal Abatement Cost Functions of the Form: $MCA_i = a_i Q_i^2 + b_i Q_i$.

Region	a_i	b_i	R^2
USA	0.0005	0.0398	0.9923
JPN	0.0155	1.8160	0.9938
EEC	0.0024	0.1503	0.9951
OOE	0.0085	-0.0986	0.9981
EET	0.0079	0.0486	0.9973
FSU	0.0023	0.0042	0.9938
EEX	0.0032	0.3029	0.9983
CHN	0.00007	0.0239	0.9992
IND	0.0015	0.0787	0.9970
DAE	0.0047	0.3774	0.9996
BRA	0.5612	8.4974	0.9997
ROW	0.0021	0.0805	0.9967

Source: Ellerman and Decaux (1998).

4 The Economic Effects of the Three Trading Schemes in 2010

How the total emissions reductions required of Annex 1 countries will be met depends on the extent to which the flexibility mechanisms will be allowed to contribute to meet their Kyoto targets. This is the area that remains to be decided by the climate change negotiators. For the purpose of illustration, we examine three trading scenarios here.

- No limits scenario: no caps are imposed on the use of all three flexibility mechanisms so that one Annex 1 country can trade as much as it wishes until it becomes more costly for the country to trade than to abate domestically;

- The EU ceilings with the however clause scenario: just as the name implies, the scenario follows the EU proposal for concrete ceilings on the use of all three flexibility mechanisms, as discussed in Section 1. For an importing country, the maximum acquisitions from all three flexibility mechanisms are allowed up to 50% of the difference between projected baseline emissions and the Kyoto targets in 2010, provided that the country can verify a similar volume of domestic abatement undertaken after 1993. We simply assume that such a verification is possible without incurring significant transaction costs. On the export side, we assume that unconstrained countries (those with hot air) would be limited to exporting only the amount of hot air, which is defined by the alternative 1 under the EU proposal; and
- No hot air scenario: trading in hot air is not allowed, indicating that any effectuated trading in GHG emissions must represent 'real' emissions reductions.

Under the above three scenarios, trading of emissions permits is assumed to take place globally among all the countries. But, under the Kyoto Protocol non-Annex 1 countries currently have no obligations to reduce their GHG emissions. Thus, in our modeling exercises, these countries are treated as if they agreed to constrain their emissions in such a manner that they are allocated permits equal to their projected baseline emissions. As such, non-Annex 1 countries only reduce their emissions by an amount equal to the number of permits they wish to sell. Following the definition of the certified credits from CDM projects, the amount of emissions reductions below the country-wide baseline trend is termed as the supply of the certified CDM credits from each non-Annex 1 countries.[2]

4.1 Contributions of the Kyoto Flexibility Mechanisms

For countries whose emissions targets specified under the Kyoto Protocol are expected to exceed their anticipated emissions requirements even in the absence of any limitation, Section 3 indicates that these countries in 2010 will have surplus assigned amounts of 105.0 MtC available for sale. Because hot air is available at zero abatement cost, hot air is assumed to be used to the maximum extent allowed, except for under the no hot air scenario. This means that, of the aggregate demand of 620.6 MtC in 2010, hot air contributes to 105.0 MtC under the no limits

[2] This implies that in our modelling exercises we treat the CDM synonymously with emissions trading as many other modellers do (Weyant, 1999). In real practice, the CDM is a project-based mechanism. Unlike homogenous permits under emissions trading, concerns about additionality and the inherent difficulty of establishing counterfactual baselines for heterogeneous CDM projects and monitoring emissions reductions below the baselines may impose high transaction costs and thereby limit the supply of CDM credits from non-Annex 1 countries (Ellerman and Decaux, 1998; US Administration, 1998).

scenario, and to 70.2 MtC under the EU ceilings with the however clause scenario. Then, the remaining demand will be met via domestic abatement actions and three flexibility mechanisms.

Inserting the aggregate magnitude of emissions reductions required and the size of hot air in 2010, the maximum allowed acquisitions and the maximum allowed transfers in 2010, and the estimated coefficients of the marginal abatement cost function for each of the above twelve regions, we solve the 12-region's marginal abatement cost-based model for determining the division of abatement actions at home and abroad using GAMS, a widely distributed nonlinear programming package (Brooke et al., 1996).

According to our calculations, even if no limits are imposed on the use of all three flexibility mechanisms, not all the emissions reductions required of an Annex 1 country will take place abroad. This is because even if no account is taken of ancillary benefits of reducing GHG emissions, it is still in the interest of any buying country to reduce its own emissions as long as the marginal costs of doing so are lower than the prevailing prices of permits. Based on the marginal costs of domestic abatement for the twelve regions, it is estimated that a reduction of 171.7 MtC, or 27.7% of the total needed emissions reductions in 2010 will be met through domestic actions of Annex 1 countries under the no limits scenario (see Table 5). This is broadly in line with the finding of the IPCC (1996) that reducing emissions by 20% from 1990 levels in developed countries within the next two or three decades can be achieved domestically through no-regrets options. When trading in hot air is not allowed, the international price of permits increases in comparison with the no limits scenario (see Table 6). As a result, the demand for permits abroad decreases, and more and more domestic abatement is undertaken. This is confirmed in Table 5, which indicates that a reduction of 203.5 MtC, or 32.8% of the total needed emissions reductions in 2010 will be met through domestic actions of Annex 1 countries under the no hot air scenario. Under the EU ceilings with the however clause scenario, the amount of abatement through domestic actions in 2010 is estimated to be 315.4 MtC, or 50.8% of the total emissions reductions required of Annex 1 countries in 2010. The reason why the Annex 1 countries as a whole abate slightly over 50% domestically under this EU ceiling scenario is that Eastern Europe abates 68.7% domestically (Zhang, 2001). This in turn is because its low marginal abatement cost allows it to benefit from exporting permits up to the extent that its domestic abatement cost equals the international price of permits.

Table 5: The Contributions of Three Flexibility Mechanisms Under the Three Trading Scenarios in 2010 (MtC).

Scenarios	Domestic actions	Hot air	Emissions trading and JI	CDM	Total supply
No limits	171.7	105.0	51.8	292.1	620.6
EU ceilings with the however clause	315.4	70.2	39.6	195.4	620.6
No hot air	203.5	0	59.6	357.5	620.6

Table 5 summarizes estimates of the contributions of three flexibility mechanisms under the three trading scenarios. It can be seen that the supply of certified CDM credits in 2010 ranges from 195.4 MtC under the EU ceilings with the however clause scenario to 292.1 MtC under the no limits scenario and to 357.5 MtC under the no hot air scenario, respectively. In comparison with the no limits scenario, the EU ceilings with the however clause restriction would reduce the size of the CDM market by 33.1%. Although such a restriction is much relaxed in terms of its impact on the size of the CDM market, in comparison with the EU ceilings without the however clause scenario under which the size of the CDM market is cut by 54.9% (Zhang, 2001), its impact on the size of the CDM market is still very severe. Of the total emissions reductions required of Annex 1 countries in 2010, the supply of the certified CDM credits accounts for 31.5% under the EU ceilings with the however clause scenario, 47.1% under the no limits scenario, and 57.6% under the no hot air scenario, respectively.

Table 6: Autarkic Marginal Abatement Costs in the No Trading Case, and Domestic Prices and the International Prices of Permits in 2010 Under the Three Trading Scenarios (at 1998 US$ per ton of carbon).

Scenarios	United States	Japan	European Union	Other OECD	International price
No emissions trading	160.1	311.8	9.1	33.4	-
No limits	9.6	9.6	9.6	9.6	9.6
EU ceilings with the however clause	46.3	126.4	6.1	6.2	5.6
No hot air	12.6	12.6	12.6	12.6	12.6

When there are no limits imposed on the use of flexibility mechanisms, the marginal cost of domestic abatement for each region equalizes, and there is no distinction between the international price and domestic prices. Based on the model, the international price of permits in 2010 is calculated to be US$ 9.6 per ton of carbon. This price is pushed up to US$ 12.6 per ton when trading in hot air is not allowed. The increase in the international price is partly because any sales for permits by the former Soviet Union are generated by no cost-free abatement undertaken to earn export permits additional to the amount of hot air, and partly because preventing trading in hot air increases the OECD countries' demand for the certified CDM credits. When the supplementarity restriction as specified under the EU proposal is imposed on the acquisitions, the purchases of permits are restricted. This will push down the market price. Thus, the international price of permits is much lower under the EU ceilings with the however clause scenario than under the no limits scenario. Moreover, because no account is taken of the differences in the marginal costs of domestic abatement among the Annex 1 countries, all the countries are required to comply with the same supplementarity rule. This could lead to a distinction between the international price of permits and domestic prices of buying countries. The lesser the extent to which it is allowed to purchase permits abroad, the higher the domestic prices, and hence the larger the distinction between the international price and domestic prices. As indicated in Table 6, because the autarkic marginal abatement cost for Japan is highest, the EU proposed restriction leads to the largest differential between the domestic price in Japan and the international price of permits. On the other hand, because the official projections of baseline GHG emissions in 2010 by most EU member countries are very close to their targets, the EU only needs to purchase a very small amount of permits to meet its targets.[3] As a result, the supplementarity restriction examined here on the EU is much less severe than on Japan and the US. Consequently, domestic price for the EU is very close to the international price of permits.

Because autarkic marginal abatement costs for the OECD countries excluding the EU are much higher than the market price of permits, trading can help these countries to lower their compliance costs by avoiding their undertaking of more costly domestic actions. The extent to which the compliance costs can be lowered depends on the relative differential between the autarkic marginal cost of the country in question and the international price of emissions permits. In percentage terms, the countries whose autarkic marginal costs are much higher than the market price will benefit more than those countries whose autarkic marginal costs are closer to the market price. As indicated in Table 7, because Japan and the US have the highest autarkic marginal abatement costs, the two countries benefit the most from trading both under the no limits scenario and under the no hot air scenario. Measured as percentages of the total abatement costs in the no trading case, the abatement costs of Japan and the US are cut by 93.1% and 85.2% under the no limits scenario, and 91.0% and 81.0% under the no hot air scenario, respectively. By contrast, because the autarkic marginal cost for the EU is very close to the market price, it achieves only small gains from trading under the

3 The low EU baseline projections are attributable in large part to internal burden sharing of the Kyoto commitments among the member countries, having incorporated the impacts of energy policies that are currently being either implemented or negotiated in response to climate change, and to the choice of base year. See Zhang (1999, 2001) for detailed discussion.

above two scenarios. Besides, the restriction on the use of flexibility mechanisms tends to lower the gain from trading. For the OECD as a whole, the reduction in abatement costs, namely, the gain from emissions trading, decreases from 86.5% under the no limits scenario to 78.4% under the EU ceilings with the however clause scenario.

Table 7: The Reductions in the Total Abatement Costs in 2010 Under the Three Trading Scenarios (%).

Scenarios	United States	Japan	European Union	Other OECD	OECD
No limits	85.2	93.1	0.2	45.3	86.5
EU ceilings with the however clause	79.8	76.5	16.3	63.9	78.4
No hot air	81.0	91.0	2.3	33.5	82.4

4.2 The Size of the CDM Market and the Shares of China and India

Multiplying the endogenously-determined, international price of permits by the supply of certified CDM credits from each non-Annex 1 country and summing over all the corresponding product, we can derive the value of the CDM market in 2010. As indicated in Table 8, our estimate ranges from US$ 1103.4 million under the EU ceilings with the however clause scenario to US$ 4512.8 million under the no hot air scenario. Subtracting their own abatement costs, the net value of the CDM market in 2010, or the net gain of non-Annex 1 countries is estimated to be in the range of US$ 603.0 million under the EU ceilings with the however clause scenario to US$ 2559.1 million under the no hot air scenario. The finding that the net gain of non-Annex 1 countries is highest when trading in hot air is not allowed indicates that to prevent trading in hot air, although in practice it is very difficult to distinguish real emissions reductions from hot air, is beneficial to the expanding of the CDM market as well as to the global climate. With respect to the geographical distribution of the CDM flows, because of a great deal of low-cost abatement opportunities available in the energy sectors of China and India and their sheer sizes of population, the two countries are expected to emerge as the dominant host countries of CDM projects. This is confirmed in Table 8, which shows that about three-quarters of the total CDM flows go to China and India, respectively. Because of relatively high abatement costs and relatively small size, the remaining four non-Annex 1 regions account for only 25% of the total flows channeled to developing countries through the CDM.

Table 8: The Value of the CDM Market and the Shares of China and India in 2010 Under the Three Trading Scenarios.

	No limits	EU ceilings without the however clause	No hot air
CDM market (million US$)	2795.6	1103.4	4512.8
of which:			
China	60.3%	60.0%	60.4%
India	15.1%	15.5%	14.9%
Net CDM market (million US$)	1565.0	603.0	2559.1
of which:			
China	59.9%	59.6%	60.1%
India	15.5%	16.0%	15.3%

Because the CDM has an important role in helping Annex 1 countries to meet their Kyoto targets at a lower overall cost, some studies have estimated the potential size of the CDM market. As indicated in Table 9, these estimates vary. Assuming the contributions from domestic abatement actions and hot air and dividing the remaining demand between emissions trading and JI among Annex 1 countries and the CDM within non-Annex 1 countries in proportion to the estimated potential of supply, Haites (1998) estimates that the size of the CDM market in 2010 ranges from 265 MtC under 50% reduction from BAU emissions scenario (under which the maximum allowed acquisitions from all three flexibility mechanisms are limited to 50% of the difference between projected BAU emissions and the Kyoto targets in 2010) to 575 MtC under the no limits scenario. The size of the market estimated by the four economic modeling studies examined ranges from 397 MtC with the OECD GREEN model (Van der Mensbrugghe, 1998) to 723 MtC with the EPPA model (Ellerman and Decaux, 1998). Austin et al. (1998) argue that such estimates derived from these global modeling exercises tend to overestimate CDM flows because, in practice, political limitations and transaction costs will probably keep CDM activity at the lower end of such estimates. Of the studies examined in Table 9, the estimate by Vrolijk (1999) is at the low end of the range. In absolute terms, our estimate is at the low to middle end. This is mainly because, as indicated in Table 9, our estimate of total emissions reductions required of Annex 1 countries, which is based on compilation of the national communications from 35 Annex 1 countries, is lower than those estimates from economic modeling studies. As discussed in Zhang (1999), the main reason is that the official projections of baseline GHG emissions in 2010 by most EU member countries are very close to their targets, thus leading to low demand for emissions reductions. In percentage terms, our estimate of the contribution of the certified CDM credits is broadly in line with other estimates. Our upper bound estimate comes from the no hot air scenario. If the supply of hot air is included as other estimates from economic modeling studies do, then our upper bound estimate will come down to 47.1% under the no limits scenario.

Table 9: Estimates of the Size of the CDM Market in 2010.

	Size of the CDM market (MtC)	Total emissions reductions required of Annex 1 countries (MtC)	Contribution of the CDM (%)
EPPA	723	1312	55
Haites	265-575	1000	27-58
G-Cubed	495	1102	45
GREEN	397	1298	31
SGM	454	1053	43
Vrolijk	67-141	669	10-21
Our projection	195-358	621	31-58

Sources: Edmonds et al. (1998); Ellerman and Decaux (1998); Haites (1998); McKibbin et al. (1999); Van der Mensbrugghe (1998); Vrolijk (1999); Zhang (2001).

Can the CDM actually produce the amount of certified credits as we estimate here? This is not that easy to answer, but experience with activities implemented jointly (AIJ) might give us some indications. The first Conference of the Parties to the UNFCCC in Berlin in April 1995 endorsed a pilot AIJ phase. During the AIJ pilot phase, emissions reductions achieved are not allowed to be credited to current national commitments of investor countries under the UNFCCC. By the time (30 June 1998) of the UNFCCC's second synthesis report on AIJ, 95 projects were listed as AIJ projects (UNFCCC, 1998). These projects are located in 24 host countries, with Africa hosting only one certified AIJ project. If all the projects were fully implemented and operating as designed, they would generate a combined GHG offset of 162 million tons of CO_2 equivalent, namely, 44 MtC equivalent, over an average lifetime of 16.5 years (UNFCCC, 1998). Translated into an annual GHG offset, it amounts to 2.7 MtC. By contrast, the projected contribution of the certified CDM credits implies as much as a 100-fold increase in this type of project-based activities. Although lack of adequate incentives for the private sector participation in AIJ project financing limits the role of the AIJ, to achieve such a substantial scale of increase in the quantity of emissions reductions, although not impossible, poses great institutional challenges for developing countries, given that most non-Annex 1 countries have not experienced an AIJ project within their own countries.

5 Conclusions

The Kyoto Protocol incorporates emissions trading, joint implementation and the clean development mechanism to help Annex 1 countries to lower the costs of meeting their Kyoto targets. To what extent their compliance costs can be lowered depends on the extent to which the flexibility mechanisms will be allowed to contribute to meet their Kyoto targets. Taking the year 2010 as representative of the first commitment period, and using a global model based on marginal abatement costs of 12 countries and regions, this paper estimates how many of the emissions reductions required of Annex 1 countries in 2010 will be met through domestic abatement actions, emissions trading and JI, and acquisitions of the certified CDM credits under the no limits scenario, under the EU ceilings with the however clause scenario, and under the no hot air scenario.

Our results suggest that, of the total emissions reductions required of Annex 1 countries in 2010, domestic actions account for 27.7% under the no limits scenario, 32.8% under the no hot air scenario, and 50.8% under the EU ceilings with the however clause scenario. The contributions of the certified CDM credits in 2010 are estimated to range from 31.5% under the EU ceilings with the however clause scenario to 47.1% under the no limits scenario and to 57.6% under the no hot air scenario. In absolute terms, the supply of the certified CDM credits in 2010 ranges from 195.4 MtC under the EU ceilings with the however clause scenario to 292.1 MtC under the no limits scenario and to 357.5 MtC under the no hot air scenario, respectively. Although it is at the low to middle end of those estimates derived from economic modeling studies, our projected contribution of the certified CDM credits implies as much as a 100-fold increase in the type of project-based activities in comparison with an annual GHG offset through the AIJ of 2.7 MtC. No doubt, to achieve such a substantial scale of increase in the quantity of emissions reductions, although not impossible, poses great institutional challenges for developing countries. With respect to the geographical distribution of the CDM flows, we found that China and India account for about three-quarters of the total flows channeled to developing countries through the CDM.

Besides, our results clearly indicate that the use of flexibility mechanisms helps to lower the OECD countries' compliance costs by avoiding their undertaking of more costly domestic actions. However, the magnitude of reductions in compliance costs differs substantially among the OECD countries. In percentage terms, the countries whose autarkic marginal abatement costs are much higher than the market price of permits will benefit more than those countries whose autarkic marginal abatement costs are closer to the market price. Furthermore, our results demonstrate that restrictions on the use of flexibility mechanisms not only reduce potential of the Annex 1 countries' efficiency gains, but also are not beneficial to developing countries because they restrict the total financial flows to developing countries under the CDM. Thus, from the perspective of husbanding the world's limited resources, the fewer the restrictions on the use of flexibility mechanisms the gains from their use are greater.

Acknowledgements

This paper formed part of the presentations at the Asian Development Bank/United Nations Environment Programme's Workshop for Policymakers on the Institutional Design of the Kyoto Protocol Cooperative Implementation Mechanisms, UN Conference Center, Bangkok, 12-14 October 1999; the International Workshop on Enhancing GHG Mitigation through International Cooperative Mechanisms in Asia, Kanagawa, Japan, 26-27 January 2000; China-US Workshop on Economic and Environmental Modeling, Beijing, 5-6 March 2000; the 3rd Biennial Conference of the European Society for Ecological Economics, Vienna, 3-6 May 2000; the 10th Annual Conference of the European Association of Environmental and Resource Economists, Crete, Greece, 30 June - 2 July 2000; the Joint Meeting of the Energy Modeling Forum and the International Energy Workshop, Stanford University, 20-22 June 2000; the 40th Congress of the European Regional Science Association, Barcelona, 29 August - 1 September 2000; the Workshop on Empirical Modeling of the Economy and the Environment, Mannheim, Germany, 26-27 June 2001; and at the International Expert Meeting on Climate Change and Sustainable Development, Seoul, 6-7 September 2001; as well as at seminars at Tsinghua University, Beijing, 17 March 2000 and East-West Center, Honolulu, 16 April 2001. The author would like to thank participants in the above Workshops and Conferences, J.R. Bhatt, A. Denny Ellerman and Adam Rose for useful discussions and comments on an earlier version of the paper. Financial support from the Asian Development Bank is gratefully acknowledged. The views expressed here are those of the author, and do not necessarily reflect the positions of the Asian Development Bank. All errors and omissions are my own.

References

Austin, D., P. Faeth, R.S. da Motta, C.E.F. Young, C. Ferraz, J. Zou, J. Li, M. Pathak, and L. Srivastava (1998), *Opportunities for Financing Sustainable Development via the CDM: A Discussion Draft.* Presented at a Side-Event of the Fourth Conference of the Parties to the UNFCCC, 7 November, Buenos Aires.

Brooke, A., D. Kendrick, and A. Meeraus (1996), *GAMS Release 2.25: A User's Guide*, GAMS Development Corporation, Washington, D.C. 20007.

Edmonds, J., C. MacCracken, R. Sands, and S. Kim (1998), *Unfinished Business: The Economics of the Kyoto Protocol*, Pacific Northwest National Laboratory, Prepared for U.S. Department of Energy.

Ellerman, A.D. and A. Decaux (1998), Analysis of Post-Kyoto CO_2 Emissions Trading Using Marginal Abatement Curves, MIT Joint Program on the Science and Policy of Global Change, *Report No. 40*, Massachusetts Institute of Technology.

Ellerman, A.D. and I.S. Wing (2000), Supplementarity: An Invitation to Monopsony?, MIT Joint Program on the Science and Policy of Global Change, *Report No. 59*, Massachusetts Institute of Technology.

European Union (1999), *Community Strategy on Climate Change: Council Conclusions*, No. 8346/99, 18 May, Brussels.

Haites, E. (1998), *Estimate of the Potential Market for Cooperative Mechanisms 2010*, Margaree Consultants Inc., Toronto, Canada.

IPCC (1996), *Climate Change 1995: Economic and Social Dimensions of Climate Change*, Contribution of Working Group III to the Second Assessment Report of the Intergovernmental Panel on Climate Change (IPCC), Cambridge.

McKibbin, W.J., M.T. Ross, R. Shackleton, and P.J. Wilcoxen (1999), *Emissions Trading, Capital Flows and the Kyoto Protocol*, Presented at the Intergovernmental Panel on Climate Change Working Group III Expert Meeting on the Economic Impacts of Annex 1 Mitigation Policies on non-Annex 1 Countries, 27-28 May, The Hague.

Tulpulé, V., S. Brown, J. Lim, C. Polidano, H. Pant, and B.S. Fisher (1998), *An Economic Assessment of the Kyoto Protocol Using the Global Trade and Environment Model*, Presented at the OECD Workshop on the Economic Modeling of Climate Change, 17-18 September, Paris.

UNFCCC (1997), *Kyoto Protocol to the United Nations Framework Convention on Climate Change* (UNFCCC), FCCC/CP/1997/L.7/Add.1, Bonn.

UNFCCC (1998), *Review of the Implementation of Commitments and of Other Provisions of the Convention. Activities Implemented Jointly: Review of Progress under the Pilot Phase (Decision 5/CP.1)*, Second Synthesis Report on Activities Implemented, United

Nations Framework Convention on Climate Change (UNFCCC), FCCC/CP/1998/2, Bonn.

US Administration (1998), *The Kyoto Protocol and the President's Policies to Address Climate Change: Administration Economic Analysis*, Washington D.C.

Van der Mensbrugghe, D. (1998), *A (Preliminary) Analysis of the Kyoto Protocol: Using the OECD GREEN Model*, Presented at the OECD Workshop on the Economic Modeling of Climate Change, 17-18 September, Paris.

Vrolijk, C. (1999), The Potential Size of the CDM, *Greenhouse Gas Emissions Trader*, No. 6, UNCTAD, Geneva, 2-4.

Weyant, J.P. (Ed., 1999), The Cost of the Kyoto Protocol: A Multi-Model Evaluation, *Energy Journal* 20 (Special Issue), 1-398.

Zhang, Z.X. (1999), *Estimating the Size of the Potential Market for All Three Flexibility Mechanisms Under the Kyoto Protocol*, Report Prepared for the Asian Development Bank under Contract TA-5592-REG.

Zhang, Z.X. (2001), An Assessment of the EU Proposal for Ceilings on the Use of Kyoto Flexibility Mechanisms, *Ecological Economics* 37 (1), 53-69.

Sharing the Burden of Carbon Abatement in the European Union

Christoph Böhringer

Centre for European Economic Research, P.O. Box 103443, 68034 Mannheim, Germany, *boehringer@zew.de*

Glenn W. Harrison

Department of Economics, Moore School of Business, University of South Carolina, Columbia, SC 29208, USA, *harrison@moore.sc.edu*

Thomas F. Rutherford

Department of Economics, University of Colorado, Boulder, CO 80309-0256, USA, *rutherford@colorado.edu*

Summary:
We evaluate the welfare implications of alternative ways in which the EU could distribute its aggregate emission reduction commitment under the Kyoto Protocol across member states. An endogenous burden sharing calculation, in which the welfare costs across member states are equalized, differs substantially from uniform proportional cutbacks as well as the specific burden sharing rule actually adopted by the EU.

Keywords: Carbon abatement, EU climate policy, burden sharing.

1 Introduction

The Third Conference of the Parties (COP3) in Kyoto, 1997, produced a climate protocol which entails quantified emission limitations for developed countries (United Nations, 1997). Within the Kyoto Protocol the European Union (EU) is committed to cut back its greenhouse gas emissions by 8% on average during the period 2008-2012 as compared to 1990 emission levels. This target will be binding on the member states of the EU as a whole, but there is no presumption that all should bear equal cuts. In fact, a key question for EU climate policy has been how to allocate reduction requirements across EU countries in order to achieve the overall EU target. The answer inevitably involves equity considerations.[1] An overall EU reduction target will only be approved by the parliaments of the EU member countries if the implied distribution of economic costs across EU regions is viewed as fair.

To what extent are policy proposals for reductions schemes consistent with equity considerations? We "translate" the burden sharing rules defined over two major abatement proposals, flat rate reduction targets and differential reduction targets based on the conclusions of the EU Council of Ministers (COM, 1999, Annex 1), into welfare impacts.

One difficulty that must be recognized in any calculation of an equitable sharing of the burden of abatement is that the underlying economy will react to the proposed allocation of the burden. That is, if "equitable" carbon limits are varied across EU members then we would expect that there would be behavioral reactions, causing the assumptions under which the initial limits were determined to be equitable to become invalid. In short, the burden sharing rule needs to be endogenous to some model of the economy in order to derive an allocation of emission reductions across EU countries which is equitable by some welfare metric. The second step in our evaluation, then, is to ask if there exists an endogenous burden sharing rule which will indeed equalize the welfare impacts across member countries. We find that such rules do exist, but are sensitive to (i) the precise degree of aversion to inequity that applies to member states, (ii) the ability of EU members to trade emission rights openly amongst themselves, and (iii) to alternative baseline energy projections which determine the emission intensity of GDP in the business-as-usual reference case.

Our main findings have to do with the distributional impacts of alternative ways in which the EU could distribute an aggregate EU reduction target across its member

---

1 The EU Council of Ministers (Environment) has agreed that "the initial distribution between Member States will be reviewed by the Council ... in any case after completion of protocol negotiations ... taking account of the principles and approaches of ... equitable burden sharing among Member States with regard to the overall emission reduction objective by the Community as a whole ..." (European Commission, 1997, paragraphs 17, 11-14).

states. In addition we investigate the potential for efficiency gains from trade with different allocations rules. The key insights from our analysis are as follows:

- Adopting the EU's assumption on the baseline development of EU economies, neither differentiated targets based on the burden sharing proposal of the EU Council of Ministers nor a flat reduction produces an equitable outcome in welfare terms. When equity is an explicit concern, Germany, France, and the Netherlands will have to accept larger cuts in emission rights than the average, Denmark and the United Kingdom fall slightly below the average, and Italy and Spain will even be allowed to increase their emission over 1990 levels. Compared to the differentiated reduction targets proposed by the EU Council of Ministers, an equitable burden sharing implies that Germany, Denmark, Italy, and the United Kingdom are allocated more emission rights whereas France, Spain, and the Netherlands will obtain less emission rights. The derived pattern for an equitable permit allocation is determined by the "comparative advantage" across countries of mitigating carbon emissions given their benchmark economic structure.
- The welfare implications of different reduction schemes are correlated with exogenous projections of GDP growth rates and improvements in carbon efficiency. These projections determine the carbon emissions in the baseline, which serve as the reference point for emission cutbacks with respect to 1990 levels. As might be expected, the overall magnitude and cross-country differences in welfare costs of abatement are closely tied to differences in reduction requirements for countries with respect to their reference emission level in 2010. The importance of the baseline projections for an equitable burden sharing scheme highlights the need for a broad consensus on the future development of the different EU economies among climate policy negotiators of the member states.
- The overall efficiency gains from allowing trading of carbon emissions rights within the EU are limited. The reason is that cross-country differences in marginal abatement costs are relatively small and are quickly "absorbed" when limited low-cost abatement options are exhausted by polluters with high abatement cost.

We draw these conclusions from simulations of a large-scale computable general equilibrium (CGE) model of EU production, trade and energy usage. The model is originally due to Böhringer, Ferris, and Rutherford (1998), and is extended to consider endogenous burden sharing. In Section 2 we provide a non-algebraic model description, in Section 3 we review the scenarios, in Section 4 the results are presented, and in Section 5 we provide some final remarks.

2 Model Structure and Parameterization

2.1 Regional and Sectoral Disaggregation

We use a static 7-region, 17-sector CGE model[2] of the EU. The regional disaggregation covers seven EU member countries which account for the largest part of EU GDP, trade and carbon emissions: Denmark (DK), Spain (ES), France (FR), Germany (GE), Italy (IT), the Netherlands (NL) and the United Kingdom (UK). All other EU and non-EU regions are represented by an aggregate Rest of the World (ROW) whose representation is reduced to import and export flows to the EU countries.

At the sectoral level the model provides sufficient detail on sector-specific differences in factor intensities, degrees of factor substitutability and price elasticities of output demand in order to trace back the structural change in industrial production induced by a policy shift. With respect to the analysis of carbon abatement policies, the sectors in the model have been carefully selected to keep the most carbon-intensive sectors in the available data as separate as possible. The energy goods identified in the model include coal (COA), gas (GAS), crude oil (CRU), refined oil products (OIL) and electricity (ELE). This disaggregation is essential in order to distinguish energy goods by carbon intensity and the degree of substitutability. In addition the model features important carbon-intensive and energy-intensive industries which are potentially those most affected by carbon abatement policies. Specifically, the remaining sectors are: agriculture (AGR), iron and steel (ORE), chemical products (CHM), non-ferrous metals (NFM), non-metallic minerals (NMM), machinery (MAC), transport equipment (TEQ), paper, pulp, and printing (PLP), wood and wood products (WOO), food processing, beverages, and tobacco (FOO), textiles and leather (TEX), transport (TRN), and other industries (Y). Table 1 provides a summary of the sectors and regions included in the model.

2.2 Factor Markets

Primary factors include labor and capital, which are assumed to be mobile across sectors within each region but not internationally mobile. In fossil fuel production part of the capital is treated as a sector-specific resource, consistent with exogenous own-price elasticities of supply. Factor markets are assumed to be perfectly competitive such that flexible prices of factors ensures market clearance.

2 See Appendix A for an algebraic summary of the model.

2.3 Production

Nested, separable constant elasticity of substitution (CES) production functions are employed to specify substitution possibilities in domestic production between capital, labor, energy, and material inputs. At the top level material inputs are used in fixed proportions, together with an aggregate of energy and a value added composite of labor and capital. The value added composite is a CES function of labor and capital. The energy aggregate is produced with a CES function of a primary energy composite and electricity. The primary energy composite is then a CES function of coal, crude oil, refined oil, and natural gas. In fossil fuel production sector-specific resources are assumed in order to constrain supply responses to appropriate elasticities.

2.4 Final Demand

Final demand of the representative consumer in each region is given as a CES composite which combines consumption of an energy aggregate and a non-energy consumption bundle. The consumption composite is a CES function of the energy composite described above and an aggregate of non-energy goods. The latter consists of the 12 non-energy goods in the model, whose substitution possibilities are given by a CES function.

2.5 Public Expenditure and Investment Demand

Government demand within each region is fixed at exogenous real levels. Public goods and services are produced with a Cobb-Douglas aggregation of commodity inputs. Savings are determined through a constant marginal propensity to save by private households, and market clearance ensures a savings-investment equality. In each region investment is a Leontief aggregation of Armington inputs, which are composed of domestic and imported commodities.

2.6 International Trade

Trade between EU regions is specified using the Armington approach to product heterogeneity, so domestic and foreign goods of the same variety are distinguished by origin. The Armington composite for a traded good is a CES function of an imported composite and domestic production for that sector. The import composite is then a CES function of an EU import composite and a ROW import. The EU composite in turn is a CES function of production from all other EU countries.

EU countries are assumed to be price-takers with respect to world market prices, so that we have perfectly elastic ROW import-supply functions and ROW export-demand functions. There is an imposed balance of payment constraint to ensure trade balance between the EU and ROW. That is, the value of imports from the ROW to the EU must equal the value of exports from the EU to the ROW after including a constant benchmark trade surplus.

2.7 Welfare Cardinalization

In order to consider the equity implications of carbon taxes on different EU countries, we employ a "cardinalized" welfare index for each representative household. We do this in order to endogenously compute "fair" distributions of emission rights across EU countries.

Specifically, we employ a cardinalization of utility which is consistent with a constant coefficient of inequality aversion, a convenient formulation widely used in welfare economics.[3] The welfare impact on region r is assessed by changes in $W_r = U_r^{1-\rho} / (1-\rho)$, where U_r is a linearly homogeneous consumption welfare index defined over the infinite horizon. Values of ρ ranging from 0 up to ∞ provide simple representations of social welfare functions ranging from Utilitarian to Leontief-Rawlsian. We specify a range of values for ρ parametrically, and trace out how our policy conclusions change as we allow for more aversion to inequality across regions.

2.8 Model Calibration: Forward Projection to 2010

An important feature of any burden-sharing rule is the extent to which it binds the economies in the future. The economic consequences of emission abatement depend to a large extent on the structural characteristics of each particular economy exhibited in the reference situation.[4] With significant (exogenous) improvements in emission intensity until 2010, for example, the Kyoto abatement target will be less stringent for the European economy. Hence emission reduction from 2010 levels toward the Kyoto target, which refers to 1990 levels, will be less costly.

In our analysis we take the most recent consistent economic and energy flow data (see Appendix B) and undertake a forward projection of the European economy

---

3 Atkinson (1970), Boadway and Bruce (1984: 277), and Layard and Walters (1978: 48).

4 The EU Council of Ministers (Environment) concludes that "... the equitable burden sharing among Member States ... will be discussed and agreed having regard ... to the aspects such as differences in starting points, economic structures and resource bases" (European Commission, 1997, paragraph 12).

based on the assumptions of an official EU study, "Energy in Europe – European Energy to 2020" (European Commission, 1996). In other words, we use the official EU projections of key economic indicators (see Table 3) to infer the structure of the EU economy for 2010, which is the reference year for our abatement counterfactuals.[5] Table 3 shows that these projections indicate an overall increase of carbon emissions of 7% between 1990 and 2010 for the EU countries included in our model. They also entail considerable variation of emission efficiency improvements, and hence percentage emission increases, across member states.

One concern with these projections is the optimistic nature of the improvements in carbon intensity that are implied. Historical trends would suggest that improvements of as much as 1% over any sustained period would be very high indeed, so implied rates in excess of 2% are extreme.[6] We regard the problem of consistent projections of the future economies of the EU to be surprisingly under-studied.[7] For present purposes we accept the numbers provided by EU Conventional Wisdom projections.[8]

3 Scenario

The EU is committed to a reduction in greenhouse gas emissions for the EU as a whole and has held that position ever since the Rio de Janeiro conference in 1992. This target is currently set at an 8% reduction of 2010 emissions compared to 1990. The policy issue here is to evaluate alternative ways in which it could distribute this overall goal within the EU.[9] We do not consider the possibility that the EU may be able to undertake "joint implementation" strategies outside the EU that allow it to meet its aggregate target by paying other non-EU countries to actually effect reductions in emissions.[10]

---

5 See Böhringer, Jensen, and Rutherford (2000: 219) for a detailed description of the baseline calibration routine.

6 A poll of 22 experts reported in Manne and Richels (1994) results in an average autonomous energy efficiency improvement value of 0.7%.

7 Böhringer, Jensen and Rutherford (1998) provide an analysis of the implications of alternative baseline energy projections on the costs of carbon abatement in the EU.

8 See European Commission (1996: 142-143) for a general description of the assumptions for the Conventional Wisdom Scenario underlying our model calibration.

9 We restrict our analysis to the abatement of carbon dioxide. CO_2 emissions are the major source of the anthropogenic greenhouse effect.

10 The possibility of joint implementation with non-EU countries might significantly change the total cost of EU abatement as well as the distributional impacts across EU member countries. Harrison and Rutherford (1998) address this issue from the perspective of global burden-sharing.

For our welfare evaluation of alternative burden-sharing rules we hold the aggregate EU cutback constant and assume that leakage[11] outside the EU is relatively unaffected by the method by which the EU achieves its reduction target. The gross benefit of abatement for a given (representative) household in each region is then constant over all burden sharing scenarios, and we can just concentrate on the "price tag" for the EU of delivering a public good called "aggregate EU abatement".

3.1 Burden Sharing Rules

We evaluate three alternative ways of allocating emission rights across coalition members.

3.1.1 DIFF: Differential Percentage Cuts

One exogenous allocation rule, called DIFF, refers to differentiated national reductions in 2010 where emission rights are distributed among EU countries according to the burden-sharing proposal of the EU Council of Ministers (COM, 1999, Annex 1). For the countries incorporated in our model these reduction[12] targets are: Denmark 21%, Germany 21%, Spain -15%, France 0%, Italy 6.5%, the Netherlands 6%, and the United Kingdom 10%. For our analysis we scale the reduction targets uniformly across countries such that the aggregate reduction of these countries together amounts to 8%. This yields the following differential cutbacks in emission rights as compared to 1990 emission levels: Denmark 16.6%, Germany 16.6%, Spain -11.8%, France 0%, Italy 5.1%, the Netherlands 4.7%, and the United Kingdom 7.9%.

3.1.2 UNIF: Uniform Proportional Cuts

An alternative exogenous allocation rule, called UNIF, uses uniform proportional cuts of 8% in all EU regions which reflects proposals for a flat rate emission reduction as stated by several parties of the Framework Convention on Climate

---

11 It is possible to make some attempt to estimate the effects of leakage in a small, open economy (Harrison and Kriström, 1998), and this idea could be applied here as well to the ROW. The idea is to assume that all net trade effects generated by some policy in the country generate emissions according to some assumed foreign emissions coefficient. The problem with this calculation in the present case is that our ROW is an amalgam of many diverse countries, and the carbon coefficients could vary enormously depending on which countries actually undertake that trade. For example, if China undertakes the trade there would be much more leakage than if Japan undertakes it. We prefer to remain explicitly agnostic on the leakage issue, leaving that to global models to account for (e.g. Harrison and Rutherford, 1998).

12 Negative percentages refer to allowed increases in emissions.

Change (United Nations, 1995). This is clearly equitable in terms of the metric of the percentage cut in abatement, but need not be equitable with respect to welfare. With this allocation rule the emissions allocations for EU members are formulated as fixed shares of EU aggregate emissions.

3.1.3 ENDOG: Endogenous Targets to Equalize Welfare Changes

The third allocation rule is endogenous, and is imaginatively called ENDOG. It shares the burden of carbon abatement differentially across EU member countries so as to equalize the welfare impact across EU members. We cardinalize utility to approximate in formal terms what might be meant by "fair", and we then let the model allocate emission rights within the EU to produce an equal welfare burden across regions. Specifically, we impose the constraint that:

$$\frac{W_r^{1-\rho}}{1-\rho} - \frac{\overline{W}_r^{1-\rho}}{1-\rho} = \Delta W^{EU}$$

where $\overline{W}_r$ is the base year level of the welfare index for region r and the parameter ρ reflects the degree of aversion to inequalities in welfare impacts across EU member states.[13] This rule constrains the burden to be shared so that it is the same for each EU country in terms of the cardinalized change in welfare.

3.2 Tradable Permits

In any of the scenarios permit prices can be interpreted as carbon tax rates which are necessary to produce an abatement outcome equivalent to the total amount of emission permits in use. Revenues from carbon permits accrue to the representative consumer in each region.[14] We evaluate two regimes with respect to the use of carbon emission permits in the EU.

3.2.1 NO_TRD: No Trade in Emission Permits Across National Borders

In this case, called NO_TRD, permit rights cannot be traded across the borders of EU member countries. The price of emission rights will typically differ across regions. In the event that a country's emissions fall below the constraint it must meet, there is no carbon tax imposed.[15]

---

13 As ρ increases, the aversion to inequality of welfare impacts increases also.

14 The shadow price of emission permits is then zero, reflecting the fact that emission rights are not scarce.

15 We do not analyse the public finance issue of how existing tax distortions, and the method of revenue recycling, affect the welfare implications of carbon abatement (see Goulder, 1995, for an overview).

3.2.2 TRD: Trade in Emission Permits Across National Borders

In this regime, denoted TRD, we allow tradable permits throughout EU regions. Trade will ensure that the prices of permit rights are equalized across all EU countries, hence marginal abatement costs will also be equalized.

4 Results

4.1 Basic Results

Table 4 presents the main results from our core scenarios. Block A shows the welfare cost in terms of percentage changes in (Hicksian) equivalent variation in real income. The aggregate welfare impacts are the composition of substitution effects (domestic and international) as well as cross-country income effects triggered by the imposition of carbon emission constraints. It is worth emphasizing that our welfare measure does not show any effects from the reduction in global carbon emissions that these policies generate. We implicitly assume that these gross benefits are separable from the welfare derived from consumption of private goods, and focus on the latter.[16] Block B shows marginal abatement costs expressed in terms of 1995 US dollars per ton of carbon abated. These costs are equivalent to the price of emission permits, or carbon tax rates which have to be levied in order to meet the exogenous reduction target. Block C reports the percentage reduction in emissions in each region and in the EU compared to 1990 levels. Block D lists the emission cutbacks for each country as compared to projected 2010 benchmark emission levels.

4.1.1 Equalizing Welfare Impacts Implies "Different Differential" Targets

In the no-trade regime the welfare cost to each EU state is equalized at 0.36% of equivalent variation in income when we allow endogenous burden-sharing (ENDOG/NO_TRD). Block C shows that an equitable burden sharing rule implies a non-uniform, differential emission right allocation, but that distribution of emission rights is quite unlike the exogenous profile DIFF. As compared to reduction targets based on the burden sharing proposal by the EU Council of Ministers, Denmark, Germany, Italy, and the United Kingdom should be endowed

16 Alternative approaches are either to specify some explicit damage function or to constrain the model to keep aggregate emission reductions constant over all scenarios. The former approach has been popular in several recent studies, despite the acknowledged lack of any hard data on the specification of the damage function.

with more emission rights whereas France, Spain, and the Netherlands should face a cut in emission rights.

To understand the nature of infra-marginal abatement costs consider the cutback in emission rights effected by the alternative allocation schemes compared to the 2010 baseline.

From Table 4D we see that equitable cutbacks of emissions across countries range from 12% (Italy) up to 21% (the Netherlands). The endogenous reduction scheme reflects differences in the benchmark economic structure across EU countries, such as variations in the carbon efficiency of the energy system, differences in the fuel mix, and trade relations with other countries which determine the aggregate welfare costs of emission abatement for any individual country. For an equitable outcome the extent of abatement has to outweigh cross-country differences in the ease of carbon substitution through fuel switching or energy savings[17], as well as the country-specific opportunities for tax-burden shifting through changes in the terms of trade.

The marginal abatement costs under ENDOG depend to a large extent on differences in the carbon intensity of different sectors across countries. For example, France faces much higher marginal abatement costs compared to Denmark to achieve an almost identical relative cutback of carbon emissions. The reason is that France is not carbon-intensive in sectors with low-cost substitution possibilities, such as electricity generation, and carbon-intensive in sectors where carbon substitution is relatively costly (e.g., final demand). Therefore, the carbon tax rates needed to produce the same relative carbon abatement are higher in France than in Denmark, which has low cost substitution possibilities in carbon-intensive sectors (e.g., power generation).

The range of marginal abatement costs under ENDOG/NO_TRD is much more narrow than those associated with DIFF/NO_TRD or UNIF_NO_TRD. The larger variation of carbon tax rates under DIFF/NO_TRD and UNIF/NO_TRD can be attributed directly to the larger differences in the required extent of abatement across countries. With DIFF/NO_TRD, Denmark, Germany, Italy, and the United Kingdom suffer the highest welfare losses because these countries have to achieve large carbon cutbacks, requiring high carbon taxes with severe distortionary effects on domestic production and consumption. When equity is considered, the first-order of business is to assign reduced emission targets (more emission rights) to Denmark, Germany, Italy, and the United Kingdom to help them mitigate their relatively worse welfare position: this is the explicit goal of the constrained optimization underlying our ENDOG calculations. Since Denmark, Germany, Italy, and the United Kingdom are reducing their abatement level, somebody else

17 For a single country carbon abatement gets increasingly more expensive with higher reduction targets when low-cost options, such as cheap fuel-switching in electricity production from coal to gas, have been exhausted.

in the EU must be increasing their abatement level in order to maintain the same aggregate EU reduction. Equalization of welfare cost across countries then implies that France, Spain, and the Netherlands have to undertake higher emission reductions.

4.1.2 Trade in Carbon Rights Improves Welfare, but Gains from Trade Are Rather Limited

EU-wide cost effectiveness of carbon abatement suggests that marginal costs of emission reduction should be equalized across sources. With respect to EU welfare, a tradable permit system which leads to uniform permit prices (or uniform carbon taxes) should be Pareto-superior to the NO_TRD cases where differences in the marginal costs of abatement are not exploited. The welfare figures in Block A confirm this economic intuition, but also indicate that the scope for efficiency gains from coordinated action is rather limited. As expected, the magnitude of the efficiency gains is positively correlated with differences in marginal abatement costs across countries. For UNIF/NO_TRD, where the relative variations in marginal abatement costs is most distinct, a shift to tradable permits produces efficiency gains of around 10% of EU-wide abatement costs. Under DIFF/NO_TRD and, ENDOG/NO_TRD, the variation of abatement costs is much smaller and so are the efficiency gains from permit trade.

The pattern of permit trade for alternative allocation rules can be explained on the basis of marginal abatement costs for the NO_TRD scenarios. Countries whose marginal abatement costs under NO_TRD are below the uniform permit price under TRD will sell permits and abate more. In turn, countries whose marginal abatement costs under NO_TRD are above the uniform permit price under TRD will buy permits and abate less. Trade in permits does not benefit all countries. The sign of the welfare change for an individual region depends on the relative strength of the associated domestic substitution effect and the terms of trade effects (i.e., international substitution effect plus income effect).[18] With respect to permit trade, the welfare effects from the domestic substitution effect are opposite to the welfare effect from the terms of trade effect, with alternating signs for the permit sellers and permit buyers. Permit sellers face higher marginal abatement costs after trade. This worsens their welfare losses due to the domestic substitution effect, but on the other hand it improves their terms of trade. The opposite applies to countries which are purchasing permit rights. Only under ENDOG_TRD will all countries benefit from emission trading since efficiency gains are, by definition, equally distributed across trading partners.

Finally, comparing entries for EU under UNIF/TRD and ENDOG/TRD in Table 4A, we find a trade-off between equity and efficiency in carbon abatement for the

---

[18] Income effects also include tax shifting and transfers through permit sales or purchases.

EU economy. A flat rate emission reduction produces different income effects, which in turn can affect the overall efficiency losses for the EU from carbon abatement due to associated changes in relative prices.

4.2 Equity Concerns

How do the results change when we vary the extent to which equity effects matter? We can analyze this question by parametrically varying the cardinalization of welfare changes in the model. The results are reported in Table 5.

Comparing the NO_TRD cases, we see that reducing the aversion to inequality of welfare impacts implies that Spain bears a larger share of the aggregate EU cutback. Spain's emission cutback as compared to 2010 baseline levels increases from 16.4% (ENDOG1) to 20.8% (ENDOG0). Conversely, if we have more of an aversion to inequality Spain has a significantly smaller cutback imposed: only 12.5% under ENDOG2. Italy is affected by varying degrees of aversion to inequality the same way as Spain, although the effect is much less pronounced.

Since Spain and Italy are bearing more of the burden under reduced inequality aversion, who is getting off with less? Denmark gains with less concern about inequality, as does Germany. France, the Netherlands, and the United Kingdom are relatively unaffected by these varying degrees of equity angst. So concerns about equity translate here into de facto transfers from Danish and German households to Spanish and Italian households. These transfers are being effected through a distortionary scheme, the distribution of emissions targets, rather than in some lumpsum form. Although it is customary and appropriate to question the real world efficacy of lumpsum transfers, the EU has necessarily[19] developed these to a fine art through creative accounting for regional disparities.

The welfare implications of varying degrees of aversion to inequality are directly related to changes in the allocation of emission rights. Whereas Denmark and Germany suffer higher welfare losses with more concern about inequality, Spain and Italy gain.

4.3 Sensitivity Analysis to Baseline Projections

The magnitude and the distribution of abatement costs for meeting the Kyoto target through alternative permit allocation rules across EU countries depends on

---

19 Due to the use of unanimity voting in many issues of EU policy. Unanimity can be more efficient than majority rule when there exists some "costless" means of making sidepayments from voter to voter. One can view the development of sidepayment mechanisms in the EU as one rational public choice response to the constraint of having to use unanimity.

the evolution of anthropogenic carbon emissions in the baseline (Business as Usual). The baseline emission level determines the extent to which emission control commitments with respect to 1990 emission levels actually constrain the economies. Carbon emissions are directly linked to the combustion of fossil fuels, so it follows that baseline assumptions about GDP growth rates, energy efficiency improvements, and the future fuel mix, matter for the reference level of carbon emissions. For our baseline projection we have used official EU data. In our sensitivity analysis we analyze the changes in welfare associated with alternative burden sharing rules keeping the evolution of EU countries rather uniform. For this purpose we take the EU average across country-specific values of key indicators (listed in Table 3) and use these average values for model calibration.

The effects of making these projections to 2010 are displayed in Table 6. By comparing these results to the effects using our original baseline (in Table 4), we see how crucial the choice of the baseline is.[20] An alternative projection of the future EU economy which abstracts from cross-country differences in growth rates substantially changes the country-specific welfare implications of the DIFF and UNIF permit allocations.

To interpret the results at the country level we have to keep in mind that the magnitude of welfare costs depends on the effective reduction requirement determined by the individual reduction target with respect to 1990 levels and the baseline emission level in 2010. The latter depends on GDP growth rates and projected improvements in CO_2 intensity. Under DIFF Denmark experiences the most severe welfare loss because a high reduction target goes together with high baseline emissions, yielding an effective reduction target of almost 35%. Table 3 indicates the reasons for the higher baseline emissions. Under uniform projections Denmark has a higher GDP growth rate and a considerably smaller improvement in CO_2 intensity as compared to the differential projections used in our core scenarios. The same reasoning applies for the United Kingdom though the increase in reduction requirement and welfare costs relative to the original baseline is much less pronounced. For France, the Netherlands and Germany the reduction requirements (as well as the implied welfare costs) remain relatively constant because the uniform projections on GDP growth and carbon intensity improvements are relatively consistent with the initial country-specific values. Spain and Italy find it easier to meet their domestic policy targets with respect to

---

20 Calibration of the model along the time path involves a two-step approximation of autonomous energy efficiency improvement factors, which scale energy demand functions to match exogenous emissions as close as possible taking into account given potential GDP growth rates, changes in the fuel mix of electricity generation, and world market prices (see Böhringer, Jensen, and Rutherford, 2000: 219). Calibrating the model to EU average values as well as to our baseline values, we set the same emission target in both case but cannot assure that both calibrations yield the same approximated overall EU emission level. This explains the difference in the overall EU emission level between the baseline and our sensitivity analysis.

1990 emission levels because smaller GDP growth rates and higher improvements in carbon intensity yield a lower reference emission level.

For a flat rate, uniform cutback of emission rights based on 1990 levels the variation of effective abatement requirements across countries, and hence the differences in welfare costs, become less pronounced. Nevertheless, the UNIF (NO_TRD) allocation scheme is still far from producing an equitable outcome.

Adjustments in the endogenous allocation of emission rights under ENDOG/NO_TRD reflect how the revised benchmark economic structure change the "comparative advantage" across countries of mitigating carbon emissions.

5 Concluding Remarks

We evaluated the welfare implications of alternative ways in which the EU could distribute an aggregate EU reduction target across its member states. We compared a uniform proportional cutback in emissions and the actual EU burden sharing agreement with an equitable allocation scheme derived from an endogenous burden sharing calculation in which the welfare cost across member states is equalized.

Our current modeling framework ignores several issues which are potentially important for the analysis of burden sharing. Lines of future research include:

- the incorporation of national energy policy constraints, such as a nuclear phase-out;
- a more detailed projection of the future economic development, in particular projections of (non price-induced) energy efficiency improvements;
- the inclusion of the other EU countries to assess the potential efficiency gains from trade in permits; and
- the inclusion of other non-EU countries that are differentiated with respect to carbon intensities to allow a more complete evaluation of "carbon leakage" and "joint implementation".

Table 1: Sectors and Regions in the Model.

Production Sectors (in brackets labels of Eurostat's R59 standard)			Regions	
Energy			1. GE	Germany
1.	COA	Coal (031, 033)	2. DK	Denmark
2.	CRU	Crude oil (071)	3. ES	Spain
3.	OIL	Refined oil (073)	4. FR	France
4.	GAS	Natural and manufactured gases (075, 098)	5. IT	Italy
5.	ELE	Electricity and steam (097, 099)	6. NL	Netherlands
			7. UK	United Kingdom
Non-Energy				
6.	AGR	Agriculture (010)		
7.	ORE	Iron and steel (135, 136)		
8.	CHM	Chemical products (170)		
9.	NFM	Non-ferrous metals (137)		
10.	NMM	Non-metallic minerals (151, 153, 155, 157)		
11.	MAC	Machinery (190, 210, 230, 250)		
12.	TEQ	Transport equipment (270, 290)		
13.	PLP	Paper, pulp, and printing (471, 473)		
14.	WOO	Wood and wood products (450)		
15.	FOO	Food processing, beverages, and tobacco (310, 330, 350, 370, 390)		
16.	TEX	Textiles and leather (410, 430)		
17.	TRN	Transport (611, 613, 617, 631, 633, 650)		
18.	ROI	Other industries (095, 490, 510, 530, 550, 570, 590, 670, 690, 710, 730, 750, 770, 790, 810, 850, 890, 930)		

Table 2: Key Elasticities Assumed in the Model.

Exports EU Region/Exports EU Region	4
Imports/Domestic Production	4
Labour-Capital/Energy	0.3
Labour/Capital	1.0
Electricity/Fuels	0.8-1.2
Fuel/Fuel	0.8

Table 3: Baseline Assumptions for 1995-2010 (EU 1996).

	GE	DK	ES	FR	IT	NL	UK	EU
A. Average GDP growth rates (in % per annum)								
	2.7	2.3	2.7	2.6	2.2	2.3	2.3	2.5
B. Average CO_2 growth rate (in %)								
	0.5	-0.3	1.0	0.5	1.1	0.4	0.3	0.6
C. Average growth rates for fossil fuel inputs to power generation (in % per annum)								
COA	-0.6	-1.8	-0.5	-7.2	1.3	-2.5	-5.5	-1.8
GAS	5.2	9.1	9.5	9.1	6.3	3.3	8.5	6.5
OIL	1.1	-1.2	0.2	-15.2	-4.0	-5.1	-0.3	-2.2
D. Implied average improvement in CO_2 intensity (in % per annum)								
	2.1	2.6	1.7	2.0	1.1	1.9	2.0	1.9
E. Total carbon emission (in million tons of CO_2)								
1990	976.3	52.0	206.5	366.7	403.6	156.5	579.8	2741.3
1995	894.9	61.4	216.2	362.7	395.1	159.7	583.0	2672.8
2010	969.8	58.3	251.7	391.9	462.4	170.0	608.0	2912.0

Table 4: Effect of Burden Sharing Rules & Tradable Emissions Permits in Base Years (2010).

	DIFF		UNIF		ENDOG	
	NO_TRD	TRD	NO_TRD	TRD	NO_TRD	TRD
A. Welfare Change as % of Income						
GE	-0.51	-0.51	-0.20	-0.26	-0.36	-0.35
DK	-0.71	-0.61	-0.39	-0.44	-0.36	-0.35
FR	-0.11	-0.13	-0.27	-0.25	-0.36	-0.35
ES	-0.11	-0.11	-0.86	-0.54	-0.36	-0.35
IT	-0.59	-0.47	-0.71	-0.52	-0.36	-0.35
NL	-0.12	-0.08	-0.15	-0.17	-0.36	-0.35
UK	-0.41	-0.41	-0.42	-0.41	-0.36	-0.35
EU	-0.37	-0.35	-0.41	-0.36	-0.36	-0.35
B. Carbon Taxes in 1995 USD per Ton						
GE	70	66	31	66	50	66
DK	90	66	55	66	51	66
FR	35	66	86	66	107	66
ES	40	66	168	66	91	66
IT	111	66	132	66	68	66
NL	42	66	54	66	86	66
UK	70	66	71	66	60	66
EU	66	66	66	66	66	66
C. Percentage Emission Cut as Compared to 1990 Levels						
GE	16.6	15.8	8.0	15.8	12.6	15.8
DK	16.6	10.8	8.0	10.8	6.8	10.8
FR	0	5.3	8.0	5.3	10.7	5.3
ES	-11.8	-6.5	8.0	-6.6	-2.2	-6.6
IT	5.1	-1.5	8.0	-1.5	-1.1	-1.5
NL	4.7	10.6	8.0	10.7	14.9	10.7
UK	7.9	7.3	8.0	7.3	6.4	7.3
EU	8.0	8.0	8.0	8.0	8.0	8.0
D. Percentage Emission Cut as Compared to 2010 Levels						
GE	16.1	15.3	7.4	15.3	12.1	15.3
DK	25.6	20.4	17.9	20.4	16.9	20.4
FR	6.5	11.4	13.9	11.4	16.4	11.4
ES	8.4	12.8	24.7	12.7	16.4	12.8
IT	17.2	11.4	19.7	11.4	11.7	11.3
NL	11.5	17.0	14.5	17.0	21.0	17.0
UK	12.2	11.6	12.3	11.6	10.8	11.6
EU	13.4	13.4	13.4	13.4	13.4	13.4

Table 5: Effects of Varying Degrees of Aversion to Inequality of Welfare Impact.

	ENDOG0		ENDOG1		ENDOG2	
	NO_TRD	TRD	NO_TRD	TRD	NO_TRD	TRD
A. Welfare Change as % of Income						
GE	-0.31	-0.30	-0.36	-0.35	-0.42	-0.40
DK	-0.26	-0.25	-0.36	-0.35	-0.51	-0.49
FR	-0.34	-0.32	-0.36	-0.35	-0.39	-0.38
ES	-0.60	-0.57	-0.36	-0.35	-0.22	-0.21
IT	-0.40	-0.38	-0.36	-0.35	-0.33	-0.32
NL	-0.37	-0.36	-0.36	-0.35	-0.35	-0.34
UK	-0.37	-0.36	-0.36	-0.35	-0.35	-0.34
EU	-0.37	-0.36	-0.36	-0.35	-0.36	-0.35
B. Carbon Taxes in 1995 USD per Ton						
GE	44	66	50	66	56	66
DK	37	66	51	66	69	66
FR	100	66	107	66	113	66
ES	129	66	91	66	65	66
IT	76	66	68	66	61	66
NL	87	66	86	66	84	66
UK	62	66	60	66	58	66
EU	66	66	66	66	66	66
C. Percentage Emission Cut as Compared to 1990 Levels						
GE	11.3	15.8	12.6	15.8	13.9	15.8
DK	2.6	10.7	6.8	10.8	11.5	10.8
FR	9.9	5.3	10.7	5.3	11.3	5.3
ES	3.3	-6.6	-2.2	-6.6	-6.8	-6.5
IT	0.1	-1.5	-1.1	-1.5	-2.4	-1.5
NL	15.3	10.7	14.9	10.7	14.5	10.7
UK	6.7	7.3	6.4	7.3	6.1	7.3
EU	8.0	8.0	8.0	8.0	8.0	8.0
D. Percentage Emission Cut as Compared to 2010 Levels						
GE	10.8	15.3	12.1	15.3	13.4	15.3
DK	13.1	20.4	16.9	20.4	21.1	20.4
FR	15.7	11.4	16.4	11.4	17.1	11.4
ES	20.8	12.7	16.4	12.8	12.5	12.8
IT	12.8	11.4	11.7	11.3	10.6	11.3
NL	21.3	17.0	21.0	17.0	20.6	17.0
UK	11.1	11.6	10.8	11.6	10.5	11.6
EU	13.4	13.4	13.4	13.4	13.4	13.4

Table 6: Sensitivity Analysis with Respect to the Baseline Assumptions.

	DIFF		UNIF		ENDOG	
	NO_TRD	TRD	NO_TRD	TRD	NO_TRD	TRD
A. Welfare Change as % of Income						
GE	-0.61	-0.58	-0.25	-0.31	-0.38	-0.37
DK	-1.83	-1.04	-1.23	-0.86	-0.38	-0.37
FR	-0.13	-0.15	-0.3	-0.27	-0.38	-0.37
ES	-0.02	0.05	-0.51	-0.39	-0.38	-0.37
IT	-0.35	-0.34	-0.44	-0.40	-0.38	-0.37
NL	-0.17	-0.11	-0.20	-0.21	-0.38	-0.37
UK	-0.60	-0.52	-0.60	-0.52	-0.38	-0.37
EU	-0.41	-0.37	-0.40	-0.37	-0.38	-0.37
B. Carbon Taxes in 1995 USD per Ton						
GE	77	67	35	67	50	67
DK	183	67	133	67	48	67
FR	39	67	89	67	107	67
ES	12	67	122	67	99	67
IT	72	67	93	67	78	67
NL	44	67	54	67	81	67
UK	91	67	91	67	61	67
EU	67	67	67	67	67	67
C. Percentage Emission Cut as Compared to 1990 Levels						
GE	16.6	14.8	8.0	14.8	11.4	14.8
DK	16.6	-6.4	8.0	-6.5	-11.9	-6.7
FR	0.0	4.9	8.0	4.9	10.3	4.9
ES	-11.8	0.1	8.0	0.0	4.9	0.0
IT	5.1	4.3	8.0	4.4	5.9	4.4
NL	4.7	11.2	8.0	11.3	14.4	11.3
UK	7.9	4.3	8.0	4.3	3.1	4.2
EU	8.0	8.0	8.0	8.0	8.0	8.0
D. Percentage Emission Cut as Compared to 2010 Levels						
GE	16.9	15.1	8.30	15.1	11.8	15.1
DK	35.4	17.6	28.8	17.6	13.4	17.4
FR	7.60	12.1	15.0	12.1	17.1	12.1
ES	2.40	12.8	19.7	12.7	17.0	12.8
IT	11.4	10.6	14.1	10.7	12.1	10.6
NL	13.3	19.2	16.2	19.2	22.1	19.2
UK	16.3	13.0	16.4	13.0	11.9	13.0
EU	13.7	13.7	13.7	13.7	13.7	13.7

Appendix A: Algebraic Model Summary of the Generic EU Model

Table 7: Summary of Prices and Output Quantities Determining an Equilibrium.

Summary of Prices	
P_{ir}	Output price of good i produced in region r
P^M_{ir}	Import price aggregate for good i imported to region r
P^E_{ir}	Composite price for aggregate energy inputs into sector i in region r (i=C for final consumption)
P^C_r	Composite price for aggregate household demand in region r.
P^G_r	Composite price for government demand in region r.
P^I_r	Composite price for investment demand in region r.
w_r	Economy-wide wage rate in region r
r_r	Rate of return for sector-specific capital inputs, sector j in region r
PCO	Price of carbon emission rights (carbon tax)

Summary of Quantity Indices	
Y_{jr}	Level of production, sector j in region r
C_r	Aggregate household consumption, region r
I_r	Aggregate investment, region r
G_r	Aggregate public output, region r

Table 8: Key Benchmark Shares, Endowment Parameters, and Elasticities.

Key Benchmark Shares, Endowment Parameters, and Elasticities	
θ_{ijr}^{M}	Import value share for sector i inputs to sector j in region r (j=G: government; j=C: household; j=I: investment)
α_{jr}^{E}	Energy input value share of KLE, sector j in region r
β_{jr}^{K}	Value shares for capital in value added of sector j in region r
a_{jr}^{KLE}	Benchmark value share of capital, labor, and energy inputs in sector j of region r
a_{ijr}	Benchmark value share of non-energy input i in sector j of region r
$\theta_{irr'}^{MM}$	Benchmark value share of region r exports in aggregate imports of good i into region r'
ε_{ijr}	Carbon emission coefficient for energy input i into sector j in region r
θ_{ijr}^{E}	Benchmark value share of energy good i in aggregate energy demand by sector j in region r (j=C: household)
$\overline{L_r}$	Aggregate labor endowment for region r
$\overline{K_r}$	Aggregate endowment of capital, region r
$CRTS_r$	Carbon emission rights endowment, region r
$\overline{G_r}$	Exogenously-specified demand for public output, region r
σ_{MM}	Elasticity of substitution between imports from different foreign countries
σ_{DM}	Elasticity of substitution between domestic and imported inputs or demands
σ_{E}	Elasticity of substitution between energy inputs

Exhaustion of Product Conditions

1. Production:

$$\pi_{jr}^{Y} = P_{jr} - \sum_{i \notin EG} a_{ijr} \left(\theta_{ijr}^{M} P_{ir}^{M^{1-\sigma_{DM}}} + \left(1-\theta_{ijr}^{M}\right) P_{ir}^{1-\sigma_{DM}} \right)^{\frac{1}{1-\sigma_{DM}}}$$

$$-a_{jr}^{KLE} \left[\alpha_{jr}^{E} P_{jr}^{E^{1-\sigma_{DM}}} + \left(1-\alpha_{jr}^{E}\right) \left(w_{r}^{(1-\beta_{jr}^{K})} r_{r}^{\beta_{jr}^{K}} \right)^{1-\sigma_{KLE}} \right]^{\frac{1}{1-\sigma_{KLE}}}$$

$$= 0$$

2. Sector-specific energy demand:

$$\pi_{jr}^{E} = P_{jr}^{E} - \left[\sum_{i \in EG} \theta_{ijr}^{E} \left\{ \left(\theta_{ijr}^{M} P_{ir}^{M^{1-\sigma_{DM}}} + (1-\theta_{ijr}^{M}) P_{ir}^{\,1-\sigma_{DM}} \right)^{\frac{1}{1-\sigma_{DM}}} + \varepsilon_{ijr} PCO \right\}^{1-\sigma_{E}} \right]^{\frac{1}{1-\sigma_{E}}}$$

$$= 0$$

3. Import demand:

$$\pi_{jr}^{M} = P_{jr}^{M} - \left(\sum_{r' \neq r} \theta_{jrr'}^{MM} P_{jr'}^{\;1-\sigma_{MM}} + \theta_{jROWr}^{MM} P_{jROW}^{M\;\;1-\sigma_{MM}} \right)^{\frac{1}{1-\sigma_{MM}}} = 0$$

4. Investment:

$$\pi_{r}^{I} = P_{r}^{I} - \sum_{i} a_{ir}^{I} \left(\theta_{iIr}^{M} P_{ir}^{M^{1-\sigma_{DM}}} + \left(1-\theta_{iIr}\right) P_{ir}^{\;1-\sigma_{DM}} \right)^{\frac{1}{1-\sigma_{DM}}} = 0$$

5. Public output:

$$\pi_{r}^{G} = P_{r}^{G} - \prod_{i} \left(\theta_{iGr}^{M} P_{ir}^{M^{1-\sigma_{DM}}} + \left(1-\theta_{iGr}^{M}\right) P_{ir}^{\;1-\sigma_{DM}} \right)^{\frac{a_{ir}^{G}}{1-\sigma_{DM}}} = 0$$

6. Household consumption demand:

$$\pi_{r}^{C} = P_{r}^{C} - \left[\theta_{Cr}^{E} P_{Cr}^{E^{1-\sigma_{EC}}} + \left(1-\theta_{Cr}^{E}\right) \left\{ \prod_{i \notin EG} \left(\theta_{iCr}^{M} P_{ir}^{M^{1-\sigma_{DM}}} + \left(1-\theta_{iCr}^{M}\right) P_{ir}^{\;1-\sigma_{DM}} \right)^{\frac{\alpha_{ir}}{1-\sigma_{DM}}} \right\} \right]^{\frac{1}{1-\sigma_{DM}}}$$

$$= 0$$

7. Household energy demand:

$$\pi_{Cr}^{E} = P_{Cr}^{E} - \left[\sum_{i \in EG} \theta_{iCr}^{E} \left\{ \left(\theta_{iCr}^{M} {P_{ir}^{M}}^{1-\sigma_{DM}} + \left(1-\theta_{iCr}^{M}\right) P_{ir}^{1-\sigma_{DM}} \right)^{\frac{1}{1-\sigma_M}} + \varepsilon_{iCr} PCO \right\}^{1-\sigma_E} \right]^{\frac{1}{1-\sigma_E}}$$

$$= 0$$

Market Clearance Conditions

8. Labour:

$$\overline{L_r} = \sum_j Y_{jr} \frac{\partial \pi_{jr}^{Y}}{\partial w_r}$$

9. Capital:

$$\sum_r \overline{K}_r = \sum_{j,r} Y_{jr} \frac{\partial \pi_{jr}^{Y}}{\partial r_r}$$

10. Output:

$$Y_{ir} = \sum_j Y_{jr} \frac{\partial \pi_{jr}^{Y}}{\partial P_{ir}} + C_r \frac{\partial \pi_r^{C}}{\partial P_{ir}} + I_r \frac{\partial \pi_r^{I}}{\partial P_{ir}}$$
$$+ G_r \frac{\partial \pi_r^{G}}{\partial P_{ir}} + \sum_{r'} M_{i'r} \frac{\partial \pi_{i'r}^{M}}{\partial P_{ir}} + M_{irROW}$$

11. Imports:

$$M_{ir} = \sum_j Y_{jr} \frac{\partial \pi_{jr}^{Y}}{\partial P_{ir}^{M}} + C_r \frac{\partial \pi_r^{C}}{\partial P_{ir}^{M}} + I_r \frac{\partial \pi_r^{I}}{\partial P_{ir}^{M}} + G_r \frac{\partial \pi_r^{G}}{\partial P_{ir}^{M}}$$

12. Balance of Payments:

$$\sum_{i,r} \overline{P_i^x} M_{irROW} = \sum_{i,r} \overline{P_i^M} M_{iROWr} + \sum_r \overline{B_r}$$

13. Final consumption demand:

$$C_r = \frac{\left(1-mps_r\right)\left(w_r \overline{L_r} + r_r \overline{K_r} + PCO \cdot CRTS_r - P_r^G \overline{G_r} - \overline{B_r}\right)}{P_r^C}$$

14. Savings:

$$I_r = \frac{mps_r\left(w_r\overline{L_r} + r_r\overline{K_r} + PCO \cdot CRTS_r - P_r^G\overline{G_r} - \overline{B_r}\right)}{P_r^I}$$

15. Carbon emission permits:

$$\sum_r CRTS_r = \sum_{j,r} Y_{jr}\frac{\partial \pi_{jr}^Y}{\partial PCO}$$

Appendix B: Benchmark Data Sources and Reconciliation

The model is based on economic transactions in a particular benchmark year (2010 in our case). Benchmark data are used to calibrate parameters of the functional forms from a given set of benchmark quantities, prices, and elasticities. Data for this model calibration stem from four different sources which need to be reconciled to yield a consistent benchmark data set.

First is the EUROSTAT (1995) input-output tables. The input-output data base covers the outputs and intermediate inputs of the 17 sectors of the model, the primary inputs (labor, profits, depreciation, and production taxes), imports, and the final demand categories (consumption, investment, government expenditures and exports).

Second is the CHELEM trade data due to WEFA (1995). The share parameters in the CES function for Armington imports across different regions are calculated using a trade matrix with bilateral trade flows.

Third is the IEA (1992) energy balances and energy prices/taxes. IEA statistics on physical energy flows and energy prices for industrial and household demands are used for the "bottom-up" calibration of energy demands and supplies as well as for the derivation of sector-specific and energy-specific CO_2 coefficients in CO_2 units per national currency units.

Fourth is the European Commission's (1996) projection on the future development of the European economy.

Reconciliation of Eurostat input-output data and trade data is an important step in constructing the base year equilibrium. For this purpose, we employ a nonlinear least squares procedure to calibrate bilateral trade flows provided by CHELEM to the intra-EU and extra-EU trade totals provided in the Eurostat input-output tables (see Böhringer, Ferris, and Rutherford, 1998, for further discussion). CHELEM does not cover all trade flows, and for the missing goods a target bilateral trade matrix representing the average over all traded goods is used.

When the monetary flows of the national input-output tables are supplemented with physical flow data on the emission-relevant fossil fuel use in production sectors and final demand, the implicit price for the same fuel varies substantially across different sectors and countries. In applied economic model-building the common approach is to ignore these differences and assume uniform prices for fuels across users. The latter assumption involves an ad hoc scaling of sector-specific and energy-specific CO_2 coefficients in order to obtain official figures on the total national carbon emissions. The problem with this shortcut is that CO_2 coefficients are then not based on actual energy flows, and the implicit marginal abatement cost curves can be significantly in error. We therefore use physical energy flows and energy prices as reported in energy statistics, rather than aggregate input-output monetary values.

Finally, we perform a forward calibration of our model to 2010 incorporating the European Commission's exogenous baseline assumptions on GDP growth, world energy prices, changes in the fossil fuel mix for electricity generation, and energy (carbon) efficiency improvements over time. Böhringer, Jensen, and Rutherford (2000) provide a detailed description of this forward projection.

References

Atkinson, A.B. (1970), On the Measurement of Inequality, *Journal of Economic Theory* 2, 244-263.

Boadway, R.W. and N. Bruce (1984), *Welfare Economics*, Cambridge.

Böhringer, C., M. Ferris, and T.F. Rutherford (1998), Alternative CO_2 Abatement Strategies for the European Union, in: Braden, J. and S. Proost (Eds.), *Climate Change, Transport, and Environmental Policy*, London, 16-47.

Böhringer, C., J. Jensen, and T.F. Rutherford (2000), Energy Market Projections and Differentiated Carbon Abatement in the European Union, in: Carraro, C. (Ed.), *Efficiency and Equity of Climate Change Policy*, Dordrecht, 199-219.

COM (1999), Commission Communication, *Preparing for Implementation of the Kyoto Protocol*, http://europa.eu.int/comm/environment/docum/99230_en.pdf.

European Commission (1997), *Meeting Document CONS/ENV/97/REV1*, General Secretariat of the Council, March 3, Brussels.

European Commission (1996), *Energy in Europe – European Energy to 2020 – A Scenario Approach*, Brussels.

Eurostat (1995), *The Input-Output Tables Database of Eurostat*, Luxembourg.

Goulder, L.H. (1995), Environmental Taxation and the 'Double Dividend': A Reader's Guide, *International Tax and Public Finance* 2, 157-183.

Harrison, G.W. and B. Kriström (1998), General Equilibrium Effects of Increasing Carbon Taxes in Sweden, in: Brannlund, R. and I. Gren (Eds.), *Green Taxes: Theory and Practice*, London.

Harrison, G.W. and T.F. Rutherford (1998), Burden Sharing, Joint Implementation, and Carbon Coalitions, in: Carraro, C. (Ed.), *International Environmental Agreements on Climate Change*, Amsterdam.

International Energy Agency (1994), *Energy Prices and Taxes / Energy Balances of OECD Countries*, Paris.

Layard, P.R.G. and A.A. Walters (1978), *Microeconomic Theory*, New York.

Manne, A. and R.G. Richels (1994), The Costs of Stabilizing Global CO_2 Emissions: A Probabilistic Analysis Based on Expert Judgements, *The Energy Journal* 15 (1), 31-56.

WEFA (1995), *Comptes Harmonisés sur les Echanges et L'Economie Mondiale*, Paris.

United Nations (1995), *Framework Convention on Climate Change, Conference of the Parties*, Berlin.

United Nations (1997), *Kyoto Protocol FCCC/CP/1997/L.7/Add1*, Kyoto.

Banking and Trade of Carbon Emission Rights: A CGE Analysis

Gunter Stephan and Georg Müller-Fürstenberger

Institute of Applied Micro-Economics, University of Bern, Gesellschafts-strasse 49, 3012 Bern, Switzerland, *gunter.stephan@vwi.unibe.ch*, *georg.mueller@vwi.unibe.ch*

Summary:
This paper analyses trading and banking of carbon emission rights. Within the framework of a modestly simple, integrated assessment model that breaks the world economy in just two regions, North and South, it can be shown: (1) There exists separability between environmental targets and the choice of instruments. Increasing the "when and where" flexibility in greenhouse gas abatement either through banking or trading of carbon emission permits or both positively affects global welfare. It has, however, almost no impact on global climate change. (2) Depending upon the choice of instruments there are significant distributional effects across regions. Both regions can improve welfare simultaneously, if carbon emission rights are traded on open international markets. But if it were feasible to bank or borrow carbon permits, then – independent of whether there is trading of carbon rights or not – the South suffers welfare losses compared to a No-Trade-No-Banking situation.

Keywords: Carbon rights, climate policy, integrated assessment, banking and trade.

1 Introduction

Does banking and trading of carbon dioxide emission permits on open international markets affect the atmospheric accumulation of greenhouse gases? Will banking and borrowing of carbon emission rights improve welfare? Does efficiency in greenhouse gas abatement depend upon whether there is interregional and/or intertemporal trading of carbon rights? Are there regional differences, if emission permits are allowed to be banked, traded or even borrowed? Analyzing these issues within the framework of an integrated assessment model of global climate change is the issue of this paper.

Banking and trade of marketable emission permits is by no means a new institution. For example the 1990 US Clean Air Act Amendments explicitly allow for trading and banking of sulfur dioxide (SO_2) emission permits. The owner of a permit is entitled to emit a certain amount, say one ton, of SO_2 in the year for which the permit is labeled. But he can also save it either to sell it or to use it in later years. Based on this experience the US Department of State has suggested to establish a similar system of tradable and bankable carbon emission permits under the Framework Convention on Global Climate Change (USDOS, 1997). The reasoning seems quite obvious. Typically it is argued that for a given environmental quality the more flexible a policy intervention can be handled by those who should be regulated the lower are the welfare losses due to the specific policy measure. Trading of carbon dioxide emission permits on open international markets increases the "where to abate" flexibility. Banking of carbon rights can provide flexibility in "when to abate" greenhouse gas emissions. Or to put it differently, efficiency in greenhouse gas abatement is promoted simply by transferring abatement activities in space and in time. This should reduce the costs of greenhouse policy.

There is, however, a major difference between SO_2 and carbon dioxide (CO_2). Global climate change is a stock damage problem. It is driven by the accumulation of CO_2 in the atmosphere and is not directly associated with the flow of emissions. Therefore, the timing of emissions and damages is not coincident as it is in the case of sulfur dioxide. For this reason environmentalists conclude that excessive banking or borrowing of permits today could cause quite drastic damages in the future. In other words, what might be beneficial today can be damaging tomorrow.

Furthermore, the theoretical literature is rather ambiguous about the welfare effects of an intertemporal permit system. For example, Biglaiser et al. (1995) show that intertemporal permit trade is not optimal. Kling and Rubin (1997) lay out that unrestricted banking and borrowing of flow pollutant rights is not granted to be socially optimal. At the same time they insist that the banning of permit banking and borrowing is also not optimal. And more recently, Leiby and Rubin (2001) have argued that banking of carbon emission rights can lead to welfare losses unless these permits are traded at the correct intertemporal exchange rate. The latter is determined by the ratio of current marginal stock damages to discounted future value of marginal stock damages less the decay rate of emissions.

Although global trade in carbon emission rights is a feature of many integrated assessment models (for example, see Nordhaus and Yang, 1996, Manne and Richels, 1995, or Bernstein et al., 1999), banking of carbon permits typically is not included in these models. This is in contrast to the present paper that explicitly incorporates trading, banking, and borrowing of carbon emission rights into a regional differentiated version of the MEDEA framework (see Stephan and Müller-Fürstenberger, 1998). Breaking the world just into North and South, it will be shown:

(1) Optimal climate policy is almost independent of the choice of policy instruments.

(2) Increasing the "when and where" flexibility in greenhouse gas abatement through trading and/or banking of carbon emission rights has virtually no impact on the development of atmospheric carbon concentration.

(3) "When and where" flexibility increases global welfare, but affects regions in quite different ways. Both North and South can improve welfare simultaneously, if carbon emission rights are traded on open international markets. If it were feasible, however, to bank or borrow carbon permits only, then the South always suffers from welfare losses compared to a No-Trade-No-Banking situation.

The rest of the paper is organized as follows. Section 2 characterizes the theoretical setting upon which the numerical exercise is based. Section 3 describes the different scenarios that will be carried out. It also provides first economic insight into the results we can expect. Section 4 discusses the major findings of our numerical analysis, and Section 5 covers concluding remarks.

2 Regional MEDEA

As the purpose of our numerical thought experiments is insight, not numbers, the theoretical framework is kept deliberately simple. To relate our results to the literature, numerical parameters from the RICE (see Norhaus and Yang, 1996), MERGE (see Manne and Richels, 1995) and MEDEA (see Stephan and Müller-Fürstenberger, 1998) integrated assessment models are taken over into our stylized-facts framework of the world economy.

There are two regions of the world. For vividness let them be called North (N) and South (S). North consists of the OECD countries including the former Soviet Union. South covers the rest of the world. Each region is represented as though it were an infinite-lived agent, maximizing the discounted logarithm of consumption.

We take a descriptive rather than a prescriptive view and suppose that both North and South discount consumption at the market rate of interest. South enjoys a higher rate of potential GDP growth than North. This immediately allows for the possibility of different rates of return on capital between the two regions (see Manne and Stephan, 1999).

Time is discrete and periods are one decade in length. For each time period, there are just two tradable commodities: a numeraire and carbon emission rights. The numeraire can be produced within each region and may be used for consumption, investment, net exports and to cover energy costs. Carbon-free energy resources such as hydro or solar are viewed as back-stop resources. They are provided at constant, but high marginal costs. Greenhouse resources such as oil, gas and coal are supplied at low but increasing costs.

Among the various greenhouse gases, carbon dioxide (CO_2) is considered as the most relevant one. The demands for carbon emissions are generated through regional consumption of greenhouse resources. Global warming is directly attributed to cumulative CO_2 emissions and affects production, but not utilities.

2.1 Climate-Economy Interaction

There are two channels through which the environment and the economy interact. One is the consumption of greenhouse resources which directly determines the flow of CO_2 emissions into the global atmosphere. The second link is provided through the concept of the "green output" by which global climate change is translated into its economic impact.

A "two-box" model is used to cumulate carbon emissions over time, and to translate them into global concentrations (for a detailed discussion, see Joos et al., 1999). In any time period t the stock of atmospheric carbon dioxide, $Q(t)$, is a function of the former one, $Q(t\text{-}1)$, and the world's past period emissions, $s(t\text{-}1)$:

$$Q(t) = \Psi Q(t\text{-}1) + \Theta s(t\text{-}1). \tag{1}$$

Ψ is the factor by which natural abatement reduces the current stock of atmospheric CO_2. Θ is the fraction of past global emissions that has accumulated in the atmosphere.

For the sake of simplicity, we neglect the thermal inertia lag between global CO_2 concentration and climate change. We also neglect the cooling effects of aerosols and the heating effects of greenhouse gases other than carbon dioxide. What we reflect, however, is that the climate effect may differ across single regions of the world.

The model is calibrated such that with zero abatement concentrations will rise from 353 ppm (the 1990 level) to 550 ppm (twice the pre-industrial level) by about 2070. This leads to damages of 3.5% of gross output in the South and 1.5% of GDP in the North. At other concentration levels, the regional damages are projected as though they were proportional to the square of the increase in concentrations over the 1990 level:

$$\Phi^r(t) = 1\text{-}[Q(t) / \Omega^r]^2. \tag{2}$$

Ω^r marks the critical value of atmospheric CO_2 concentration. At that level of perturbation, production in region r = N(orth), S(outh) is reduced to zero.

$\Phi^r(t)$ is the so-called environmental loss factor. It indicates economic damages induced by global climate change in region r = N,S. The corresponding economic costs are measured in terms of forgone GDP. I.e., if the atmospheric stock of carbon dioxide is raised to levels $Q(t)$ above pre-industrial atmospheric carbon, then in region r the productivity of factors is reduced such that only $\Phi^r(t)$ percent of the original gross production is at the region's disposal.

2.2 Production, Emissions, and Abatement

Principally, there are two ways to reduce CO_2 emissions. One is to replace greenhouse fuels by carbon-free energy inputs. A second option is to uncouple economic growth from fossil fuel consumption by increasing the energy efficiency and by substituting capital for energy. To capture both possibilities, the regional production possibilities are represented through a nested constant elasticity of substitution (CES) production function[1]:

$$y^r(t) = [\beta_1(l^r(t)^{\alpha} k^r(t)^{1-\alpha})^{\varepsilon} + \beta_2(e^r(t))^{\varepsilon}]^{1/\varepsilon}. \tag{3}$$

Capital $k^r(t)$, labor $l^r(t)$ and energy inputs $e^r(t)$ together produce the conventional (i.e., without climate effects) output, $y^r(t)$. β_1 and β_2 are CES-coefficients derived from base year data, and ε is the CES elasticity of substitution between capital/labor and energy.

Substitution between capital and labor is described by a Cobb-Douglas formulation where α is the corresponding parameter. Total energy inputs into regional production,

$$e^r(t) = f^r(t) + n^r(t), \tag{4}$$

are the sum of flows of fossil fuels $f^r(t)$ and of back-stop energy resources $n^r(t)$.

2.3 Material Balance of Produced Goods

Climate change negatively affects the productivity of the regional economies (see equation 2). Only the fraction $\Phi^r(t)$ of conventional gross output $y^r(t)$ is still at their disposal. Within each region r, "green output" $\Phi^r(t)y^r(t)$ can be consumed, invested into conventional capital formation, or used to supply either greenhouse resources or carbon-free energy.

Energy supply costs are measured in units of gross production. Marginal costs b^r of carbon-free energy are constant, but approximately four times as high as the

---

1 For better readability parameters do not carry regional indices.

costs of greenhouse resources in the initial year. Marginal costs $a^r(t)$ of greenhouse resources increase over time, depending upon extraction in prior periods.

To prevent an excessively rapid market penetration once renewable resources become competitive, it is assumed that after an energy supply system is installed, it is quasi-irreversible and cannot be abandoned immediately. With this formulation there is the possibility that market prices of energy temporarily overshoot the marginal costs of the renewable resources.

Regional output is considered as numeraire that can be traded internationally. Therefore, if $x^r(t)$ denotes net imports, $c^r(t)$ consumption, and $i^r(t)$ investment in conventional capital, for each period t

$$x^r(t) + \Phi^r(t)y^r(t) \geq c^r(t) + i^r(t) + a^r(t)f^r(t) + b^r n^r(t) \tag{5}$$

is the material balance of commodities produced and traded in region r. Finally, since net imports have to balance out in each period t, condition

$$x^N(t) + x^S(t) = 0 \tag{6}$$

has to be obeyed globally.

2.4 Intertemporal Decisions

At each point of time the society has control over two capital stocks, the endowment of physical capital on the one hand and the environmental capital on the other. The environmental capital stock is determined by the atmospheric CO_2 concentration. As such it depends upon abatement activities and – as will be discussed below – upon the way, climate policy is organized internationally.

The regional endowment $k^r(t)$ in physical capital depends upon investment activities, $i^r(t\text{-}1)$, the former capital stock, $k^r(t\text{-}1)$, and the capital depreciation at rate υ^r

$$k^r(t) = (1\text{-}\upsilon^r)\, k^r(t\text{-}1) + i^r(t\text{-}1). \tag{7}$$

At first glance, the natural approach to the economics of global climate change would be to employ an overlapping generations model. It was shown, however, that under reasonable assumptions both an infinitely-lived agent framework and an overlapping generations model will identify the same greenhouse policies as being efficient (see Stephan et al., 1997). Therefore, without loss of generality it can be supposed that for striking an optimal balance between consumption, physical investment and greenhouse gas abatement regions follow a Ramsey path.

Let δ be the social discount rate, then consumption, production, investment into physical capital and greenhouse gas abatement are determined in each region r =N, S, as if a policy maker has maximized the discounted sum of the logarithm of consumption, $c^r(t)$

$$W^r = \Sigma_t \delta^{-t} \ln(c^r(t)). \tag{8}$$

If there is no capital mobility and no investment into greenhouse gas abatement, both regions would develop independently. But if the regions agree to cooperate on greenhouse abatement, prices, supplies and demands are generated through a multi-region multi-period general equilibrium model. Solutions are obtained via Rutherford's sequential joint maximization method – a specialization of the Negishi approach (see Rutherford, 1999).

3 Banking and Trade

Throughout this paper it is supposed that all regions contribute to the solution of the global climate problem. Voluntarily they bind themselves to an international agreement on greenhouse gas abatement through implementing a system of carbon emission permits. Now, how could an institutional regulation look like that coordinates climate policy internationally? One idea might be to assume that such an agreement separates between the issue of allocating the climate resources optimally and the issue of distributing emission rights across regions (for a discussion, see Manne, 1999). That is, a credible inter-governmental agency sets and implements global CO_2 emission targets, $w(t)$, for each period t such that global climate resources are used optimally. Then emission shares $m^r(t)$ are assigned to each single region r through international negotiations. Roughly, this corresponds to the greenhouse gas emission trading proposal of the US Department of State. Note that our calculations are based upon the assumption that North's share in carbon rights will decline over a period of fifty years from 45% to 20%. This reflects a transition from grand-fathering to per-capita allocation (see Figure 4).

Given these presumptions, four "flexibility" scenarios come to mind:

(1) No-Trade-No-Banking (NT&NB),

(2) Trade but No-Banking (T&NB),

(3) No-Trade but Banking (NT&B), and finally,

(4) Trade and Banking (T&B).

Without trade, banking, and borrowing of carbon emission permits (NT&NB), each region r would be allowed to consume per period its own endowment,

$m^r(t)w(t)$, of carbon dioxide emissions only. This represents the situation of lowest "when and where" flexibility. Optimality in global greenhouse gas abatement is assured, but – as we expect – in an inefficient way, hence at highest economic costs.

International trade of carbon rights grants that the marginal costs of abatement are equalized across regions. This is a necessary requirement for Pareto-efficiency in climate policy. Consequently, compared to the NT&NB scenario both regions should gain from increasing the "where to abate" flexibility.

If emission permits can freely be transferred between regions as long as the world's restrictions in carbon endowment are obeyed, then the T&NB scenario has to satisfy

$$[w(t)-s^{N}(t)-s^{S}(t)] \geq 0 \tag{9}$$

for each period t.[2] And since each region r can generate additional income by selling carbon emission rights, it must be assured that each region stays on it's intertemporal budget constraint. This means, net imports have to be financed through counter trade of carbon rights and vice versa.

"When" flexibility in greenhouse gas abatement is given if the regional economies were allowed to allocate freely their endowment of carbon emission rights over time. Therefore, if trading of emission permits is not feasible in addition, a region r has to fulfill

$$\Sigma_t [m^{r}(t)w(t)-s^{r}(t)] = 0. \tag{10}$$

I.e., regional economies may save and borrow carbon emission permits, but they are not allowed to sell or buy them.

At first glance one might expect that banking and borrowing of carbon emission rights might have a similar effect on welfare as trade has. Since carbon abatement activities are now allocated efficiently over time, costs of abatement should be reduced and welfare should be positively affected. However, there could be two countervailing effects on regional welfare.

First, since costs of abatement are borne early but benefits do not accrue until the distant future, at least the North has an incentive to borrow carbon emission rights. This might lead to what environmentalists fear – a change of the time profile of carbon emissions with higher atmospheric carbon concentration in the future, hence higher ecological damages and – as a consequence – additional welfare losses (see Section 2.1).

---

2 $s^{r}(t)$, r = N, S denotes the CO_2 emissions of region r in period t.

Second, since the global climate has the properties of a public good, the South is affected by the North's banking and borrowing decision. In other words, if the North is allowed to operate banking and borrowing of carbon rights independently, this might create an intertemporal external effect on the southern economies. For an optimal solution it would be necessary, however, to internalize these effects. This is possible only, if the South compensates the delay in greenhouse gas abatement by the North through increasing his abatement activities in the near distant future. Again this can imply welfare losses for the southern economies.

Full flexibility in greenhouse gas abatement both in space and in time is given, if carbon emissions rights may be banked and borrowed, bought and sold on internationally open markets (T&B). Instead of regional and annual restrictions on CO_2 emissions, the world now has to obey a single intertemporal constraint:

$$\sum_t [w(t) - s^N(t) - s^S(t)] = 0. \qquad (11)$$

Based on conventional wisdom we expect that full inter- and intra-temporal flexibility in greenhouse gas abatement promotes economic efficiency, but can create the same problem as mentioned for the NT&B scenario.

4 Computational Experiments

Parameters for the computational experiments are benchmarked against 1990 data (for details, see Appendix). To reduce end-of-time-horizon effects, results are reported until 2150 but computations are carried out until 2200.

4.1 Carbon Emissions and Atmospheric Accumulation

Banking allows nations to decide individually and independently when to exercise their carbon emission rights. Does such a loss of control over the timing of emissions lead to what environmentalists fear – excessive borrowing or banking of carbon emission permits at early periods such that the time profile of atmospheric carbon will be changed significantly?

As Figure 1 shows the answer to this question is no. Increasing the "when-and-where" flexibility can even force lower atmospheric carbon concentrations, in particular in the middle of the time horizon. An explanation could be that the rate at which emissions accumulate is unimportant as long as emissions are small compared to the atmospheric carbon stock and as long as the total allowable amount of emissions is not exceeded. In any case, it is optimal to stabilize the atmospheric concentration below 550 ppm. This corresponds roughly to doubling the pre-industrial level.

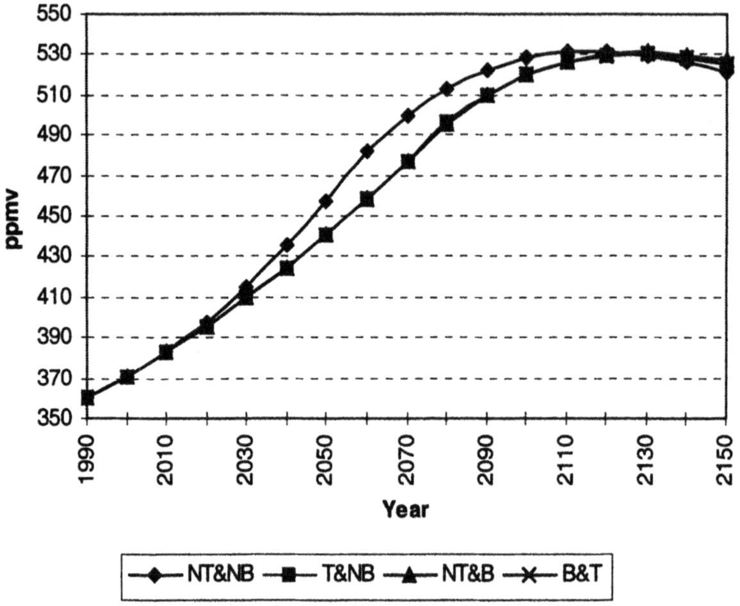

Figure 1: Atmospheric Carbon.

Although banking and trade of carbon rights does not significantly affect the atmospheric accumulation of carbon dioxide, regional emissions notably change. As Figure 2 illustrates, increasing flexibility stipulates CO_2 emissions in the northern hemisphere. If the option exists to buy carbon rights from the South, the North takes this option to reduce abatement activities and costs. If there is "when" flexibility, the North is motivated to borrow from itself carbon rights and to postpone abatement, just as was expected in Section 3. Since banking without trade imposes an adding-up constraint on total regional emissions (see equation 10), increasing emissions in early periods must be compensated by lowering them by the end of the time horizon, unless there is international trade in carbon permits. This explains the differences between a pure banking and the trade scenarios. Note that adding "when" flexibility to an international trade setting induces no changes in emission pattern.

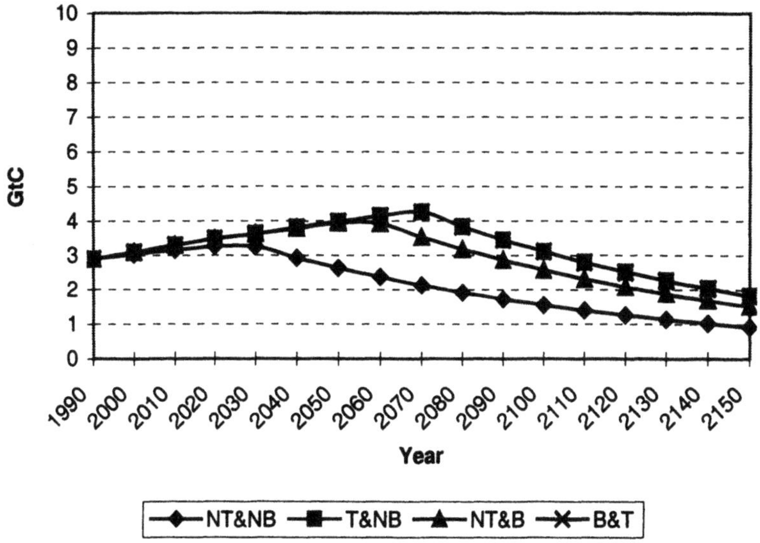

Figure 2: North's Carbon Emissions in GtC.

A similar invariance result is observed for the South (see Figure 3). Once international trade in carbon rights is allowed, banking does not induce further changes in emissions. With respect to trading of carbon permits either with or without banking, the South exhibits quite the opposite pattern from the North. The South trades a significant share of its emission rights against imports in numeraire from the North. Therefore, domestic abatement activities have to be extended. And if banking were allowed, the South will save carbon permits for the first century. Afterwards, emissions coincide with those in a No-Trade-No-Banking (NT&NB) scenario.

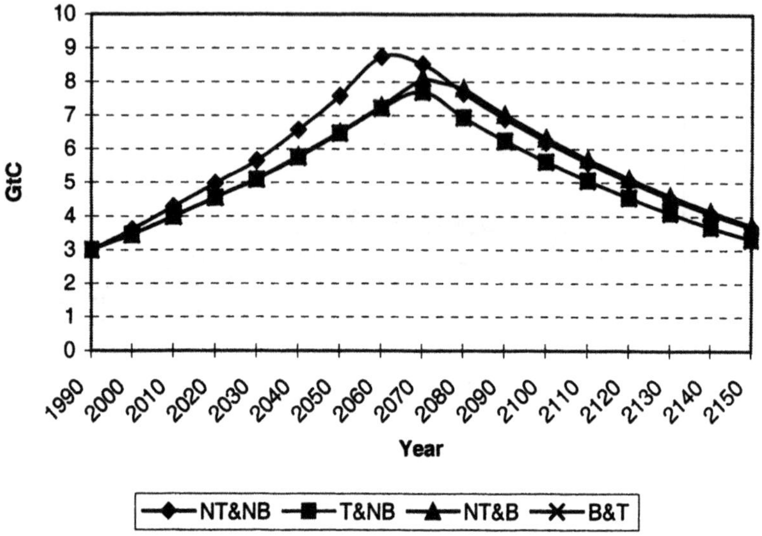

Figure 3: South's Carbon Emissions in GtC.

Summing up, our simulations suggest: If carbon emission rights are already traded internationally, then allowing for banking does not impose additional costs in terms of climate change. Moreover, in terms of atmospheric carbon concentration, differences between pure trade (T&NB), pure banking (NT&B) as well as banking and trade of carbon rights (T&B) can almost be neglected.

4.2 Economic Effects

As we expect, increasing the flexibility in greenhouse gas abatement – either trough trading or banking and borrowing of carbon emission permits or both – has a positive impact on welfare (see Table 1).

Table 1: Global and Regional Welfare Effects.

	NT&NB	T&NB	NT&B	T&B
Total Welfare	5.7698	5.7759	5.7738	5.7740
Welfare North	3.2017	3.2044	3.2059	3.2065
Welfare South	2.5681	2.5715	2.5679	2.5675

So far the calculations are not surprising. What surprises, however, is:

(1) Differences in aggregated and discounted welfare are almost negligible.

(2) Banking is less favorable than trade, even if combined with trade.

(3) There are obvious differences between regions depending upon the choice of instruments.

Both regions will gain simultaneously in terms of discounted welfare, if it is allowed to trade carbon emission permits internationally. If banking of carbon permits were feasible, then only the North will be better off compared to a No-Trade-No-Banking (NT&NB) as well as the pure trade (T&NB) situation. The South loses welfare under such circumstances and – equally important – total welfare is smaller with banking and borrowing than in a situation where trading of carbon permits is feasible.

The reasoning for this outcome was already given in Section 3. If banking and borrowing of carbon emission permits were allowed, then without international coordination the North can use this opportunity to delay abatement activities. However, by reducing costs of carbon abatement, he negatively affects the development of the world climate and imposes an intertemporal external effect on the South. Since the economic costs of climate change are higher in the South than in the North (see Section 2.1) the South has to counteract for an optimal solution by extending its abatement activities in early periods.

So far the question has not been answered why the effects of different flexibility regimes on aggregated welfare are almost negligible. Figure 4 suggests an answer. As mentioned in Section 3, emission shares are attributed exogenously to the single regions. Figure 4 identifies four different allocation schemes. GRAND represents an allocation of shares according to the so called grand-fathering rule. PCAP means that emission shares are attributed on an equal per-capita basis. PARETO would be a distribution of carbon rights that supports a Pareto-efficient solution of the global climate problem without the necessity of trade in carbon emission rights. ZEW is the exogenously given allocation of rights as supposed in this paper. Incidentally ZEW is close to the Pareto-efficient allocations as Figure 4 indicates. It is no surprise therefore that increasing flexibility can exhibit small changes only.

Within our representative agent framework, regional welfare is directly related to the logarithm of consumption (see equation 9). Therefore, per capita consumption might be used as proxy for regional welfare. For evaluating how different stages of flexibility in greenhouse gas abatement affect regional welfare, per capita consumption of the T&B scenario is taken as base-line. For better visibility, T&B per capita consumption is set equal to one and differences in welfare are then expressed as percent deviation in consumption from the base-line.

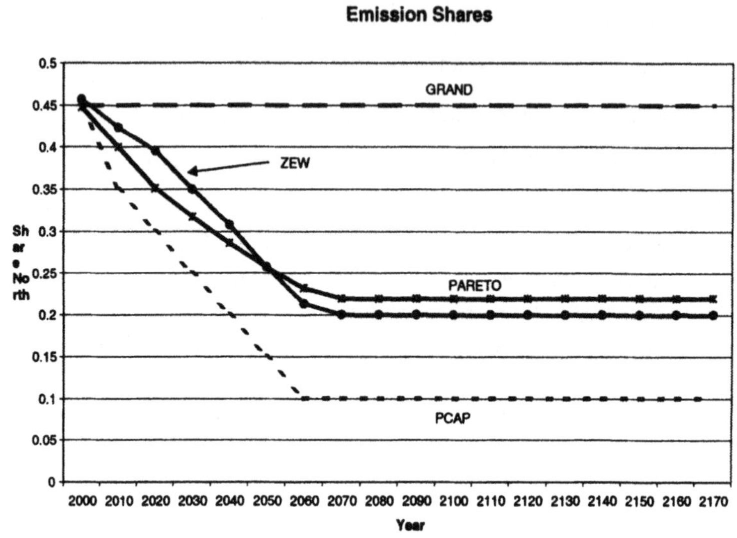

Figure 4: Exogenously Determined Emission Shares.

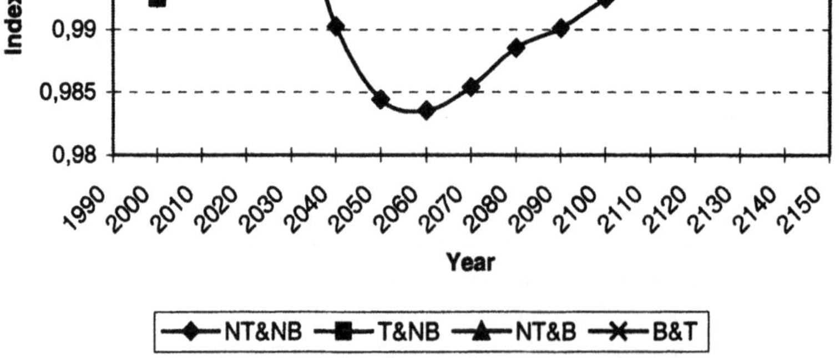

Figure 5: North's Consumption Relative to Banking-and-Trade Scenario.

On the one hand economists' conventional wisdom tells that welfare is positively correlated to "where to abate" flexibility. Trading carbon emission rights on an open international market implies identical marginal costs of abatement across all regions – a requirement that is necessary for efficiency in greenhouse policy. On the other hand the theoretical literature (for example, see Leiby and Rubin, 1998) suggests that welfare will decrease with "when to abate" flexibility, unless carbon emission rights are traded at the correct intertemporal exchange rate.

Our results do not completely support the last conjecture, but clearly identify winners and losers from "where and when" flexibility (see Figures 5 and 6). For the next century, No-Trade-No-Banking (NT&NB) is the worst case for both regions. For latter periods the results look a bit different. In any case flexibility works mainly in favor of the northern region, which gains mostly from postponing abatement either by trade or by banking.

If there were only one type of flexibility at choice, the North would prefer "when" over "where" flexibility in greenhouse gas abatement. To it the option to exercise a major part of its own carbon rights just at the beginning of the next century is superior to buying carbon rights from the South.

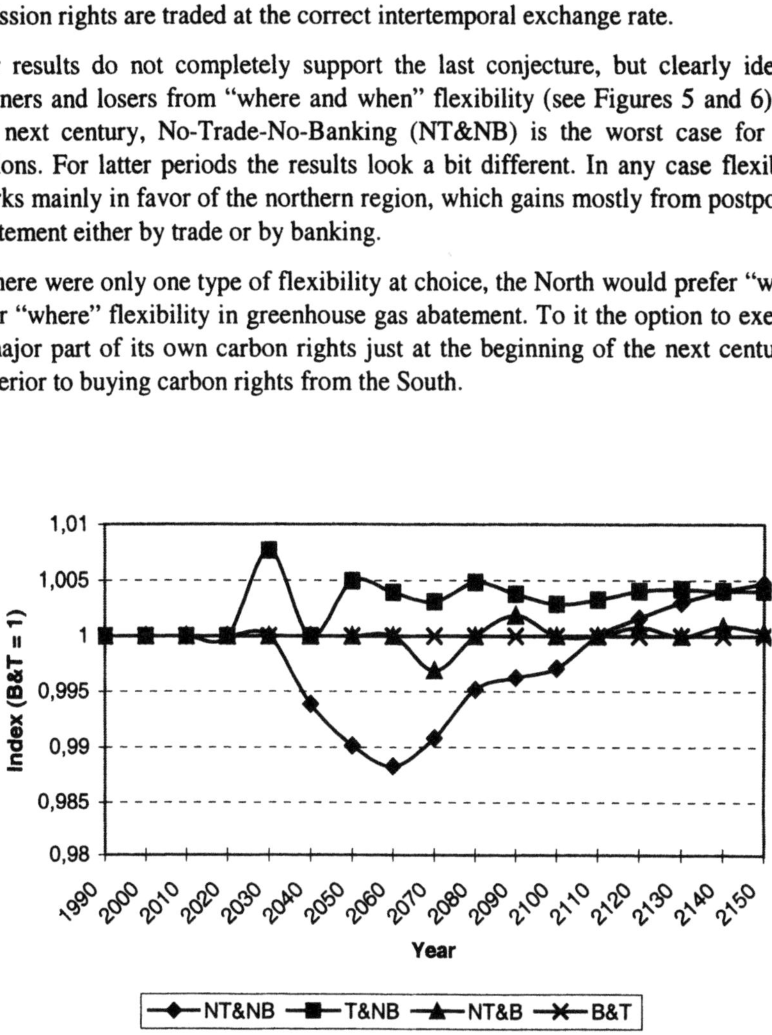

Figure 6: South's Consumption Relative to Banking-and-Trade Scenario.

From South's perspective, trade and banking (T&B) is the least preferred scenario. If the South has to vote for at least one type of flexibility, it would prefer international trading of carbon rights. This is due to potential gains from selling carbon rights to North. Banking decreases North's demand for carbon rights since it offers a substitute to trade. This worsens the terms-of-trade, imposing welfare losses on the South. Therefore, a trade only (T&NB) scenario is preferred to a trade and banking (T&B) scenario.

5 Conclusions

In summary, this analysis reveals two important results. First, contrary to the theoretical literature, increasing flexibility in greenhouse gas abatement has no negative impact on the global climate. The accumulation of the atmospheric carbon stock is almost independent of whether there is trading and banking of carbon rights or not. This implies that there exists separability between ecological targets and the choice of policy instruments.

Second, our paper shows that there exists a conflict between North and South with respect to "when and where" flexibility in greenhouse gas abatement. North is best off in a regime that allows both banking and trade of carbon emission rights. The worst outcome would happen to the North if there were no banking and no trade in carbon rights at all. For the South just the opposite is observed. South is best off in a pure trade scenario, but banking and trade of carbon emission rights is a bad situation for it even compared with a no flexibility scenario. Trading of carbon emission rights is a compromise. Both regions will gain welfare under such circumstances.

References

Bernstein, P.M., W.D. Montgomery, and T.F. Rutherford (1999), Global Impacts of the Kyoto Agreement: Results from the MS-MRT Model, *Resource and Energy Economics* 21, 375-413.

Biglaiser, G., K.J. Horowitz, and J. Quiggin (1995), Dynamic Pollution Regulation, *Journal of Regulatory Economics* 8, 33-44.

Joos, F., G. Müller-Fürstenberger, and G. Stephan (1999), Correcting the Carbon Cycle Representation: How Important Is It for the Economics of Climate Change?, *Environmental Modeling and Assessment* 4, 133-140.

Kling, C. and J. Rubin (1997), Bankable Permits for the Control of Environmental Pollution, *Journal of Public Economics* 64, 101-115.

Leiby, P. and J. Rubin (2001), Bankable Permits for the Control of Stock and Flow Pollutants: Optimal Intertemporal Greenhouse Gas Emission Trading, *Environmental and Resource Economics* 19, 229-256.

Manne, A.S. (1999), Equity, Efficiency and Discounting, in: Portney, P. and J. Weyant (Eds.), *Discounting and Integenerational Equity*, Washington D.C., 111-130.

Manne, A.S. and R.G. Richels (1995), The Greenhouse Debate, Economic Efficiency, Burden Sharing and Hedging Strategies, *The Energy Journal* 16, 1-37.

Manne, A.S. and G. Stephan (1999), Climate-Change Policies and International Rate-Of-Return Differentials, *Energy Policy* 27, 309-316.

Nordhaus, W.D. and Z. Yang (1996), A Regional Dynamic General-Equilibrium Model of Alternative Climate-Change Strategies, *American Economic Review* 86, 741-765.

Rutherford, T.F. (1999), Sequential Joint Maximization, in: Weyant, J. (Ed.), *Energy and Environmental Policy Modeling*, Dordrecht, 139-176.

Stephan, G. and G. Müller-Fürstenberger (1998), Discounting and the Economic Costs of Altruism in Greenhouse Gas Abatement, *Kyklos* 51, 321-338.

Stephan, G., G. Müller-Fürstenberger, and P. Previdoli (1997), Overlapping Generations or Infinitely-Lived Agents: Intergenerational Altruism and the Economics of Global Warming, *Environmental and Resource Economics* 10, 27-40.

Appendix

Data are taken from Manne and Stephan (1999) and Stephan and Müller-Fürstenberger (1998). Key economic data are listed below.

Table 2: Basic Economic Data.

Data	North	South
GDP 1990 (US$ trillions)	16.3	4.5
Capital-GDP ratio	2.5	3.0
Energy consumption (Exajoules)	176	168
Elasticity of substitution energy-value added	0.2	0.35
Carbon emissions (GtC)	2.9	3.0
Economic damage due to double pre-industrial CO_2 (%)	1.5	3.5
Utility discount rate	0.03	0.03

Table 3: Potential GDP Growth Rates.

	North	South
1990-2120	1.5	3.5
2130	1.5	3.25
2140	1.5	3
2150	1.5	2.75
2160	1.5	2.5
2170	1.5	2.25
2180	1.5	2
2190	1.5	1.75
2200	1.5	1.5

Cost-Efficiency Methodology for the Selection of New Car Emission Standards in Europe*

Zeger Degraeve

London Business School, Sussex Place – Regent's Park, London NW1 4SA, UK, *zdegraeve@lbs.ac.uk*

Stef Proost

Faculty of Economics, KULeuven, Naamse Straat 69, 3000 Leuven, Belgium, *stef.proost@econ.kuleuven.ac.be*

Gunther Wuyts

Faculty of Economics, KULeuven, Naamse Straat 69, 3000 Leuven, Belgium, *gunther.wuyts@econ.kuleuven.ac.be*

Summary:
In the Auto-Oil Programme, the European Commission looks for emission limits for cars such that the urban air quality targets are reached at minimum cost. This optimization problem was solved by Degraeve et al. (1998). In this paper we deal with two methodological problems in this cost efficiency approach. We study first what is known as the overachievement problem in cost-effectiveness analysis. In a pure cost-efficiency approach, there is a tendency to understate the merits of federal regulatory measures: Because these measures are uniform they will always do more than required in some regions. We prove this and show how this problem can be solved using minimum information on the benefits of environmental improvements. The second problem we study is the implementation problem of local measures. From a European wide perspective, it may be cost-efficient that some regions take local measures but this is not necessary in the interest of these regions when there is transfrontier pollution. When this behavioral constraint is taken into account, the cost efficient bundle will change. We show how these two considerations affect the selection of optimal emission standards for cars in Europe.

---

* We thank the workshop organizers, participants and in particular Olaf Hölzer for his detailed comments on this paper.

1 Introduction

The transport sector has been up to now the most important source of emission for conventional pollutants (CO, VOC, NO_2, PM) in urban areas. It is not only the most important source of emissions in quantity terms but also the most damaging source per unit of emission because of the high concentrations of emissions at low height and the high population density at the places of emissions. It is accepted that 1 ton of emissions emitted at ground level could be 5 to 10 times more damaging than 1 ton emitted at higher height by central heating, industry or power stations. Therefore transport emissions have much more local health effects than the emissions of other sectors.

Achieving an acceptable air quality in urban areas is one of the priorities of the European environmental policy. The EU has up to now relied mainly on emission standards for cars and trucks and on emission standards for other sources to achieve this objective. In this paper we will analyze more in detail the emission by the transport sector.

The first directives of the Commission implementing the catalytic converter standards in 1991 and 1994 (91/441/EEC and 94/12/EC) were criticized heavily by industry. The complaint was that neither the benefits nor the costs of the standards had been assessed. The Auto-Oil Programme (abbreviated as AOP), an assessment exercise of the Commission in collaboration with the automobile industry and the oil industry, has innovated in terms of assessment methodology. In the AOP-I Programme (1992-1996), an explicit cost-effectiveness approach has been followed. The basic premise was that emission standards on new cars as well as new fuel qualifications are only justified if they are the cheapest way to reach urban air quality targets in EU regions. The complexity of the relation between emission reductions and urban air quality was taken into account via atmospheric air pollution models and there was an effort to make car emission standards compete against other policy instruments as there are fuel qualifications, gasoline taxes etc. The European Parliament did not follow the recommendations of the AOP-I but opted for more stringent car emission and fuel standards. One of the reasons might have been the unbalanced composition of the AOP-I group where the regulated industry was overrepresented.

The second AOP-II study that started in 1997 takes into account a wider range of policy measures and relies on improved models and data. CO_2 emissions are in principle not an issue in the Auto-Oil Programme.

In the Auto-Oil Programme, the European Commission looks for emission limits for cars such that the urban air quality targets are reached at minimum cost. This is a complex exercise because of two difficulties. The first is the presence of economies of scale in car manufacturing. This means that emission limits cannot be differentiated regionally and that additional local measures can become interesting to reach the air quality limits. The second difficulty is the presence of transfrontier pollution. Part of the pollution originating in one region ends up damaging other regions in Europe. Selecting emission standards for cars is therefore a complex exercise in which local transport and other measures are traded off against European wide measures on car emission abatement and fuel quality.

This optimization problem was solved by Degraeve et al. (1998). They combined information on the cost of different measures and information from air quality models into a cost minimization framework. It was found that a balanced combination of traditional car emission regulation measures and locally diversified transport measures can achieve the objectives at much lower overall cost. This first exercise has been the basis for the European emission regulations for new cars from 2005 onwards. It was followed by a new more complete cost-effectiveness exercise (called AOP-II).

In this paper we discuss some of the methodological difficulties encountered in this exercise and show how the choice of methodology affects the results.

We study first what is known as the overachievement problem in cost-effectiveness analysis. In a pure cost-efficiency approach, there is a tendency to understate the merits of federal regulatory measures: because these measures are uniform they will always do more than required in some regions. We prove this and show how this problem can be solved using minimum information on the benefits of environmental improvements. In fact this problem is a variant of the problem solved by Oates et al. (1989). The second problem we study is the implementation problem of local measures. From a European wide perspective, it may be cost-efficient that some regions take local measures but this is not necessary in the interest of these regions. This element needs to be integrated in the cost-effectiveness methodology and affects the choice of the optimal bundle. We show how these two considerations affect the selection of optimal emission standards for cars.

Section 2 presents the current AOP-II cost-efficiency methodology. Section 3 discusses the overachievement problem. Section 4 treats the implementation problem of local measures. The last section illustrates both problems numerically. The data, optimization procedure and effects on the selection of the policy bundle are briefly explained. We conclude with some caveats.

2 The Current Cost-Effectiveness Methodology

In the Auto-Oil Program of the European Commission, the objective of the analysis is to select emission standards for cars and fuel quality regulations such that in all urban areas, the ambient air quality targets (CO, Benzene, NO_2, PM10, Ozone) are met at the lowest cost. Therefore the cost and efficiency of emission standards for cars have to be compared with measures on fuel quality, inspection and maintenance programs, transport demand measures, fiscal measures and emission reduction in non-transport sectors. This is done simultaneously for the period 2000-2020 and for all pollutants considered.

This is a very ambitious goal. The approach followed can best be understood using Figure 1. One can imagine different types of measures to improve the urban air quality. We concentrate first on transport measures. Different categories of measures are illustrated at the bottom of Figure 1. They include vehicle emission standards, fuel quality standards, inspection and maintenance measures, transport demand management and economic instruments (including taxes etc.).

The potentially effective measures are fed into a transport model (one for each urban area and country considered) in order to determine its overall effect on emissions and its welfare cost. This is necessary because the welfare cost and emission reduction effect of transport demand measures requires working through its effects on all transport markets. As an example consider lower public transport fares. This will increase bus use and bus emissions but will decrease car use and car emissions. The model used for this computation is the TREMOVE transport model (DRI-KULeuven, 1999). The output of this step is a list of measures with, for each area studied, the net emission reduction effect for that area and the gross welfare cost of the emission reduction.

The gross welfare cost of each policy measure consists of three elements: the change in consumer surplus on the transport markets expressed in generalized cost terms[1], the change in producer surplus plus the change in tax revenue valued at the marginal cost of public funds. A broad welfare concept was needed in order to assess on the same basis the variety of policy measures to be considered. The welfare concept is gross because it does not include the air pollution benefits.

The next step is the selection of the least-cost mix of measures to achieve the target ambient air quality in the different urban areas in Europe. This is done via an optimization model (LEUVEN II, Degraeve et al., 1998) that combines information on emission reduction and costs for each potential policy in the transport sector and in other sectors for each emission region with air quality modeling relations that translate emission reductions in each emission region into changes in ambient air quality in all European urban areas. This problem is solved simultaneously for all selected urban areas, all pollutants and for the whole period 2000-2020.

The mathematical model used in this second step is set out in detail in Appendix 1. Three factors are responsible for the complexity of this model. First, there is the

---

1 The generalised cost of a trip or a kilometre on a transport market equals the sum of the resource cost, the time cost and the taxes paid. This price concept allows to take into account the behavioural response to changes in congestion levels via changes in the time costs.

interaction between different policy measures (the total effect of one measure can be smaller than the sum of two measures). Next, there is the large number of combinations of measures that can be used. The different measures are considered as discrete steps so that the model is an integer optimization problem. Finally there is the interaction between the different urban problems due to the transboundary effects of ozone and the uniformity constraint for the global measures.

In the two next sections, we discuss the two methodological problems in the present approach.

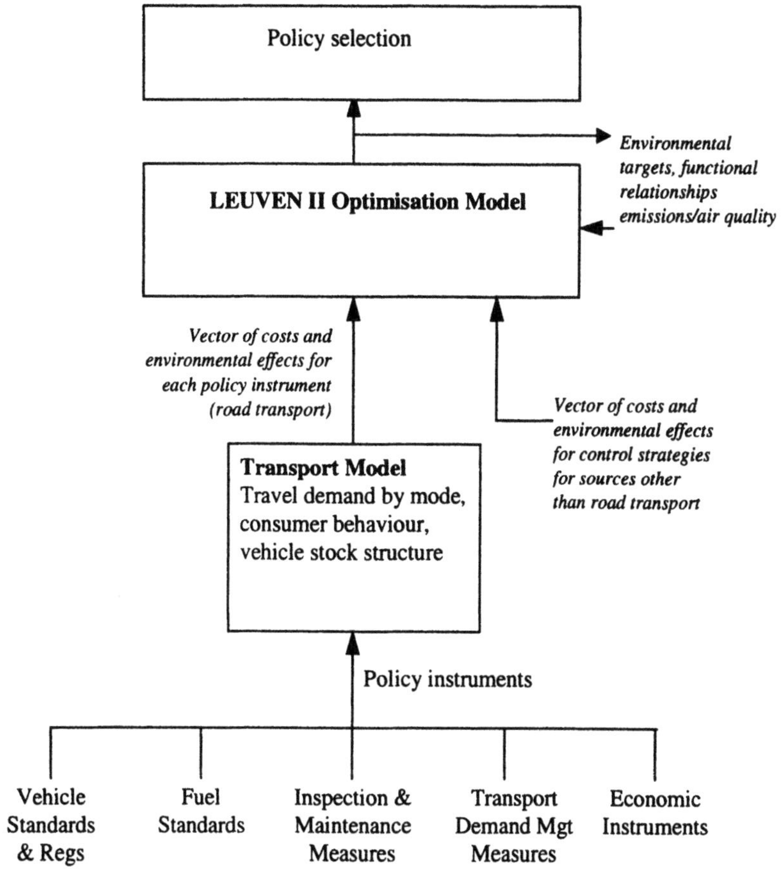

Figure 1: Structure of the Auto-Oil II Cost Effectiveness Analysis.

3 The Overachievement Problem

In a cost-effectiveness analysis such as AOP-II, overachievements of targets cannot be neglected. Apart from the benefits of a cleaner environment, neglecting may also lead to solutions that are biased against the use of global (or federal) measures. Emission limits for cars are such a global measure. We will now first develop a simple example to illustrate the type of problem that has to be solved.

Consider 2 regions (A and B) and 2 types of measures (x and y). The intensity at which both types of measures are introduced in region A is denoted by x^A and y^A, while x^B and y^B represent the intensities in region B. The cost of introducing measures x and y in region A is given by a cost function $C^A(x^A, y^A)$. For region B, a similar notation is used. Ambient air quality objectives are expressed as reductions in concentration compared to a given baseline and are denoted by Q^A and Q^B for regions A and B respectively. The diffusion of pollutants is formulated via constant coefficients t_{AB} relating the emissions in A to the reduction in ambient concentration in B.

The cost minimization problem to be solved is then:

$$\min_{x^A,x^B,y^A,y^B} \; C^A\left(x^A,y^A\right)+C^B\left(x^B,y^B\right) \tag{1}$$

such that

$$t_{AA}\left(x^A+y^A\right)+t_{BA}\left(x^B+y^B\right)\geq Q^A$$

$$t_{AB}\left(x^A+y^A\right)+t_{BB}\left(x^B+y^B\right)\geq Q^B$$

$$x^A,x^B,y^A,y^B\geq 0$$

The Auto-Oil methodology carefully avoids the use of monetary damage estimates to compute optimal emission reductions. It is well known that air pollution damage estimates cannot be determined with precision (for an illustration see the discussion between Krupnick et al. (1991) and Hall et al. (1992) or the ranges in the Extern-E results (Bickel et al., 1997). This explains why in policy discussions like Auto-Oil one prefers cost-effectiveness studies that use an absolute ambient air quality target.

In problems with many pollutants and policy instruments that are common to several regions, there will often be overachievement for some pollutants and some regions. Some of the stakeholders in the Auto-Oil debate have pointed to the possible bias against the use of uniform European instruments when overachievement is present. The idea is that European wide emission standards for cars and European wide oil quality standards tend to generate overachievement due to their lack of flexibility. Not attributing any value to this overachievement leads to a bias against uniform measures.

In a pure cost-effectiveness analysis overachievement receives no benefit at all and this is not very logical. Giving a small benefit to overachievement may change the optimal mix of policy measures.

We consider a simplified example to discuss this issue. We assume that there is no transboundary pollution and that cost functions are strictly convex and continuous. There are three types of policy measures: local measures for each of the regions A and B: x^A and x^B and a uniform or global measure y that affects both regions. By global measure we mean that this measure has to be introduced in both regions at the same intensity. Measures of type y are typically regulatory emission standards for cars or some oil quality standards. The cost-effectiveness problem (1) can then be rewritten as:

$$\min_{x^A, x^B, y} \; C^A\left(x^A\right)+C^B\left(x^B\right)+C\left(y\right) \tag{2}$$

such that

$$x^A + y \geq Q^A$$
$$x^B + y \geq Q^B$$

The first-order conditions for this problem are:

$$\begin{aligned} x^A &\geq 0 \quad \frac{\partial C^A}{\partial x^A} - \lambda^A \geq 0 \\ x^B &\geq 0 \quad \frac{\partial C^B}{\partial x^B} - \lambda^B \geq 0 \\ y &\geq 0 \quad \frac{\partial C}{\partial y} - \lambda^A - \lambda^B \geq 0 \end{aligned} \tag{3}$$

If at the optimal solution $\bar{x}^A$, y, $\bar{x}^B > 0$ then there is no overachievement because if there was an overachievement, one could always decrease total costs by reducing the local measure intensity and this contradicts the optimality of $\bar{x}^A$, y, $\bar{x}^B > 0$. On the other hand, if at the optimal solution $\bar{x}^A$, $y > 0, \bar{x}^B = 0$ then overachievement is possible in B.

We will now show that overachievement and its treatment can have important impacts on the selection of the least cost bundle. This type of problems is present in many cost-effectiveness analyses. Oates et al. (1989) examined the relative efficiency of standards and tradable permits for a cost-effectiveness problem in which certain air quality targets had to be met. They found that standards tend to give rise to more overachievement. When this overachievement is not valued, one will overestimate the advantage of tradable permits. Here we analyze a problem not related to the choice of instruments but the choice of measures that are more or less flexible.

What impact can be expected from valuing overachievement? If we give credits s^A and s^B to overachievement of the ambient air quality objectives in region *A* and *B* and add deduct these credits from the objective function in (2), we obtain new first-order conditions for our previous optimization problem:

$$
\begin{aligned}
x^A \geq 0 \quad & \frac{\partial C^A}{\partial x^A} - \lambda^A - s^A \geq 0 \\
y \geq 0 \quad & \frac{\partial C}{\partial y} - \lambda^A - s^A - \lambda^B - s^B \geq 0 \\
x^B \geq 0 \quad & \frac{\partial C^B}{\partial x^B} - \lambda^B - s^B \geq 0
\end{aligned}
\tag{4}
$$

Assume you start with s^A, $s^B = 0$ and x^A, $y > 0$ but $y^B = 0$ and overachievement in region *B*. In this case, increasing the credits for overachievement s^A and s^B is likely to improve the ranking of measure *y* and could lead to more overachievement. To see this, check what happens when s^A or s^B are increased starting from 0. With s^A increasing nothing changes because there was initially no overachievement so as long as s^A is smaller than the marginal cost $\lambda^A > 0$ of achieving the air quality target in *A* there is no interest in increasing x^A. Consider now a small increase in s^B. The optimal use of measure x^B is not affected as long as the credit for overachievement is smaller than the marginal cost of measure *B*. But the use of measure *y* will be increased as:

$$
\frac{\partial c}{\partial y} - s_B = \lambda^A = \frac{\partial c^A}{\partial x^A}
\tag{5}
$$

so that measure *y* has become more interesting to use than measure x^A.

We see that the ranking between global measures and local measures is twisted again more to global measures when credits are given to overachievement. Introducing benefit measures for pollution reduction matters for the selection of instruments.

Of course we could also discuss the problem of underachievement. Reaching the target air quality in some regions like Athens may be impossible in the short term. It is important to foresee this possibility by using a maximum estimate for the cost of underachievement. Combining benefits and fines would bring us closely to the stepwise incentive function advocated by Roberts and Spence (1976) to solve the regulation problem with imperfect knowledge about abatement costs.

The optimization problem that is presented in the appendix allows for over- and underachievement of the air quality targets and takes both possibilities into account when determining the best policy mix for the different regions. In the objective function (A.1), the second and the third sum give the total benefit for overachievement. This is the amount by which the targets for the different (local and regional) pollutants are exceeded in the different regions multiplied by the credit that is given to such an averachievement. The amounts are determined in the constraints (A.2) and (A.3) and are given by sro_{ij}. The sign of these benefits in the objective function is of course negative since a benefit is a negative cost. The fines for underachievement are represented by the last two sums in the objective function (A.1). The interpretation is very similar to the one of the overachievements, except for the signs since an underachievement is treated as an extra cost.

4 The Implementation Problem of Local Measures

In this section, we will discuss a second problem-type that can be encountered in an analysis such as the Auto-Oil Programme. In the selection of the optimal mix, one compares local and European measures using the same welfare cost concept. However ultimately, the optimal mix of measures has to be implemented and this can give rise to some problems as the local policy maker may have a different objective function than the European decision-maker. The optimization model selects the optimal policy mix from a European-wide point of view, but if the European policy maker cannot enforce the regions to implement this strategy, and more specifically the local measures, it is possible that these regions have an incentive to deviate from this optimal mix. We will again illustrate this problem by means of a simple example and afterwards, we will discuss how it is tackled in the LEUVEN II-model.

Let us return to our model (2) in the previous section. Here, it was assumed that there was no transboundary pollution. If we introduce this type of pollution, we can rewrite the model to obtain the optimal mix of uniform measures y and local measures x in the following way:

$$\min_{x^A, x^B, y} \; C^A\left(x^A\right)+C^B\left(x^B\right)+C\left(y\right) \tag{6}$$

such that

$$t_{AA}\left(x^A + y\right)+t_{BA}\left(x^B + y\right) \geq Q^A$$
$$t_{AB}\left(x^A + y\right)+t_{BB}\left(x^B + y\right) \geq Q^B$$
$$x^A, x^B, y \geq 0$$

The first-order conditions of this problem are:

$$\begin{cases} \dfrac{\partial C^A}{\partial x^A} - (\lambda^A t_{AA} + \lambda^B t_{AB}) = 0 \\ \dfrac{\partial C^B}{\partial x^B} - (\lambda^B t_{BB} + \lambda^A t_{BA}) = 0 \\ \dfrac{\partial C(y)}{\partial y} - \lambda^A (t_{AA} + t_{BA}) - \lambda^B (t_{AB} + t_{BB}) = 0 \end{cases} \tag{7}$$

Let $\bar{x}^A$, $\bar{x}^B$, $\bar{y}$ denote the optimal solution. This type of solution has been used in the Auto-Oil decision process. The problem is that the local measures (x) will be decided upon by local governments that use a different objective function. To see this consider the problem solved by the local government A: to choose x^A for given $\bar{y}$ and given $\bar{\bar{x}}^B \neq \bar{x}^B$ and where BEN^A represents the benefits of emission reduction in region A:

$$\underset{x^A}{\text{Min}}\ C^A(x^A) - BEN^A\left(t_{AA}\left(x^A + \bar{y}\right) + t_{BA}\left(\bar{\bar{x}}^B + \bar{y}\right)\right) \tag{8}$$

The interior optimum of this is given by:

$$\frac{\partial C^A}{\partial x^A} - \frac{\partial BEN^A}{\partial Q^A} t_{AA} = 0 \tag{9}$$

We need to compare this with:

$$\frac{\partial C^A}{\partial x^A} - \bar{\lambda}^A t_{AA} - \bar{\lambda}^B t_{AB} = 0 \tag{10}$$

when one central government decides on the local measures too.

Assume that the local marginal benefits of emission reduction are constant. We see that the local government will not implement sufficient local measures because it does not have the incentives to do so. The optimal mix selected with this methodology is only optimal in as far as the European policy maker can indeed correct the local government behavior. If this is not so, it may be better to rely less on local measures and use more central measures.

In the optimization model (detailed in Appendix), we solved the implementation problem in a different way. It proved impossible to fully model the two tier problem of the European government, so we try to mimic the problem by imposing that a certain percentage of the air quality targets has to be reached by global measures because of lack of control on the implementation of local measures. These constraints can be found in constraints (A.6) and (A.7) in the appendix. The intuition for these extra constraints is that we have argued above that if the implementation problem of local measures exists, it may be optimal to select more global measures.

5 How Important Is the Choice of Cost-Efficiency Methodology – A Numerical Evaluation

5.1 Scope

The choice of methodology is numerically illustrated for the Auto-Oil I problem, since there were no sufficient data available from AOP II. In AOP-I, seven European regions (Athens, Köln, Den Haag, London, Lyon, Madrid, and Milano) and four pollutants were incorporated in the analysis, which concerns one representative year (2010). Remark that we have one transboundary pollutant (Ozone) and three local pollutants. The goal of the model is then to determine which measure should be introduced in each region to obtain the required air quality objectives at least cost. In order to provide as much flexibility as possible, the model can choose between a wide variety of measures in the transport sector. Also the non-transport sector (e.g. industry or combustion) has been included, such that the model can freely determine the contribution of each sector.

5.2 Data

From the mathematical model in the appendix, the data requirements can be determined. First, data on the cost of introducing a measure in a region is needed. Also, the environmental effectiveness of that measure in a region on a pollutant has to be specified. Furthermore, the model demands the required air quality targets and data on emission concentration relationships. Finally, benefits for overachievement and fines for underachievement have to be quantified.

The costs of introducing sets of measures in a region[2] are based on a welfare cost concept that not only comprises technical costs such as the costs of fitting pollution abatement equipment to vehicles, but also the costs to consumers of behavioral adjustments induced by policy measures, e.g. reductions in mobility. Hence, these costs can be interpreted as total costs to society. The cost of a bundle (see Appendix for a definition of the different concepts) is simply the sum of the costs of all sets of control measures in that bundle.

The effectiveness of a set of control measures is defined in terms of emission reductions. In this paper, we use a specific formula to compute the effectiveness of a bundle, the LEUVEN II-model uses a more general and flexible way, based on a tree-structure. We refer to the appendix for more details.

Also air quality targets need to be specified, i.e. the percentage reductions in urban emissions as well as in ozone concentrations that are required to bring down air pollution to acceptable levels. Both in the USA and in the EU the basic approach to air quality problems consists of requiring the attainment of uniform air quality standards. Obviously, the first step is the definition of these standards. In the EU air quality standards are based on World Health Organization guidelines and are set at uniform levels across the EU for concentrations of pollutants at different places during different time intervals. These levels are generally thresholds above which health risks become serious or significant damage to vegetation and agriculture occurs.

Although air quality is expressed in terms of concentrations, policies can only affect emissions and this implies that relations between emissions and concentrations will have to be incorporated in the analysis. With the exception of ozone these relations are linear, e.g. a 1% reduction in NOx emissions, brings down NOx concentrations by 1%. For ozone, these relations are highly nonlinear, but – over certain ranges – they can be approximated by linear functions which link reductions in regional VOC and NOx emissions in region i to changes in

---

2 The objectives and costs of policy measures in a region are assumed to be representative of the countries in which the region is situated. This means, for example, that the total costs of a fuel measure introduced in the Köln region consists of the per vehicle cost (annual mileage times cost per liter) multiplied by the size of the entire German vehicle fleet. Implicitly, therefore, the analysis covers the following countries: Greece, Germany, The Netherlands, United Kingdom, France, Spain, and Italy. Clearly, this procedure assumes a significant degree of averaging within countries. Several consequences arise from this assumption. First, although there is no guarantee that the air quality objectives for a pollutant are reached everywhere in a particular country, it nevertheless seems reasonable to assume that large parts of the country satisfy the standard since, the objectives have been formulated for regions, where, air quality problems are relatively severe. Secondly, to the extent that local measures can only be introduced in those parts of a country where they are really necessary, their "true" costs are likely to be overestimated because they would only have to be introduced in non-attainment areas.

ozone levels in region *j*. This, in turn, is summarized in two so-called "blame" matrices (one for regional NOx and one for regional VOC) the coefficients of which give the percentage reduction in ozone levels in a particular region due to a 1% reduction in emissions of the corresponding precursor in another region.

Based on the relations above, expected economic growth as well as measures already scheduled in a variety of sectors, a forecast has been made in the context of the Auto/Oil Programme of "baseline" emissions and air quality levels for the year 2010 in seven representative European regions (European Commission 1996). A comparison of those levels with the air quality standards gives a set of required reductions in emissions.

The approach taken for ozone is, however, different. The ozone reduction objectives represent percentage reductions in forecast ozone AOT90 (Accumulative exposure Over a Threshold) figures. AOT90 is defined as the sum of hourly ozone concentrations exceeding 90 parts per million during the months May, June and July. The AOT90 values are calculated for the year 2010 with the EMEP MSC-W model of the Swedish Environmental Research Institute (Simpson 1991, update 1995). The reduction target imposed for AOT90 in our analysis, results from keeping excess ozone levels below 5 ppb. Our ozone target can be interpreted as a required reduction in peak ozone concentrations that occur because of man made emissions from transportation and nontransport sources.

Finally, benefits of overachieving certain targets are considered as well. These benefits are defined as the 'reward' for a certain region of overachieving its targets for a certain pollutant. An example of such benefits can be found in a benefit study by IVM et al. (1997). In our paper, we will illustrate the importance of correctly computing these benefits by varying the level.

As said before, a wide variety of control measures is included. In the appendix it is explained that these control measures are put together in sets, which contains zero, one or more control measures. These sets are then classified in categories. In Table 1, an overview is given of the categories and the sets of control measures that are included in the simulations. A broad range of categories is taken into account, as well in the Transport sector as in the Non-Transport sector (Non Transp). Since the focus of AOP-I was on the transport sector, this sector has been split up further. The first subdivision, Technical Standards is broken down in diesel cars (TED) and gasoline cars (TEG)[3]. Furthermore, we consider different fuel standards (FUELS) that lead to emission reductions for an unchanged engine and inspection and maintenance schemes (IM). These inspection and maintenance schedules can take the form of remote road sensing or the installation of on board diagnostics. Also non-technical control measures are included. More specifically,

---

3 These technical standards are engineering packages that achieve increasing percentages of emission reduction for cars and light duty vehicles and this for the different pollutants considered.

we look at public transport subsidies (PUB) of up to 40% reductions in fares, scrappage schemes (SCR) and structured circulation taxes (SM) that relate the circulation tax to the average yearly emission of certain pollutants for that type of car. The sets in the two categories of Technical Standards (TED and TEG) only contain global control measures and are indicated in bold in the table. All control measures in the other categories are assumed to be local. Measures in the non technical category contain measures for the emissions of power plants and other large combustion installations.

Table 1: Categories and Sets of Control Measures.

Transport							Non Transp
Technical Standards			Non Technical				
TED	TEG	FUELS	IM	PUB	SM	SCR	
TED00	TEG00	F00	IM00	PUB00	SM00	SCR00	NT00
TED01	TEG01	F04	IM01	PUB01	SM03	SCR01	NT01
TED03	TEG03	F05	IM02	PUB02	SM05	SCR02	NT02
TED04	TEG04	F06					NT04
TED06	TEG06	F08					NT06
TED07	TEG07	F09					NT08
TED08	TEG08	F10					NT10
TED09	TEG10	F14					
TED10	TEG11						
TED12	TEG12						
TED13	TEG14						
TED14	TEG15						
TED16							
TED17							

5.3 Results

In Table 2 in the appendix, the results for different optimization runs of the model can be found. From the third column onwards, each column gives the results of a different run. Within each column, we first present a general overview of the results of that scenario. Here, one can find the gross costs, this is the total EU-wide cost of introducing the control measures net of benefits or fines. Also the total benefits and fines and the total net costs, defined as the gross cost minus the benefits are given. In the lower columns we report results for each of the 7 regions: the costs, the composition of the bundle chosen, and the environmental effectiveness of that bundle are specified.

The first scenario in the table is called the base case because we will compare all other scenarios with this one. In the base case, we follow the standard AOP2 methodology, the benefits for overachievement are equal to zero for each pollutant, while the fines for underachievement are infinity in all cases. From Table 2 follows that the EU-wide costs for implementing the control measures necessary for attaining the required targets in the different regions are equal to 4629.143 million Euro. In the remainder of the column, further details about the specific choice for each region can be found. The results should be interpreted as follows. For the region of Athens, the gross cost of the bundle introduced in that region is 251.19 million Euro. In the next part, the composition of the bundle selected is given. We can see that from the category FUELS, the set F04 is chosen and this set has a cost of 15.2 Million Euro. The set IM03 was picked from IM which has a cost of 7.26 million Euro and so on. In the third part, the achieved effectiveness for each pollutant is shown. Remark that for Ozone, this is not only the effectiveness of the bundle selected by this region, but also the sum of the effect that other regions have on the level of Ozone in this region. We can see that in Athens a reduction of 60.72% was achieved for Ozone (to be compared with the required 45%) , 35.12% for Regional NOx and so on. The results for the other six regions should be interpreted in the same way as just explained. If we compare the achieved effectiveness in the different regions with the required targets for each region which are presented in the second column, we can detect the constraints that are binding. These are the Ozone-constraint in Den Haag and the constraints for Urban NOx in Athens, Madrid and Milano. Later on in the discussion, it will be quite often these urban air quality constraints that drive the results obtained.

A first extension of the base case is to allocate benefits when a region overachieves its required targets. We consider three cases: a benefit of first 20 Euro and then 30 Euro per percent per capita overachievement of the target for Urban NOx and a benefit of 5 Euro per percent per capita overachievement of the targets for Urban NOx and Ozone. The results can be found in Table 2 in the columns "Ben NOx Urb 20", "Ben NOx Urb 30" and "Ben NOx Urb Ozone 5" respectively. It is trivial that the net total costs are smaller compared with the base case. Remark, however, that the gross costs are higher in the three cases. This is an indication that stronger control measures have been chosen (a more expensive control measure has also a greater effect). Secondly, in most cases, the amount of overachievement of a target increases for the pollutant(s) with a benefit, while for the pollutants without a benefit the picture is mixed. Although it holds for most control measures that if they have more effect on one pollutant, they also are more effective for the other ones, the selection of control measures is apparently more aimed at the pollutants with a benefit. Thirdly, also the composition of the bundles changes, as well for the part of local control measures as for the part of global control measures. When we increase the benefit for Urban NOx from 20 to 30 Euro, it is mainly the choice of the global control measures that changes, the local control measures selected remain more or less the same. Finally, remark that even a relative small benefit of 5 Euro has a quite large impact on the bundles selected, this shows the importance of incorporating benefits for overachievement in the analysis.

In contrast with the case of benefits, one can include fines for underachievement in the analysis, hereby allowing a region not to satisfy completely all targets for the different pollutants. In the columns "Fines NOx Urb Ozone 50" and "Fines NOx Urb Ozone 100" of Table 2 we introduced a fine of 50 and 100 Euro respectively per percent underachievement per capita of the target for Urban NOx and Ozone. A first important point is that the net costs are both smaller than in the base case. In case of a fine of 50 Euro, we find lower gross costs but a higher fine than when the fine is 100. This is intuitive since a lower fine gives a higher incentive for a region not to introduce a (relatively) expensive control measure to reach a target, but to pay the (relatively lower) fine instead. Furthermore, the overachievement decreases in almost all cases. Also the composition of the bundles that are chosen by the regions change. When a fine of 50 Euro is imposed, less strong global control measures are selected than in the base case. Furthermore, we find an underachievement of a target in three cases: Ozone in Den Haag and Urban NOx in Madrid and Milano. Recall that these were also cases in which the constraint for the specific pollutant and region was binding in the base case. With a fine of 100 Euro, only an underachievement is found for Ozone in Den Haag, all other targets are satisfied. However, the cost (4286.927 million Euro) is quite a bit lower than in the base case (4629.143 million Euro). The conclusion is that when we allow for the possibility of underachievement of targets by means of fines, we can achieve a cheaper solution. The reason for this is that when a region has huge difficulties in reaching its targets, it might have to introduce strong control measures and hence also strong global control measures. The other regions then have to implement these global control measures as well. This means that the EU-cost is increased. When we allow a region to undersatisfy (slightly) a target, it might be possible that the overall solution changes only marginally in terms of achieved effectiveness, but is less expensive. Of course, an important parameter is the level of the fine. A fine of 200 Euro for Urban NOx and Ozone e.g. results again in the same solution as in the base case. It is clear that also other types of fines, for instance non-linear schemes could be examined as well.

In the last column we tackle the problem of the decentralization of local measures. As has been explained before, we mimic the problem by imposing that a minimum of the reduction efforts has to be realized via global control measures. In this way, we get an idea of the consequences of this problem. When we set this minimum effort at 10%, total costs increase significantly and of course more global control measures are introduced into the optimal bundle.

6 Conclusion

In this paper we examined the cost-efficiency methodology to decide on new emission standards for cars and fuel quality standards in Europe. We detected two important problems that could lead to a bias against European wide measures: the overachievement problem and the problem of decentralization of local measures. We showed how the first problem can be solved by attributing benefits to overachievement and the second problem can be handled by using explicitly the behavior of the local government as constraint in the problem.

We demonstrated numerically that both problems are important in the Auto-oil set up and that a correct solution of the two problems can lead to important welfare gains.

There are many caveats in this analysis. One interesting extension is the transformation of the constraint on global measures into a weaker constraint where the weakening implies giving up part of the economies of scale of the global measure.

References

Bickel, P., S. Schmid, W. Krewitt and R. Friedrich (eds) (1997), *External Costs of Transport in ExternE, Final Report*, IER, Stuttgart.

Degraeve Z. and G.J. Koopman (1998), Achieving Air Quality Standards in the European Union, *Operations Research* 46(5), 643-654.

Degraeve, Z., G. Koopman, C. Denis and L. Teunen (1998), Deriving and Selecting Policy Instruments to Meet Air Quality Standards in the European Union, in: Proost, S. and J. Braden, *Climate Change, Transport and Environmental Policy – Empirical Applications in a Federal System*, Edward Elgar.

DRI-KULeuven (1999), *The TREMOVE Model, Auto-Oil Pprogramme*, Commission of the European Communities.

European Commission (1996), *Air Quality Studies for the Auto/Oil Programme, Report of Subgroup 2, DG XI*, Brussels, Belgium.

Hall, J., A.M. Winer, M. Kleinman, F. Lurman, S. Dcolome and V. Brajer (1992), Valuing the Benefits of Clean Air, *Science* 255, 812-17.

IVM, NILU, IIASA (1997), *Economic Valuation of Air Quality Targets for Sulphut Dioxide, Nitrogen Dioxide, Fine Particulate Matter and Lead, Final report*, European Commission, DGXI.

Krupnick, A. and P.R. Portney (1991), Controlling Urban Air Pollution : A Benefit-Cost Assessment, *Science* 252, 522-528.

Oates, W.E., P.R. Portney and A.M. McGartland (1989), The Net Benefits of Incentive Based Regulation: A Case Study of Environmental Standard Setting, *American Economic Review* 79(5), 431-53.

Roberts, M.J. and M. Spence (1976), Effluent Charges and Licenses Under Uncertainty, *Journal of Public Economics* 5.

Simpson, D. (1991, Update 1995), Long Period Modeling of Photochemical Oxidants in Europe, *Cooperative Programme for Monitoring and Evaluation of the Long Range Transmission of Air Pollutants in Europe*, Swedish Environmental Research Institute, EMEP MSC-W Report 2/91.

Appendix 1: Formal Description of the LEUVEN II Model

Definitions

In this section, we present key definitions and discuss the conceptual modeling framework. We define a control measure to be an instrument capable of reducing emissions. Instruments can be combined to form sets of control measures. We classify sets of control measures in a table such as Table 1. So as a result, we consider 8 categories of control measures. For each of those, a finite list of sets of control measures was developed. An example of such lists can be found in Table 1. Each set of control measures is characterized by its cost and its effectiveness, i.e. its emission reduction potential, expressed in terms of the percentage reduction in economy-wide emissions forecast in 2010 if no additional policy action were taken. The use of the discrete sets of control measures is not restrictive for our modeling purposes: the same instrument such as a reduction in public transport fares but applied at a different intensity, e.g. a 10% or a 40% reduction, is considered to be a different type of measure.

A control measure is also classified according to its geographical scope. Global control measures must be introduced simultaneously throughout the European Union. They are indicated in bold in Table 1. So in our analysis they are restricted to the technical control measures. Of course, other categories can contain global control measures as well. We assume that control measures from the remaining categories are local. Local measures can be tailored to a specific regional situation and can hence be introduced at various intensities throughout the EU.

We specify the sets and parameters presented above by the following notation:

R: Set of regions, index j,

P: Set of pollutants, index p,

P_{local}: Set of local pollutants, index l,

M: Set of all control measures,

C: Set of categories of control measures,

C_t: Set of control measures of category t, $t = 1, 2, \ldots, |C|$, index i_t,

$v^t_{i_t j}$: Cost of introducing set of control measures i_t in region j, $t = 1, 2, \ldots, |C|$, $\forall\, i_t \in C_t, \forall\, j \in R$,

$e^t_{i_t jl}$: Effectiveness of set of control measures i_t for the abatement of pollutant l in region j, $t = 1, 2, \ldots, |C|$, $\forall\, i_t \in C_t, \forall\, j \in R, \forall\, l \in P_{local}$.

d_{pj}: Required abatement of pollutant l in region j as a percentage of emissions level in 2010, $\forall\, p \in P, \forall\, j \in C$.

The problem for each region consists of selecting sets of control measures from the 8 categories C_t, $t = 1, 2, \ldots, |C|$, taking exactly one set from each category, to form a bundle that would achieve the required emissions reduction at least cost. In Table 1, the empty set is also included as the first element (being x00) in each category allowing for the possibility not to choose a set of control measures from the corresponding category. We then introduce the following notation:

B: Set of bundles, index b,

$B^t_{i_t}$: Subset of bundles containing set of control measures i_t, $t = 1, 2$, $\forall\, i_t \in C_t$.

$z_{bj} = 1$: If bundle b will be introduced in region j, 0, otherwise, $\forall\, b \in B, \forall\, j \in R$,

f_{bj}: Cost of bundle b in region j, $\forall\, b \in B, \forall\, j \in R$,

ne_{bjl}: Net effectiveness of bundle b for the abatement of pollutant l in region j, $\forall\, b \in B, \forall\, j \in R, \forall\, l \in P_{local}$.

The cost of the bundle is simply the sum of the cost of the sets of control measures selected. Since the interaction effects with respect to costs are negligible, this additive functional form is acceptable:

$$f_{bj} = \sum_{t=1}^{|C|} v^t_{i_t j} \qquad \forall b \in B, \forall j \in R$$

The effectiveness of the bundle cannot likewise be computed as there potentially exist interaction effects among some of the categories. This requires the effectiveness of bundles to be partly endogenous. In some cases there are synergy effects implying that the simultaneous introduction of two sets of control measures leads to an abatement of emissions that is larger than the sum of the emissions reduction potential of the individual sets. This will be the case when sets of control measures are complementary. However, generally, the joint impact will be lower than the sum of the two emissions reduction. Nevertheless, we have to properly take the cumulative effectiveness of combining several sets of control measures into account. This is because a second set of control measures can only be applied to the emissions level remaining after the first set has had its effect. The effectiveness of a bundle is computed as follows:

$$e_{(i_1 \ldots i_7)jl} = 100\left(1 \;-\left(1 - e^{1}_{i_1 jl} - e^{2}_{i_2 jl}\right)\prod_{l=3}^{7}\left(1 - e^{l}_{i_l jl}\right)\right)$$

Its net effectiveness then amounts to:

$$ne_{bjl} = e_{(i_1 \ldots i_7)jl} + e^{8}_{i_9 jl} \qquad \forall b \in B, \forall j \in R, \forall l \in P_{local}$$

The Treatment of Ozone

Ozone can be treated in different ways. The most ideal way would be to have exact relationships between ozone and its precursors. If these are not available, an approximation could be applied by using blame matrices, as was done in the LEUVEN I model. A final possibility is to translate the targets for ozone into targets for its precursors. In the remainder of the text, we explain the second approach. The third approach is similar, only the constraints concerning ozone disappear and constraints for the precursors, analogous to the ones for the other local pollutants, are added.

So, the approach taken in the LEUVEN II model is to use blame matrices, one for each precursor of ozone (typically these are regional NOx regional VOC). The coefficients of these matrices give the percentage reduction in ozone levels in a particular region due to a 1% reduction in emissions of the corresponding precursor in another region. The elements are denoted in the following way:

a_{kjl}: Entries of the blame matrix, impact of reducing the ozone precursor l in region k on the reduction of ozone in region j, $\forall\, k, j \in R$, $l = VOC, NOx$.

The effect of a bundle b on ozone in region j is then equal to:

$$\sum_{l=VOC,NOx} \sum_{k=1}^{|R|} a_{kjl}\left(ne_{bkl}\, z_{bk}\right)$$

So, we take the net effectiveness of a bundle *b* on the precursors for ozone (regional *NOx* and *VOC*) in region *k*, we account for the effect of region *k* on region *j* by multiplying with the appropriate element in the blame matrix and we sum over all regions *k* to get the effect on region *j*.

Benefits or Fines for Deviating from the Emission Ceilings

In the model as it would be formulated at the moment, the constraints on the achievement of the required targets will be satisfied as tightly as possible. The reason for this is that a region has no incentive to do better than the target since overachieving certain targets only results in a higher cost. The exception on this is ozone, because the model takes account interactions between a certain region and the other regions. In order to provide an incentive for the different regions to oversatisfy some targets, we will attach a 'reward' to an overachievement. This reward can have a number of causes, since an overachievement implies a cleaner environment, which in turn can lead to smaller medical expenditures, less damage from pollution and so on.

In addition to the benefits of a cleaner environment, there is another, conceptual reason to include rewards for overachievement in the analysis. In an extended exercise like the Auto-Oil II Programme, in which a number of pollutants and regions are included, there will be often an overachievement of the targets for certain pollutants in certain regions. This might imply a bias against the use of uniform, global control measures. The main idea is that global control measures, such as European wide standards, tend to generate an overachievement due to their lack of flexibility.

On the other hand, it is also possible that a region or a country cannot satisfy its targets for one or more local pollutants (recall that for ozone a region or country can be helped out by another region). If one solves the model, this will of course result in an infeasible solution. By introducing fines if a region or country does not satisfy its targets, we can overcome this problem. Introducing fines means that a region or country will balance the fine that has to be paid versus the cost of the control measures that are necessary to achieve the target. This means that when fines are low, a region or country might choose not to reach its targets although it in principle could, because the fines for underachievement are lower than the costs for reaching the targets. This shows that the importance of setting the level of the fine. In the next section we will introduce benefits and fines in the formal model.

Now that we have argued why it is interesting to take into account benefits of overachievement and fines for underachievement of targets, we will now discuss more formally what are the consequences for the LEUVEN II-model. In a first step, we need to determine by how much a target is over- or undersatisfied. This is done by using so-called slack variables. We define a slack variable as the absolute value of the difference between the target and the effectiveness achieved. Hence, it is clear that we need slack variables for the constraints on the required emission abatements for the regions as well as for the required national emission abatement. We will denote them in the following way.

sro_{pj}: Slack variable for overachievement of regional requirement, $\forall\, p \in P$, $\forall\, j \in R$,

sru_{pj}: Slack variable for underachievement of regional requirement, $\forall\, p \in P$, $\forall\, j \in R$.

Let us finally introduce the notation for the benefits and fines:

br_{pj}: Benefit of overachievement of a regional target, $\forall\, p \in P$, $\forall\, j \in R$,

fr_{pj}: Fine of underachievement of a regional target, $\forall\, p \in P$, $\forall\, j \in R$.

The Implementation Problem of Local Measures

The LEUVEN II-model determines the mix of local and global control measures that is most cost-effective from a European-wide point of view. In doing so, we assume that the local authorities implement the measures that are chosen. However, a bundle that is the most cost-effective from a European-wide perspective, is not necessarily so for a region. This means that local authorities might have incentives to select other local measures than the ones in the 'optimal' bundle. Intuitively, we can expect that the local government will not implement sufficient local control measures because it does not have the incentives to do so. Therefore, the optimal mix selected in the Auto Oil Programme is only optimal in as far as the European policy maker can indeed correct the local government behavior. If this is not so, one will tend to favor again more the European wide, global control measures.

The same arguments could of course be made on the national level, since also countries might have incentives not to implement the optimal bundle from a European point of view. However, we will only discuss the case of local authorities (this means at the level of the regions), the case for countries is very similar to this case.

In the discussion above, it was argued that a consequence of the implementation problem of local control measures is that this has an influence on the mix between global and local control measures. Hence, we will use this result to introduce the implementation problem formally into the LEUVEN II-model. More specifically, we add a constraint that states that a certain percentage of a target needs to be achieved by global control measures. More formally, let us introduce the following notation:

gne_{bjl}: Net effectiveness of bundle b, achieved by the global control measures in bundle b, for the abatement of pollutant l in region j, $\forall\, b \in B$, $\forall\, j \in R$, $\forall\, l \in P_{local}$,

g_{lj}: Percentage of the required abatement of pollutant l in region j to be achieved by global control measures, $\forall\, l \in P_{local}$, $\forall\, j \in C$.

The Formal Model

The formal model can be formulated as follows:

(1) Minimize total cost of introducing bundles;

$$Min \quad \sum_{j=1}^{|R|}\sum_{b=1}^{|B|} f_{bj}\, z_{bj} - \sum_{l\in P_l}\sum_{j=1}^{|R|} br_{lj}\, sro_{lj} - \sum_{j=1}^{|R|} br_{O3j}\, sro_{O3j} + \sum_{l\in P_l}\sum_{j=1}^{|R|} fr_{lj}\, sru_{lj} + \sum_{j=1}^{|R|} fr_{O3j}\, sru_{O3j} \qquad \text{(A.1)}$$

Subject to

(2) Achieve the required regional emissions abatement;

$$\sum_{b=1}^{|B|} ne_{bjl}\, z_{bj} - sro_{lj} + sru_{lj} = d_{lj} \qquad \forall l \in P_{local}, \forall j \in R \qquad \text{(A.2)}$$

(3) Achieve the required ozone abatement;

$$\sum_{l=VOC,NOx}\ \sum_{k=1}^{|R|} a_{kjl}\left(\sum_{b=1}^{|B|} ne_{bkl}\, z_{bk}\right) - sro_{O3j} + sru_{O3j} = d_{O3j} \qquad \forall j \in R \qquad \text{(A.3)}$$

(4) Choose one bundle for each region;

$$\sum_{b=1}^{|B|} z_{bj} = 1 \qquad \forall j \in R \qquad \text{(A.4)}$$

(5) Global control measures have to be introduced throughout the EU;

$$\sum_{b=1}^{|B_{i_t}^t|} z_{bj} = \sum_{b=1}^{|B_{i_t}^t|} z_{bj+1} \qquad t = 1,2, \forall i_t \in C_t, j = 1,\ldots,|R|-1 \tag{A.5}$$

(6) Achieve the required share of global control measures;

$$\sum_{b=1}^{|B|} gne_{bjl}\, z_{bj} \geq g_{lj} d_{lj} \qquad \forall l \in P_{local}, \forall j \in R \tag{A.6}$$

(7)

$$\sum_{l=VOC,NOx} \sum_{k=1}^{|R|} a_{kjl} \left(\sum_{b=1}^{|B|} gne_{bkl}\, z_{bk} \right) \geq g_{O3j} d_{O3j} \qquad \forall j \in R \tag{A.7}$$

(8) Define the variables 0/1;

$$z_{bj} \in \{0,1\} \qquad \forall b \in B, \forall j \in R \tag{A.8}$$

A difficulty with this formulation is that for realistic problem dimensions, the number of variables z_{bj} that needs to be included in the model will be extremely large (this is $|R|^*|B|$, and $|B|$ increases exponentially). For the column generation approach that was used to tackle this difficulty, we refer to Degraeve and Koopman (1998).

Appendix 2: Results

Table 2 contains the results of the different runs of the model. We refer to the main text of this paper for a discussion of the results.

Table 2: Results

	Targets	Base Case		Ben NOx Urb 20		Ben NOx Urb 30		Ben NOx Urb Ozone 5		Fines NOx Urb Ozone 50		Fines NOx Urb Ozone100		Share gl. measures 10%	
Overview															
Total Gross Cost:			4629.14		4999.31		5349.31		5462.68		2459.74		3690.90		5581.49
Total Benefit:			0.00		2599.94		4355.17		3412.71		0.00		0.00		0.00
Total Fine:			0.00		0.00		0.00		0.00		1183.59		596.03		0.00
Total Net Cost:			4629.14		2399.36		994.14		2049.97		3643.32		4286.93		5581.49
ATHENS															
Costs															
Gross Cost			251.19		342.56		360.18		217.35		296.92		251.19		238.38
Total Benefit			0.00		114.43		240.77		182.24		0.00		0.00		0.00
Total Fine			0.00		0.00		0.00		0.00		0.00		0.00		0.00
Net cost			251.19		228.13		119.41		35.10		296.92		251.19		238.38
Bundle															
Composition															
Category		Set	Cost	Set	Cost	Set	Cost	Set	Cost	Set	Cost	Set	Cost	Set	Cost
FUELS		F04	15.20	F05	21.64	F05	21.64	F04	15.20	F00	0.00	F04	15.20	F00	0.00
IM		IM03	7.26	IM03	7.26	IM03	7.26	IM00	0.00	IM03	7.26	IM03	7.26	IM03	7.26
Non Transp		NT04	51.60	NT06	125.83	NT06	125.83	NT06	125.83	NT06	125.83	NT04	51.60	NT06	125.83
PUB		PUB01	-3.98	PUB01	-3.98	PUB01	-3.98	PUB01	-3.98	PUB01	-3.98	PUB01	-3.98	PUB01	-3.98
SCR		SCR02	20.20	SCR02	20.20	SCR02	20.20	SCR02	20.20	SCR02	20.20	SCR02	20.20	SCR02	20.20
SM		SM05	139.20	SM05	139.20	SM05	139.20	SM03	14.40	SM05	139.20	SM05	139.20	SM03	14.40
TED		TED08	13.41	TED06	0.12	TED13	13.84	TED08	13.41	TED06	0.12	TED08	13.41	TED09	42.38
TEG		TEG04	8.30	TEG07	32.29	TEG08	36.18	TEG07	32.29	TEG04	8.30	TEG04	8.30	TEG07	32.29
Effectiveness															
Pollutant	Required		Achieved		Achieved		Achieved		Achieved		Achieved		Achieved		Achieved
Ozone	45		60.72		81.24		82.63		81.36		79.07		60.72		82.07
RegNOx	0		35.12		46.42		47.44		46.16		44.93		35.12		46.50
RegVOC	30		38.57		42.35		46.80		37.08		37.01		38.57		36.33
UrbanNOx	41.25		41.44		46.97		49.28		41.34		41.50		41.44		41.54

Table 2: continued.

	Targets	Base Case		Ben NOx Urb 20		Ben NOx Urb 30		Ben NOx Urb Ozone 5		Fines NOx Urb Ozone 50		Fines NOx Urb Ozone100		Share gl. measures 10%	
DEN HAAG															
Costs															
Gross Cost			500.54		762.57		525.92		605.68		154.22		465.78		363.90
Total Benefit			0.00		1040.25		1622.56		269.53		0.00		0.00		0.00
Total Fine			0.00		0.00		0.00		0.00		784.21		596.03		0.00
Net cost			500.54		-277.68		-1096.64		336.15		938.43		1061.80		363.90
Bundle															
Composition															
Category		Set	Cost	Set	Cost	Set	Cost	Set	Cost	Set	Cost	Set	Cost	Set	Cost
FUELS		F05	34.28	F05	34.28	F05	34.28	F05	34.28	F00	0.00	F00	0.00	F00	0.00
IM		IM02	0.48	IM02	0.48	IM02	0.48	IM02	0.48	IM00	0.00	IM00	0.00	IM00	0.00
Non Transp		NT08	397.23	NT08	397.23	NT06	113.03	NT08	397.23	NT06	113.03	NT08	397.23	NT06	113.03
PUB		PUB01	-5.88	PUB01	-5.88	PUB01	-5.88	PUB01	-5.88	PUB01	-5.88	PUB01	-5.88	PUB01	-5.88
SCR		SCR00	0.00	SCR00	0.00	SCR00	0.00	SCR00	0.00	SCR00	0.00	SCR00	0.00	SCR00	0.00
SM		SM03	21.25	SM05	205.48	SM05	205.48	SM03	21.25	SM03	21.25	SM03	21.25	SM03	21.25
TED		TED08	43.60	TED06	16.25	TED13	58.22	TED08	43.60	TED06	16.25	TED08	43.60	TED09	120.79
TEG		TEG04	9.57	TEG07	114.71	TEG08	120.30	TEG07	114.71	TEG04	9.57	TEG04	9.57	TEG07	114.71
Effectiveness															
Pollutant	Required		Achieved		Achieved		Achieved		Achieved		Achieved		Achieved		Achieved
Ozone	67.5		68.47		67.91		68.63		72.09		51.82		61.54		68.42
RegNOx	0		46.86		52.29		47.78		53.55		33.38		45.79		46.99
RegVOC	0		63.81		70.67		51.77		66.33		43.93		62.78		46.22
UrbanNOx	3.75		41.39		55.76		57.84		53.07		32.31		39.41		52.76

Table 2: continued.

	Targets	Base Case		Ben NOx Urb 20		Ben NOx Urb 30		Ben NOx Urb Ozone 5		Fines NOx Urb Ozone 50		Fines NOx Urb Ozone100		Share gl. measures 10%	
KOLN															
Costs															
Gross Cost			0.00		1004.74		1280.54		1238.86		507.65		743.27		1562.27
Total Benefit			0.00		696.47		1128.68		332.18		0.00		0.00		0.00
Total Fine			0.00		0.00		0.00		0.00		0.00		0.00		0.00
Net cost			1305.02		308.28		151.86		906.68		507.65		743.27		1562.27
Bundle Composition															
Category		Set	Cost	Set	Cost	Set	Cost	Set	Cost	Set	Cost	Set	Cost	Set	Cost
FUELS		F00	0.00	F00	0.00	F00	0.00	F00	0.00	F00	0.00	F00	0.00	F00	0.00
IM		IM03	69.12	IM02	1.51	IM02	1.51	IM03	69.12	IM00	0.00	IM03	69.12	IM00	0.00
Non Transp		NT06	937.91	NT04	376.15	NT04	376.15	NT04	376.15	NT04	376.15	NT04	376.15	NT04	376.15
PUB		PUB01	-31.18	PUB01	-31.18	PUB01	-31.18	PUB01	-31.18	PUB01	-31.18	PUB01	-31.18	PUB01	-31.18
SCR		SCR00	0.00	SCR00	0.00	SCR00	0.00	SCR00	0.00	SCR00	0.00	SCR00	0.00	SCR00	0.00
SM		SM03	112.75	SM03	112.75	SM03	112.75	SM03	112.75	SM03	112.75	SM03	112.75	SM03	112.75
TED		TED08	194.93	TED06	28.43	TED13	292.00	TED08	194.93	TED06	28.43	TED08	194.93	TED09	587.46
TEG		TEG04	21.50	TEG07	517.09	TEG08	529.32	TEG07	517.09	TEG04	21.50	TEG04	21.50	TEG07	517.09
Effectiveness															
Pollutant	Required		Achieved		Achieved		Achieved		Achieved		Achieved		Achieved		Achieved
Ozone	67.5		96.96		91.75		94.04		96.88		74.86		84.49		94.45
RegNOx	0		42.81		39.33		44.07		43.75		31.48		35.98		46.20
RegVOC	0		38.56		36.11		36.44		37.45		32.77		34.19		36.18
UrbanNOx	18.75		44.32		53.57		56.37		55.80		39.02		41.52		56.59

Table 2: continued.

	Targets	Base Case		Ben NOx Urb 20		Ben NOx Urb 30		Ben NOx Urb Ozone 5		Fines NOx Urb Ozone 50		Fines NOx Urb Ozone100		Share gl. measures 10%	
LONDON															
Costs															
Gross Cost			357.29		689.98		900.95		839.59		311.74		357.29		968.83
Total Benefit			0.00		451.89		833.30		310.98		0.00		0.00		0.00
Total Fine			0.00		0.00		0.00		0.00		0.00		0.00		0.00
Net cost			357.29		238.10		67.64		528.62		311.74		357.29		968.83
Bundle															
Composition															
Category		Set	Cost	Set	Cost	Set	Cost	Set	Cost	Set	Cost	Set	Cost	Set	Cost
FUELS		F00	0.00	F00	0.00	F00	0.00	F00	0.00	F00	0.00	F00	0.00	F00	0.00
IM		IM00	0.00	IM00	0.00	IM00	0.00	IM00	0.00	IM00	0.00	IM00	0.00	IM00	0.00
Non Transp		NT04	172.07	NT02	68.01	NT04	172.07	NT04	172.07	NT04	172.07	NT04	172.07	NT04	172.07
PUB		PUB01	-22.43	PUB01	-22.43	PUB01	-22.43	PUB01	-22.43	PUB01	-22.43	PUB01	-22.43	PUB01	-22.43
SCR		SCR00	0.00	SCR00	0.00	SCR00	0.00	SCR00	0.00	SCR00	0.00	SCR00	0.00	SCR00	0.00
SM		SM03	81.11	SM03	81.11	SM03	81.11	SM03	81.11	SM03	81.11	SM03	81.11	SM03	81.11
TED		TED08	73.97	TED06	28.41	TED13	105.43	TED08	73.97	TED06	28.41	TED08	73.97	TED09	203.20
TEG		TEG04	52.58	TEG07	534.88	TEG08	564.77	TEG07	534.88	TEG04	52.58	TEG04	52.58	TEG07	534.88
Effectiveness															
Pollutant	Required		Achieved		Achieved		Achieved		Achieved		Achieved		Achieved		Achieved
Ozone	67.5		81.60		89.78		104.86		103.11		73.89		79.26		107.57
RegNOx	0		35.50		40.73		47.11		46.45		33.63		35.50		48.02
RegVOC	0		29.50		24.80		34.17		33.95		29.55		29.50		33.95
UrbanNOx	30		39.06		52.59		57.78		56.58		37.07		39.06		58.43

Table 2: continued.

	Targets	Base Case		Ben NOx Urb 20		Ben NOx Urb 30		Ben NOx Urb Ozone 5		Fines NOx Urb Ozone 50		Fines NOx Urb Ozone100		Share gl. measures 10%	
LYON															
Costs			728.99		751.23		870.85		823.43		257.55		387.27		841.91
Gross Cost			0.00		268.04		526.82		466.49		0.00		0.00		0.00
Total Benefit			0.00		0.00		0.00		0.00		0.00		0.00		0.00
Total Fine			728.99		483.19		344.03		356.94		257.55		387.27		841.91
Net cost															
Bundle															
Composition															
Category		Set	Cost	Set	Cost	Set	Cost	Set	Cost	Set	Cost	Set	Cost	Set	Cost
FUELS		F04	94.05	F00	0.00	F00	0.00	F04	94.05	F00	0.00	F00	0.00	F00	0.00
IM		IM02	4.66	IM02	4.66	IM02	4.66	IM02	4.66	IM00	0.00	IM00	0.00	IM00	0.00
Non Transp		NT06	427.25	NT04	184.23	NT04	184.23	NT04	184.23	NT02	32.66	NT04	184.23	NT04	184.23
PUB		PUB01	-22.22	PUB01	-22.22	PUB01	-22.22	PUB01	-22.22	PUB01	-22.22	PUB01	-22.22	PUB01	-22.22
SCR		SCR00	0.00	SCR00	0.00	SCR00	0.00	SCR00	0.00	SCR00	0.00	SCR00	0.00	SCR00	0.00
SM		SM03	80.35	SM03	80.35	SM03	80.35	SM03	80.35	SM03	80.35	SM03	80.35	SM03	80.35
TED		TED08	78.42	TED06	100.27	TED13	182.07	TED08	78.42	TED06	100.27	TED08	78.42	TED09	195.61
TEG		TEG04	66.49	TEG07	403.94	TEG08	441.77	TEG07	403.94	TEG04	66.49	TEG04	66.49	TEG07	403.94
Effectiveness															
Pollutant	Required		Achieved		Achieved		Achieved		Achieved		Achieved		Achieved		Achieved
Ozone	67.5		143.12		135.65		140.87		144.79		105.58		123.00		141.89
RegNOx	0		36.14		32.17		37.99		36.85		21.63		29.19		39.75
RegVOC	3.75		39.55		37.03		36.54		38.29		23.25		31.36		35.29
UrbanNOx	30		39.83		43.40		47.56		46.01		31.12		34.39		46.96

Table 2: continued.

	Targets	Base Case		Ben NOx Urb 20		Ben NOx Urb 30		Ben NOx Urb Ozone 5		Fines NOx Urb Ozone 50		Fines NOx Urb Ozone100		Share gl. measures 10%	
MADRID															
Costs															
Gross Cost			0.00		485.36		526.75		623.94		354.34		843.13		672.63
Total Benefit			0.00		2.05		1.90		422.73		0.00		0.00		0.00
Total Fine			0.00		0.00		0.00		0.00		223.92		0.00		0.00
Net cost			843.13		483.31		524.85		201.21		578.27		843.13		672.63
Bundle															
Composition															
Category		Set	Cost	Set	Cost	Set	Cost	Set	Cost	Set	Cost	Set	Cost	Set	Cost
FUELS		F00	0.00	F00	0.00	F00	0.00	F04	75.04	F00	0.00	F00	0.00	F04	75.04
IM		IM02	0.92	IM02	0.92	IM03	26.10	IM03	26.10	IM03	26.10	IM02	0.92	IM02	0.92
Non Transp		NT06	128.66	NT06	128.66	NT06	128.66	NT06	128.66	NT06	128.66	NT06	128.66	NT06	128.66
PUB		PUB01	-15.17	PUB01	-15.17	PUB01	-15.17	PUB01	-15.17	PUB01	-15.17	PUB01	-15.17	PUB01	-15.17
SCR		SCR02	99.93	SCR02	99.93	SCR01	16.49	SCR02	99.93	SCR02	99.93	SCR02	99.93	SCR01	16.49
SM		SM05	530.46	SM03	54.86	SM03	54.86	SM03	54.86	SM03	54.86	SM05	530.46	SM03	54.86
TED		TED08	90.71	TED06	52.35	TED13	147.96	TED08	90.71	TED06	52.35	TED08	90.71	TED09	248.03
TEG		TEG04	7.62	TEG07	163.81	TEG08	167.87	TEG07	163.81	TEG04	7.62	TEG04	7.62	TEG07	163.81
Effectiveness															
Pollutant	Required		Achieved		Achieved		Achieved		Achieved		Achieved		Achieved		Achieved
Ozone	0		84.52		78.19		81.36		84.26		64.33		77.91		81.37
RegNOx	0		34.05		33.61		35.06		36.06		31.11		34.05		36.48
RegVOC	22.5		40.75		37.05		38.79		40.50		34.88		40.75		38.77
UrbanNOx	37.5		37.54		37.60		37.56		37.79		33.02		37.54		37.62

Table 2: continued.

	Targets	Base Case		Ben NOx Urb 20		Ben NOx Urb 30		Ben NOx Urb Ozone 5		Fines NOx Urb Ozone 50		Fines NOx Urb Ozone100		Share gl. measures 10%	
MILANO															
Costs															
Gross Cost			642.98		962.86		884.13		1113.83		577.31		642.98		933.56
Total Benefit			0.00		26.81		1.15		1428.55		0.00		0.00		0.00
Total Fine			0.00		0.00		0.00		0.00		175.46		0.00		0.00
Net cost			642.98		936.04		882.98		-314.72		752.77		642.98		933.56
Bundle Composition															
Category		Set	Cost	Set	Cost	Set	Cost	Set	Cost	Set	Cost	Set	Cost	Set	Cost
FUELS		F00	0.00	F00	0.00	F00	0.00	F04	84.34	F00	0.00	F00	0.00	F00	0.00
IM		IM00	0.00	IM02	0.96	IM03	64.05	IM02	0.96	IM02	0.96	IM00	0.00	IM00	0.00
Non Transp		NT06	463.64	NT06	463.64	NT04	171.18	NT06	463.64	NT06	463.64	NT06	463.64	NT04	171.18
PUB		PUB01	-22.43	PUB01	-22.43	PUB01	-22.43	PUB01	-22.43	PUB01	-22.43	PUB01	-22.43	PUB01	-22.43
SCR		SCR00	0.00	SCR00	0.00	SCR00	0.00	SCR00	0.00	SCR00	0.00	SCR00	0.00	SCR00	0.00
SM		SM03	81.11	SM03	81.11	SM03	81.11	SM03	81.11	SM03	81.11	SM03	81.11	SM03	81.11
TED		TED08	111.36	TED06	44.73	TED13	190.30	TED08	111.36	TED06	44.73	TED08	111.36	TED09	308.85
TEG		TEG04	9.31	TEG07	394.85	TEG08	399.92	TEG07	394.85	TEG04	9.31	TEG04	9.31	TEG07	394.85
Effectiveness															
Pollutant	Required		Achieved		Achieved		Achieved		Achieved		Achieved		Achieved		Achieved
Ozone	60		291.45		301.32		277.12		339.58		268.38		285.42		276.18
RegNOx	0		37.10		36.67		35.31		41.78		32.76		37.10		37.70
RegVOC	22.5		39.51		42.95		37.33		48.53		39.57		39.51		35.37
UrbanNOx	37.5		37.77		38.84		37.54		43.63		33.99		37.77		39.51

Commitment and Time Consistency of Environmental Policy and Incentives for Adoption and R&D

Till Requate

Interdisciplinary Institute for Environmental Economics, University of Heidelberg, Bergheimer Strasse 20, 69115 Heidelberg, Germany,
requate@bwl.uni-kiel.de

Summary:
In this paper we survey recent developments on the incentives through environmental policy instruments to adopt advanced abatement technology. We further investigate repercussions on R&D. First we study the case where the regulator makes long-term commitments to policy levels and does not anticipate the arrival of new technology. We show that taxes provide stronger incentives than permits, auctioned and free permits offer identical incentives, and standards may give stronger incentives than permits. Second, we investigate scenarios where the regulator anticipates new technologies. We show that with taxes and permits the regulator can induce first-best outcomes if he moves after firms have invested, whereas this does not always hold if he moves first. Third, we consider a model where a polluting downstream industry is regulated either by emission taxes or by tradable permits. A separate monopolistic or duopolistic upstream industry engages in R&D and, in case of R&D success, sells an advanced abatement technology to the downstream firms. We study three different timings of environmental policy, ex post taxation (or issuing permits), ex interim commitment to a tax rate (a quota of permits) after observing R&D success but before adoption, and finally ex ante commitment before and independent of R&D success. We show that ex interim commitment always dominates ex post environmental policy. Moreover, ex interim second best taxation dominates ex interim second best optimal permit policy. There is no unique ranking, however, between ex ante and ex interim commitment. Finally we sketch ways to generalize the model to upstream duopoly.

Keywords: Emission taxes, auctioned permits, grandfathering, emission standards, technology adoption, R&D, timing, commitment.

1 Introduction

This paper addresses some recent developments regarding the incentives of environmental policy instruments to spur both R&D and the adoption of new technology. It is well known that among the wide array of pollution control instruments, economists prefer those which provide incentives through prices rather than through command and control. The main advantage of market-based instruments such as emission taxes, subsidies on abatement, and different regimes of tradable permits is their (static) cost efficiency. However, it has been pointed out early by Kneese and Schulze (1975: 38) that "over the long haul, perhaps the most important single criterion on which to judge environmental policies is the extent to which they spur new technology toward the efficient conservation of the environment." It took, however, quite a while until researchers on a theoretical level tried to answer the question, which policy provides the best incentives to adopt new, less polluting technology, but also which policy gives incentives to engage in R&D, i.e. to develop more environmentally friendly technology.

In the 1980s Downing and White (1987), Malueg (1989), Milliman and Prince (1989) and later on Jung et al. (1996) made the first attempts to rank environmental policy instruments, in particular, taxes, subsidies, auctioned permits, free permits, and emission standards with respect to their incentives to adopt less polluting technology. Their results, to be reviewed below, have been challenged by Requate and Unold (2001, 2002) who argue that not the aggregate cost savings of the whole industry but the incentive of a single firm in equilibrium is crucial to judge the performance of the above mentioned policy instruments.

Although Downing and White talk about "innovation", and Milliman and Prince use the notion of "technical change", it is mainly incentives for "adoption" of existing technology which have been the subject of investigation by those authors.

In this paper we critically discuss the arguments put forward by Milliman and Prince and Jung et al. We further survey the main results from Requate and Unold (2001, 2002) who take into account equilibrium considerations. We show that taxes lead to overinvestment if the regulator has committed to a tax policy which is optimal with respect to the conventional technology. Permits by contrast lead to underinvestment since some firms can free-ride on the investment of other firms. We also show that taxes and permits lead to first-best outcomes if the regulator anticipates adoption of new technology and decides on the level of his policy after adoption.

Whereas in those models the new technology was assumed to be available already, in the second part of this paper we focus on the incentives to engage in research and development of the new technology by a monopolistic (or duopolistic) innovator. We show that the incentives both to engage in R&D and to price the new technology depend crucially on commitment and timing of environmental policy. In particular we distinguish between ex ante commitment, ex interim commitment and ex post optimal regulation. In the latter case the regulator moves after observing both R&D success and the rate of adoption. Under ex interim regulation the regulator observes whether or not R&D has been successful, but commits to his tax rate or quota of permits, respectively, however before the pricing and adoption of the new technology. Under ex ante regulation the regulator commits to the level of his policy before R&D activity.

More recently other authors have included R&D into the analysis. Among those there are two strands of literature. One road goes along the industrial organization methodology, using concepts of game theory to analyze strategic behavior in equilibrium. Another road follows the methodology of endogenous growth theory. The latter is still in the very beginning with respect to determining optimal policies including the solution of problems such as commitment, timing and time-consistency. Hence we will not pursue that approach in this paper. Among the industrial organization (IO) models the papers by Biglaiser and Horowitz (1995), Parry (1995), Denicolo (1999), and Fisher et al. (2001) are closest to our approach. Biglaiser and Horowitz consider a model where the regulated polluting firms can engage in R&D themselves. As in our model this technology can be sold to other firms. However, Biglaiser and Horowitz consider ex post regulation only. Denicolo, like our paper compares ex post regulation to commitment, and finds that some commitment can raise the incentives to innovate. Parry finds that the second best optimal tax falls short of marginal damage, whereas in our model the second best optimal tax under ex interim commitment exceeds marginal social damage. In chapter 5 of his dissertation Unold (1999), and more recently Requate (2001) take up some ideas of those models to highlight the interplay between R&D, pricing of new technology, and finally adoption by firms which do not engage in R&D themselves.

This paper is organized as follows. In Section 2, we discuss the traditional literature on ranking environmental policy instruments to provide incentives for adoption of new technology. We also summarize the results of the alternative approach developed in Requate and Unold (2001, 2002). In Section 3 we will focus on the R&D incentives for a monopolistic firm with symmetric firms to adopt the new technology. In Section 4 we sketch some extensions of that model. The final section contains some concluding remarks.

2 Adoption

As mentioned above, Downing and White (1987), Malueg (1989), Milliman and Prince (1989) and, more recently, Jung et al. (1996) have attempted to rank environmental policy instruments with respect to their incentives to spur the adoption of advanced, low-polluting technologies. Notably Milliman and Prince (1989) compare pollution control instruments such as taxes, auctioned permits, free (or grandfathered) permits, subsidies, and emission standards at firm level. More recently, Jung et al. (1996) extended this analysis to a heterogeneous industry by investigating the incentive to adopt new technology on an aggregate level. Both papers come to the conclusion that the different control regimes can be ranked in the order

(1) auctioned permits,

(2) taxes and subsidies,

(3) free permits,

(4) emission standards.

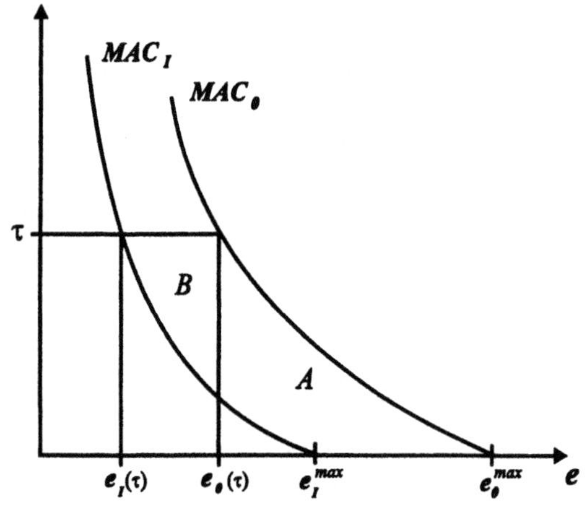

Figure 1: Single Firm's Investment Incentive Under Taxes.

2.1 The Alleged Argument

The argument by Milliman and Prince (1989) and Jung et al. (1996) is outlined in Figure 1. MAC_0 denotes the marginal abatement cost curve of a representative firm running a conventional abatement technology, whereas MAC_I is the marginal abatement cost curve after adoption of some advanced abatement technology. Let e_i^{max} (i = 0, 1) denote the largest possible emission level a firm would choose under a laissez-faire policy. It is natural to assume $e_I^{max} < e_0^{max}$.

Let us first consider pollution control by an emission tax. If a particular firm switches from conventional to advanced technology, its savings in variable abatement cost plus tax payments is given by the area $A + B$. If, as is usually the case, installing a new technology causes a fixed set-up cost $F > 0$, it will be profitable for the firm to invest in the new technology if and only if $F < A + B$.[1]

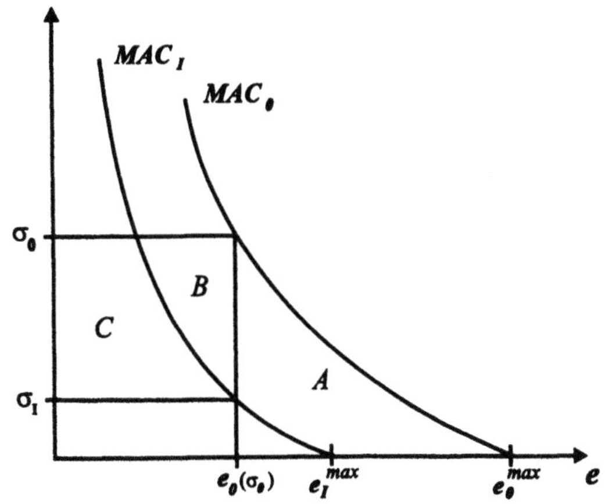

Figure 2: Single Firm's Total Cost Reduction if All Firms Invest Under Permits.

Now consider pollution control by auctioning off permits in each period, say, a year. Let σ_0 denote the original price for permits (see Figure 2) before the advanced technology is available. Assume first that only one small firm has access to that new technology. Then the incentive for this single firm to install the new technology is the same as under taxes, provided that the price for permits is not affected by the firm's decision to switch technology (this was correctly observed by Milliman and Prince). Assume now that all the firms have access to the new technology. Then the price for permits will fall if a considerable number of firms adopts the new technology. Jung et al. assume such a price, say σ_I, to be exogenously given (see Figure 2). Then indeed, each firm that has adopted the new technology has lower variable costs than in a situation where no other firm has adopted the new technology. The cost difference is equal to the area $A + B + C$ depicted in Figure 2. This cost difference, however, has nothing to do with a single firm's incentive to adopt the new technology. In other words, the criterion for whether or not to adopt the new technology is not determined by the inequality $F < A + B + C$. To see this, let us assume that there are a large number of firms n of similar size. Each firm is small compared to the entire industry.

1 In dynamic framework, F must be interpreted as the annuity of the fixed cost per period. In this case, the criterion $F < A + B$ is conditional on the tax remaining constant for a sufficiently long time.

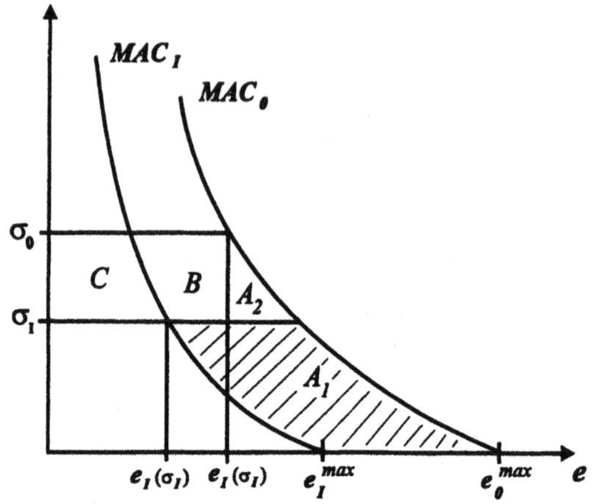

Figure 3: Single Firm's Investment Incentive Under Permits.

As before, let σ_0 denote the price before any firm has adopted the new technology, and let σ_l denote the price after all the firms have adopted the new technology. Assume now that n–1 firms have adopted the new technology and the last firm has not yet done so. Since the last firm's impact on the price can be neglected, the price for permits is now approximately σ_l. When will the last firm adopt the new technology? The answer is given by Figure 3. It will do so if $F < A_1$, which is smaller than $A_1 + A_2 + B = A + B$. Now if $A_1 < F < A + B$ holds, the last firm would adopt the new technology under taxes but would not do so under permits. Hence the incentive to invest under permits must be lower than under taxes.

Assume further that $A_1 << F << A + B$ holds, i.e. F is considerably greater than A_1 but considerably smaller than $A + B$. Then even the n–1st firm will probably not invest in the new technology. On the other hand, the first firm has an obvious incentive to adopt the new technology if no other, or a very small number of other firms will adopt the new technology. Given these considerations, it should have become clear by now that it is necessary to determine endogenously the number of firms adopting the new technology.

Assuming that all the firms will adopt the new technology – as both Milliman and Prince and Jung et al. do – is tantamount to assuming the fixed set-up cost to be sufficiently low, i.e. $F << A_1$. But if we assume that ex ante all the firms will adopt the new technology, it is no longer of interest to ask which policy instrument will provide a higher incentive to adopt the new technology.

Requate and Unold (2001, 2002), by contrast, argue that the ranking derived by that methodology relies basically on calculating the aggregate cost savings achieved by an industry-wide adoption of new technology. On the assumption that all firms adopt the new technology, Milliman and Prince and Jung et al. calculate the total costs for industry before and after adoption of the new technology and then compare the differences. These aggregate cost savings, however, do not reveal a single firm's incentive to adopt a new technology in equilibrium. This approach ignores the fact that a single firm can free-ride on a decreasing permit price caused by other firms' investment. To put it differently, equilibrium aspects with respect to the firms' investment decisions have been neglected completely, in particular for the permit regimes.

By contrast, the analysis by Requate and Unold reveals that equilibrium aspects are crucial for the ranking of the policy instruments. They investigate how many firms will adopt the new technology in equilibrium. By doing so, they are able to establish that the criterion deciding whether one policy instrument will provide higher incentives to invest than another instrument is its success in inducing more firms to invest. To highlight the argument we present briefly a model framework.

2.2 A Model Framework

Consider a competitive industry consisting of n small firms where industry size is assumed to be exogenous. All the firms emit a homogenous pollutant which can be abated. The firms' abatement cost functions C_i satisfy the usual properties: $C_i(e_i) > 0$ for $e_i < e_i^{max}$, and $C_i(e_i) = 0$ for $e_i \geq e_i^{max}$, where e_i^{max} denotes the maximal emission level under laisser-faire. Moreover, marginal abatement costs are positive in the relevant area, i.e. $-C_i'(e_i) > 0$ for $e_i < e_i^{max}$, and decreasing with more emissions, i.e. $C_i''(e_i) > 0$ for $e_i < e_i^{max}$. Since the product market is assumed to be competitive, decisions about output need not be modeled explicitly. Those decisions are implicitly accounted for in the abatement cost functions.

In the following we assume that initially all the firms are alike, thus having the same abatement cost curve C_0. However, the firms can invest in an advanced abatement technology leading to a lower marginal abatement cost curve C_I (see again Figure 1) with

$$-C_I'(e) < -C_0'(e) \quad \text{for all} \quad e < e_0^{max}.$$

Buying and installing the new technology involves a fixed cost $F > 0$.

Assume initially that the regulator can choose between setting a tax τ or issuing a number of tradable permits L where σ denotes the market price for permits. In the following let x denote a tax or a permit price. Then cost minimization on the firm side requires

$$-C_0'(e_0) = x \tag{1}$$

$$-C_I'(e_I) = x \tag{2}$$

for firms with conventional and new technology, respectively. Then a typical firm decides to adopt the new technology if

$$C_0\left(e_0(p)\right) + xe_0(\tau) - C_l\left(e_l(p)\right) - xe_l(p) - F > 0. \tag{3}$$

Firms are indifferent about adopting the new technology if

$$C_0\left(e_0(p)\right) + xe_0(p) = C_l\left(e_l(p)\right) + xe_l(p) + F. \tag{4}$$

Inequality (3) and equation (4) are the driving forces behind the argument. Assume in the following that the regulator has made a long term commitment to either a tax-rate or an aggregate quantity of permits being issued.

Consider first the case of taxation. Then there exists a tax rate $\tilde{\tau}$ such that, if $\tau < \tilde{\tau}$, no firm will adopt the new technology, and we will obtain the highest possible emission level $\overline{E} = ne_0$, while if $\tau > \tilde{\tau}$, all firms will adopt the new technology and we obtain the lowest possible emission level $\underline{E} = n\tilde{e}_l$. If $\tau = \tilde{\tau}$, the number of firms that adopt is undetermined and the total emission level is in $[\underline{E}, \overline{E}]$. But this also implies that if firms with old and firms with new technology are in the market at the same time, the tax (or the permit price) must be equal to $\tilde{\tau}$.

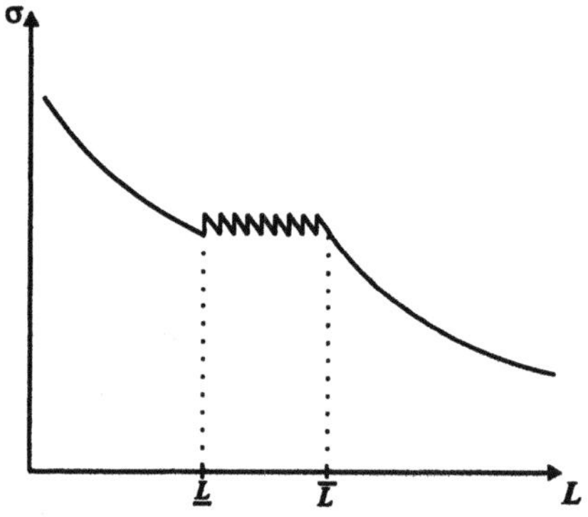

Figure 4: Quantity/Price Schedule for Permits (Few Firms).

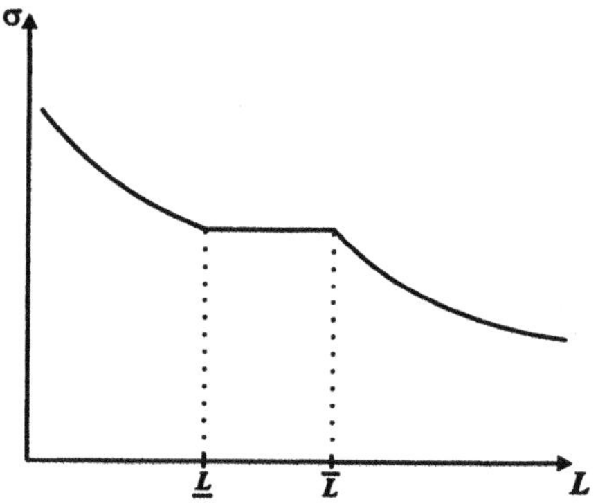

Figure 5: Quantity/Price Schedule for Many Firms.

Now assume the industry is regulated by auctioning off permits (we treat free permits in Section 2.2.1). Then, in contrast to taxes, we always obtain a unique equilibrium number of investing firms. Moreover, for many different aggregate quantities of permits both types of technologies will exist in the market.

The permit quantity price schedule is displayed in Figure 4. If the market is sufficiently large, the equilibrium price schedule is almost a flat line on some interval $L \in [\underline{L}, \overline{L}]$, see Figure 5.

2.2.1 Grandfathering

According to Malueg, Milliman and Prince, and Jung et al., free permits provide lower incentives to adopt new technology than auctioned permits. The argument they propose is that innovating makes firms' permits depreciate. However, innovating depreciates the firms' permits anyway, irrespective of whether these permits are auctioned off or distributed for free. In contrast to what Jung et al. claim, the original permit price is completely irrelevant with regard to the incentive to adopt a new technology in equilibrium. To see this more clearly, let $\hat{e}_i = \hat{e}$ be the firm's (identical) initial endowment of permits. As the firms are alike before innovation, there will be no trade before the new technology is available. Since the advanced technology leads to lower marginal abatement costs for each emission level, the investors must be sellers and the non-investors must be buyers of permits. In an equilibrium with partial adoption, firms must be indifferent about adopting the new technology or not, i.e.

$$C_0(e_0) + \sigma \cdot \left[e_0 - \hat{e}\right] = C_I(e_I) - \sigma \cdot \left[\hat{e} - e_I\right] + F. \tag{5}$$

We see that $\sigma \cdot \hat{e}$ cancels out. Thus (5) is equivalent to (4) and we see immediately that the incentive to adopt a new technology is the same for free (grandfathered) as for auctioned permits.

2.2.2 Uniform Emission Standards

Finally, consider a uniform emission standard $\bar{e}$, sometimes referred to as "command and control policy". Under such a standard, which means that a firm is not allowed to emit more than $\bar{e}$ units of the pollutant, the firm will be indifferent between staying with the conventional technology and adopting the advanced technology if

$$C_0(\bar{e}) - C_I(\bar{e}) - F = 0. \tag{6}$$

Since $-C_0'(\bar{e}) > -C_I'(\bar{e})$, the LHS of (6) decreases in $\bar{e}$. This means that the cost advantage of the new technology decreases as the emission standard is relaxed. This, however, means that there exists a standard $\bar{\bar{e}}$ (depending on F) such that no firm will adopt the new technology if $\bar{e} > \bar{\bar{e}}$, but all firms will adopt the new technology if $\bar{e} < \bar{\bar{e}}$.

2.3 Ranking Under the Standard and Charges Approach

We now briefly summarize the incentives provided by the different policy instruments if the regulator has decided to enforce an aggregate emission standard $\widehat{E}$ by either charging taxes, issuing a corresponding number of permits, paying subsidies on abatement, or setting a uniform emission standard. Following both papers, Milliman and Prince (1989) and Jung et al. (1996) we assume that the regulator has implemented his policy before the advanced technology was available and has committed himself to her policy for a sufficiently long time.

Proposition 1 (Taxes): Assume the regulator has set taxes in order to achieve an aggregate emission target $\widehat{E}$.

- Then for all emission targets $\widehat{E} > \overline{E}$ no firm adopts the new technology.
- For all targets $\widehat{E} < \overline{E}$ all the firms adopt the new technology and the new aggregate emission level E is lower than $\widehat{E}$.
- For target $\widehat{E} = \overline{E}$, the number of firms adopting the new technology is ambiguous. However, no firm has a positive incentive to adopt the new technology.

Proposition 2 (Permits): Assume the regulator has issued permits in order to achieve an aggregate emission target $\widehat{E}$.

- Then for all emissions targets $\widehat{E} > \overline{E}$ no firm adopts the new technology.
- For all targets $\widehat{E} < \underline{E}$ all the firms adopt the new technology.
- For all targets $\underline{E} < \widehat{E} < \overline{E}$, a number n_l with $0 < n_l < n$ of firms adopt the new technology.

Finally we consider emission standards:

Proposition 3 (Standards): For each F there is an aggregate emission target $\widehat{\widehat{E}}$ with $\underline{E} < \widehat{\widehat{E}} < \overline{E}$ such that

- for all aggregate emissions targets $\widehat{E} > \widehat{\widehat{E}}$ no firm adopts the new technology;
- for all aggregate emissions targets $\widehat{E} < \widehat{\widehat{E}}$ all the firms adopt the new technology;
- for $\widehat{E} = \widehat{\widehat{E}}$ the number of firms adopting the new technology is ambiguous. However, there is no positive incentive to adopt the new technology.

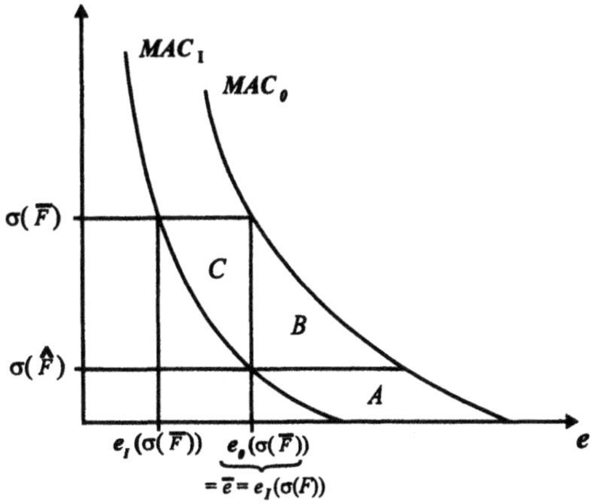

Figure 6: Comparison in Incentives by Permits and Standards, the Case $F = \overline{F}$.

The intuition for the last result can be drawn from Figure 6. If $\hat{E} < \overline{E}$ but $\hat{E}$ is sufficiently close to $\overline{E}$, then the standard $\bar{e} = \hat{E}/n$ is close to $\tilde{e}_0$ satisfying $-C_0'(\tilde{e}_0) = \tilde{\sigma}$. In equilibrium on the permit market, firms are indifferent between adopting and not adopting the new technology. Hence $F = C_0(\tilde{e}_0) - C_I(\tilde{e}_I) + \tilde{\sigma}[\tilde{e}_0 - \tilde{e}_I] = A + B + C$. On the other hand, the incentive to adopt under emission standard $\bar{e}$ is given by $F = C_0(\bar{e}) - C_I(\bar{e}) = A + B < A + B + C = F$. The opposite is true if $\hat{E}$ greater than but sufficiently close to $\underline{E}$. Since the cost difference under standards, $C_0(\bar{e}) - C_I(\bar{e})$, is continuous and decreasing in $\bar{e}$, there exists an aggregate standard $\hat{\hat{E}}$ such that all the firms invest for $\hat{E} < \hat{\hat{E}}$ and no firm invests for $\hat{E} > \hat{\hat{E}}$.

Table 1: Degree of Adoption Under Different Policies.

Range of F (Case $\underline{F} < \hat{\hat{F}}$)	Degree of Adoption Under Taxes or Subsidies	Auctioned Permits or Grandfathering	Command and Control
$F \leq \hat{F}$	complete (optimal)	complete (optimal)	complete (optimal)
$F \in (\hat{F}, \underline{F})$	complete (optimal)	underinvestment	complete (optimal)
$F \in (\underline{F}, \hat{\hat{F}})$	overinvestment	underinvestment	overinvestment
$F \in (\hat{\hat{F}}, \overline{F})$	overinvestment	underinvestment	underinvestment
$F \geq \overline{F}$	no adoption (optimal)	no adoption (optimal)	no adoption (optimal)

The results from Propositions (1)-(3) are summarized in Table 1. We see that for a wide range of emission targets, all the instruments are equivalent. In particular, if the aggregate target is sufficiently tight, all the firms adopt, if it is sufficiently loose, no firm adopts, irrespective of the instrument chosen by the regulator. If the aggregate emission target is tighter but not too tight, taxes and subsidies can be ranked over permits, and permits over command and control. For targets $\hat{E} \in (\underline{E}, \hat{\hat{E}})$, however, our result turns the traditional wisdom of the literature completely upside down. Taxes, subsidies and standards are to be ranked top. Permits, auctioned or grandfathered, are ranked lowest. But does this mean that taxes in general have to be ranked above permits from a social viewpoint? This question cannot be answered without knowing the optimal number of firms which should adopt the new technology. To have a criterion for optimality, however, we have to take into account the social damage from pollution which we have not considered so far. The optimal number of firms which should adopt the new technology clearly depends on the fixed set-up costs. It is shown in Requate and Unold (2002) that no firm should adopt the new technology if the fixed costs are sufficiently high, that all the firms should adopt the new technology if the fixed costs is sufficiently low, and that partial adoption is optimal for an intermediate range of fixed costs, say $[\underline{F}, \overline{F}]$. On that interval the optimal number of adopting firms monotonically decreases as the fixed costs go up.

In order to make a judgement about whether some policy instruments provide too little – or possibly too strong – incentives to adopt a cleaner technology, we have to be more specific about the behavior of the regulator. Following Milliman and Prince (1989) we first assume that the regulator does not anticipate the new technology and has made a commitment to the level of his policy instrument for a sufficiently long period of time. Such a commitment was made by the Clean Air Act Amendments 1990 with respect to the total supply of permits. In some European countries commitments to certain tax rates were made independently of technological development. We further assume that the regulator has set his policy to the optimal level with respect to the old technology. For this purpose let τ_0 denote the optimal tax rate before the new technology was available, L_0 the corresponding optimal amount of permits, and $\overline{e}_0 = L_0 / n$ the corresponding uniform emission standard. Then we obtain the following result.

Proposition 4: If the previous policy has been set to the optimal level, given the old technology, then

(1) **under taxes** there will be no adoption for $F > \overline{F}$, complete adoption for $F < \overline{F}$, whereas for $F = \overline{F}$ the number of firms which adopt the new technology

is ambiguous. However, there is no positive incentive for any firm to adopt the new technology, i.e. non-adoption is an equilibrium.

(2) **under permits** there exists a fixed cost $\hat{F}$ such that there will be no adoption for $F > \bar{F}$, there will be complete adoption for $F < \hat{F}$, and the number of adopting firms is less than optimal for $F \in (\hat{F}, \bar{F})$.

(3) **under standards** there exists a fixed cost $\hat{\hat{F}}$ with $\bar{F} \geq \hat{\hat{F}} > \hat{F}$ such that no firm adopts for $F > \hat{\hat{F}}$, all the firms adopt for $F < \hat{\hat{F}}$, and for $F = \hat{\hat{F}}$ the number of firms adopting the new technology is not uniquely determined. However, there is no positive incentive for any firm to adopt the new technology. Whether $\hat{\hat{F}}$ is smaller or greater than $\underline{F}$ is ambiguous.

Part (1) of the result rests mainly on our assumption that the regulator has committed herself to maintaining the tax rate at the level τ_0 for a time horizon sufficiently long to guarantee amortization of the investment. Only for one particular tax rate are firms indifferent about adopting and not adopting. This "switching tax rate", however, depends on the fixed cost. An important consequence of our result is that for $F \in (\underline{F}, \bar{F})$ too many firms will adopt the new technology under taxes. In other words, for $F \in (\underline{F}, \bar{F})$ taxes induce overinvestment.

The intuition for part (2) is that the more firms adopt the new technology, the lower the price for permits. This induces some firms to free-ride on decreasing permit prices and thus reduces the incentive to adopt the new technology. Since the aggregate supply for permits is not adjusted by assumption, the incentive to adopt is suboptimal.

Figure 6 illustrates the intuition for part (3). If the fixed cost of investment is equal to $\bar{F}$ and the original standard was equal to $\bar{e} = \bar{E} / n$, then the incentive to invest for the first firm is represented by the area $A + B + C$ whereas under a standard the cost saving corresponds to the smaller area $A + B$. By continuity, some firms will invest in equilibrium for F slightly smaller than $\bar{F}$ under a permit regime, whereas no firm will do so under a standard. A converse argument works for F slightly greater than $\underline{F}$. In that case, some firms will invest under permits in equilibrium, whereas all firms want to invest under a standard.

The results of Proposition 4 are summarized in Figure 7. Curve A denotes the socially optimal number of firms, curve B denotes the number of firms that invest under permits, curve C under taxes, and curve D under standards. The results stand in sharp contrast to both the claims of Milliman and Prince and those of Jung et al. For a considerable range of parameters both taxes and standards induce more firms to adopt a new technology than permits do. Although environmentalists may prefer taxes and standards for this reason, from an economic – and thus from a broader social point of view – we cannot rank taxes and standards above permits in general. It will depend on parameters which of those instruments will lead to the lowest welfare loss compared to the first-best allocation. As Requate and Unold (2001) show all the results hold for asymmetric firms.

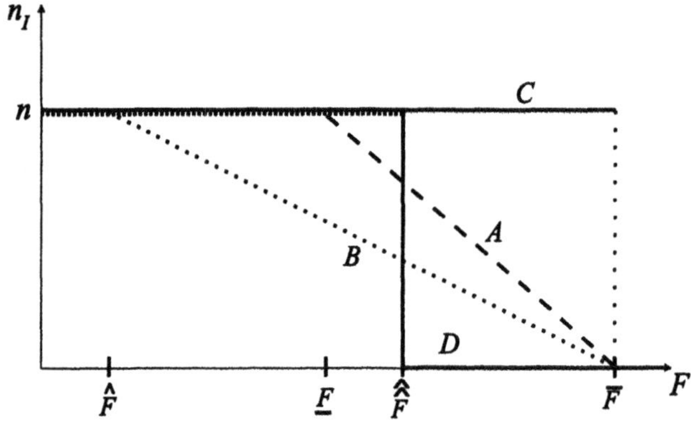

Figure 7: Number of Firms Investing in the Advanced Abatement Technology.

2.4 Optimal Policy in Anticipation of the New Technology

So far we have assumed that the regulator does not know or anticipate the new technology and thus adheres to the originally optimal policy level for a long period of time. In this section we assume that the regulator either knows the new technology before it is adopted, or learns about it after it has been adopted. We investigate two different scenarios referred to as timing B and C. (In Section 3, we study a further timing referred to as timing A).

Timing B (Ex ante optimal policy): In the first stage, the regulator makes a commitment to both the choice and the level of his policy instrument. In the second stage the firms invest.

Timing C (Ex post optimal policy): The regulator announces his policy instrument. Then the firms invest. Thereafter, the regulator observes both the number of firms adopting the new technology and the cost of the new technology, and then decides about the level of his policy instrument (Milliman and Prince refer to the latter as ratcheting).

Ex ante Optimal Policy: Consider first the case where the regulator is the first to move. Then the firms observe the new tax rate or the total supply of permits, and then decide whether or not to adopt the new technology. Denote by L_I the optimal number of permits if the new technology is available. Then we obtain the following result:

Proposition 5 (Ex ante optimal policy): If the welfare-maximizing regulator moves first, then

- under **permits** there is a unique subgame perfect equilibrium where the regulator issues the socially optimal number of permits L_l in the first stage. In the second stage, the socially optimal number of firms adopt the new technology.
- Under **taxes** for both cases, $F \leq \underline{F}$ and $F \geq \overline{F}$, there exists a unique subgame perfect equilibrium leading to a first-best outcome. If $F \leq \underline{F}$, the regulator sets $\tau = \underline{MD}^*$ in the first stage, and all the firms adopt the new technology in the second stage. If $F \geq \overline{F}$, the regulator sets $\tau = \tau_0 = \overline{MD}^*$ in the first stage, and no firms adopt. For $F \in (\underline{F}, \overline{F})$ there are multiple equilibria. One of these equilibria is efficient but there are also many inefficient equilibria with too little and too much investment.

In the case of permits, the result is driven by the fact that the permit price decreases if more firms adopt the new technology, whereas the tax is unaffected by the firms' behavior through the very timing of the game. Under taxes first-best outcomes can easily be achieved if either no adoption or complete adoption is socially optimal. A problem arises for the regulator if partial adoption is socially optimal, i.e. for $F \in (\underline{F}, \overline{F})$. For each F a particular number of firms should invest for an optimal outcome. This would require a lot of coordination between firms. Since firms, however, are indifferent between investing and not investing if the tax equals the socially optimal marginal damage – which is a necessary condition for optimality – inefficient outcomes are likely to occur in this case. Thus permits are better than taxes because with permits there is a unique equilibrium that implements the efficient allocation.

Requate and Unold (2001) study a version of this model with a continuum of asymmetric firms. In that case no interval of fixed costs exists leading to multiple equilibria. In that case we get first-best outcomes under taxes for all parameters and thus equivalence of taxes and permits.

Ex post Optimal Policy: We now consider the case where the regulator moves immediately after firms have adopted the new technology. Here things look very simple:

Proposition 6 (Ex post optimal policy): If the regulator sets her optimal policy after observing the number of investing firms, then under both taxes and permits there is a unique subgame perfect equilibrium where the optimal number of firms invests, and the regulator sets the first-best tax rate, or issues the optimal number of permits, respectively. Standards, by contrast, induce only second best allocations if partial adoption of the new technology is optimal.

The intuition behind the result is that under the tax regime the regulator can now mimic the permit market. Moving second, he can set the optimal price for pollution relative to the number of firms which have adopted the new technology. Since the firms anticipate this, the optimal number of firms will invest in the first stage. If fewer than the optimal number of firms invested in the first stage, the marginal damage would be higher than optimal. Hence the regulator's response would result in setting a tax rate higher than socially optimal (or in issuing a corresponding number of permits). But then it would not have been optimal for the non-investing firms to refrain from investing, and an allocation with less than the optimal number of firms investing would not have been an equilibrium in the first stage of the game. If, by contrast, more than the optimal number of firms invested in the first stage, the regulator's response would result in a tax rate (or the corresponding number of permits) lower than socially optimal. But then some firms will be better off by not having invested. Thus if the regulator can react immediately after investment and the firms anticipate this, the first-best outcome is a subgame perfect equilibrium under the two market-based instruments.

As mentioned above, no inefficient equilibria exist under asymmetry of firms. As an important consequence we obtain:

Corollary 7: If firms are asymmetric, taxes and permits are equivalent and lead to first-best adoption of new technology independently of who moves first the regulator or the firms.

Under standards first-best outcomes cannot be achieved through decentralized investment decisions. This even holds true for both types of timings. The reason is that if partial investment were optimal, different emission levels are optimal for firms with old and new technology, respectively. By its very nature, however, a uniform pollution standard requires uniform emission levels for both types of firms.

3 R&D

Thus far we have considered scenarios where the new technology was already given and firms had to decide whether or not to adopt it. In this section we will look at an R&D process before firms adopt the new technology. We will assume that there is an R&D-monopolist, called the innovator, who sells the new technology to the polluting firms, referred to as downstream firms. We assume that the innovator has to invest some R&D effort $R(y)$ in order to discover the new technology with a probability y. The R&D cost function satisfies $R'(y) > 0$, $R'' > 0$, and $\lim_{y \to 1} R(y) = \infty$. Besides R&D costs the innovator faces constant marginal costs c to produce one unit of the new technology. The new technology is described by an abatement cost function $C_I(e_I)$, which satisfies the same properties as in the last section.

The separation of the R&D sector from the a regulated polluting sector which may adopt new technology but does not develop it by itself is well supported by empirical evidence. Lanjouw and Mody (1996) found that 81% of all innovations in air emissions cleaning was developed by machinery industry but only 5% have been used by the same sector. For water cleaning technology we obtain 83% of innovations by the machinery industry but only 2% being used by the same sector, and for energy production from renewable resources the numbers are 85% and 8%, respectively.

We will consider three different timings of environmental policy:

- Timing A: Ex ante commitment before R&D success;
- Timing B: Ex interim commitment after observing R&D success;
- Timing C: Ex post optimal (time consistent) setting of taxes or permits.

Before we study the behavior of firms we briefly derive socially optimal allocations. The total social cost is given by

$$TSC = ySC_I + (1-y)SC_0 + R(y)$$

where $SC_0 = nC_0(e_0) + D(ne_0)$ is the social cost net R&D costs if R&D was not successful, and $SC_I = n_IC_I(e_I) + (n - n_I)C_0(e_0) + D(E)$, with $E = n_I e_I + (n - n_I)e_0$, is the social cost after R&D success.

Optimal emissions are given by the well known condition

$$-C_i'(e_i) = D'(E)$$

for $i = 0, I$.

The optimal number of adopting firms is given by $n_I = 0$ if $F \geq \overline{F}$ and $n_I = n$ for $F \leq \underline{F}$, and by

$$C_I(e_I) + F + c - C_0(e_0) = D'(E)(e_0 - e_I)$$

for $F \in (\underline{F}, \overline{F})$.

The socially optimal probability of R&D success, associated with an optimal level of R&D effort, is given by

$$SC_0 - SC_I = R'(y)$$

3.1 The Firms' Behavior

3.1.1 Behavior of Downstream Firms

We now describe the behavior of the firms starting with the downstream firms. Given any tax rate τ or permit price σ the firms set emissions such that marginal abatement costs equal the price for pollution $x = \tau, \sigma$, i.e. $-C'_i(e_i) = x$, for $i = 0, I$. The solution is denoted by $e_i(x)$. We make the additional assumption

$$\frac{d}{dx}[e_0(x) - e_I(x)] < 0, \tag{7}$$

which requires that the difference between emissions of the old and the new technology decreases if the price of emissions rises. This assumption is consistent with physical evidence according to which by the entropy law the marginal abatement cost goes to infinity if emissions go to zero.

The Investment Decision: We now turn to the downstream firms' decision to adopt the new technology. We use p to denote the license price charged by the innovator for one unit of the new technology, and x to denote the tax rate or the permit price, respectively. The downstream firm decides to invest if

$$C_I(e_I) + F + p + xe_I \leq C_0(e_0) + xe_0 \tag{8}$$

where we assume for convenience that the firms invest if (8) holds with equality.

The Innovator's Decisions: If the innovator was successful he decides about the price, or the number of units to be sold, respectively. If the regulator has set the tax rate before the innovator decides on price or output, the demand is given by:

$$n_I(p) = n \quad \text{if} \quad p \leq \Delta C(\tau) - F$$

$$n_I(p) = 0 \quad \text{if} \quad p > \Delta C(\tau) - F$$

where $\Delta C(\tau) = C_0(e_0) - C_I(e_I) + \tau(e_0 - e_I)$ is the cost saving of the variable cost.

If the regulator has issued a number of permits L, the innovator must take into account that selling more units of the new technology decreases the demand for permits, and thus causes the permit price to fall. This in turn provides less incentives to adopt the new technology. Thus the profit function can be written as

$$\Pi(n_I) = \left[\tilde{p}(\sigma(L, n_I)) - c\right] n_I$$

where

$$\tilde{p}(\sigma(L,n_I)) = C_0(e_0) - C_I(e_I) - F + \sigma(L,n_I)[e_0 - e_I]$$

is the downstream firms' marginal willingness to pay, and $\sigma(L, n_I)$ is the permit price determined by the market clearing condition

$$L = n_I e_I - (n - n_I)e_0 \tag{9}$$

Note that e_0 and e_I are also functions of $\sigma(L, n_I)$.

The first-order condition of the innovator is now given by

$$\tilde{p}'\sigma_{n_I} + p - c = 0.$$

By differentiating (9) it is easy to show that $\sigma_{n_I} < 0$ which is quite intuitive.

Our first observation now is that pre-commitment to a tax rate before the innovator decides on output leads either to complete or to no adoption of the new technology, whereas pre-commitment to a certain aggregate quantity of permits may induce the innovator to choose an interior solution of his profit maximization problem.

Finally the innovator decides on the level of R&D effort represented by the R&D success y. His objective function is given by $y\Pi - R(y)$, where Π is his profit under either policy (taxes, permits, different timings).

The first-order condition is then simply given by

$$R'(y) = \Pi.$$

3.2 The Regulator's Problem

We are now ready to study the social planner's problem. We start with timing C where the regulator moves after having observed both R&D success and the degree of adoption of the new technology. This is also the only time consistent policy.

3.2.1 Ex Post Optimal Environmental Regulation

If the regulator moves after R&D success and after adoption of new technology, we may consider her as a slave to the innovator since the optimal tax rate, or the optimal number of permits, respectively, depends on the number of adopting firms. Thus the innovator will take into account that he can influence the regulator's tax rate, or the number of permits, respectively. In case of a tax policy, the regulator sets the tax rate according to the following rule:

$$\tau = D'(E)$$

where $E = n_I e_I(\tau) + (n - n_I) e_0(\tau)$. Alternatively he can issue a number of permits according to

$$-C_0'(e_0(\sigma)) = -C_I'(e_I(\sigma)) = D'(L).$$

Once n_I is given, the aggregate marginal abatement cost curve is given, and the two policies are equivalent. Hence we can restrict attention to the case of taxes. It is useful to note that

$$\frac{d\tau}{dn_I} = -\frac{D'(E)\cdot\left(e_0(\tau) - e_I(\tau)\right)}{1 - D''(E)\dfrac{\partial E}{\partial \tau}} < 0$$

since $e'_i(\tau) < 0$ for $i = 0, I$ and thus $\dfrac{\partial E}{\partial \tau} < 0$.

(Similarly one can show that $\dfrac{dL}{dn_I} < 0$.)

Hence the price set by the successful innovator can be written as

$$p(n_I) = C_0(e_0) - C_I(e_I) - F + \tau(n_I)\left[e_0 - e_I\right]$$

where $e_i = e_i(\tau(n_I))$. Since the firms choose their emissions such that their marginal abatement costs are equal to the tax rate, we obtain

$$\frac{dp(n_I)}{dn_I} = \tau'(n_I)\left[e_0 - e_I\right] < 0.$$

The innovator's profit can thus be written as

$$\Pi(n_I) = \left[p(n_I) - c\right]n_I.$$

The first-order condition is determined by the typical monopoly rule:

$$p'(n_I)n_I + p(n_I) - c = 0.$$

The behavior of the monopolist thus leads to under-provision of new technology, as was expected. Furthermore one can show that the innovator's profit is smaller than the social value of innovation.[2] Thus R&D effort under timing C is less than socially optimal.

3.2.2 Ex Interim Optimal Environmental Regulation

Let us now consider the case of ex interim regulation. This means, that the regulator observes whether or not the innovator was successful. In case of R&D success, he commits to a tax rate or a number of permits before the innovator sets the price for his new technology and before the downstream firms adopt that technology. We start with the case of taxes:

Taxes: Recall that for any adoption cost F, there exists a tax rate $\tilde{\tau}(F)$ such that

$$C_0\left(e_0(\tau)\right)-C_I\left(e_I(\tau)\right)-F-c+\tau\left[e_0(\tau)-e_I(\tau)\right]>0 \qquad (10)$$

if and only if $\tau>\tilde{\tau}(F)$. Note that the left hand side of (10) is also the maximal price which the innovator can charge in order to induce the downstream firms to invest. Since the firms are symmetric, at any tax rate τ and any price p, either all the firms want to invest, or no firm wants to invest. This observation gives rise to the following result:

Proposition 8: Consider timing B under a tax policy.

- If $F \geq \overline{F}$, the regulator sets $\tau = D'(ne_0^*)$, the innovator offers nothing, and no firms adopts the new technology, which is socially optimal.
- For $F \in (\underline{F}, \overline{F})$ the regulator sets $\tau = \tilde{\tau}(F) - \epsilon$ if F is sufficiently close to $\overline{F}$; and $\tau = \tilde{\tau}(F) + \epsilon$ if F sufficiently close to $\underline{F}$. In the first case no downstream firm adopts the new technology, in the latter case all the firms adopt the new technology.
- For $F \leq \underline{F}$, the regulator sets $\tau = \tilde{\tau}(F)$, and all the firms adopt the new technology.

Note, however, that unless $F \leq \underline{F}$, the regulator chooses a tax rate at which the innovator makes no profit. Anticipating this in the first stage of the game the innovator will not engage in any R&D.

Permits: In this case, after observing R&D success, the government issues a number of permits L. Then the innovator sets the number of units n_I that he wants to sell. Thus the permit price σ can be written as a function of L and n_I. Hence, if the regulator can make a positive profit, this is given by $\Pi(n_I) = [p(n_I) - c]\, n_I$ where $p = C_0(e_0) - C_I(e_I) - F + \sigma(L, n_I)\,[e_0 - e_I]$. The first-order condition is then

---

2 One can show this along the lines of the proof of Proposition 6.8 of Unold (2000).

given by $p - c + p'(n_I)\, n_I = 0$. The regulator now minimizes the social cost with respect to L.

$$SC(L) = n_I\left[C_I(e_I) + F + c\right] + \left(n - n_I\right)C_0(e_0) + S(L)$$

If $F > \overline{F}$, we obtain $L = E_0^*$, and the innovator would face no demand even for $p = c$. For $F < \overline{F}$, the first-order condition of the regulator is given by

$$SC'(L) = \left[C_I(e_I) + F + c - C_0(e_0)\right]\frac{\partial n_I}{\partial L} + S'(L) \tag{11}$$

$$+ n_I C_I'(e_I) e_I'(\sigma)\sigma_L + (n - n_I) C_0'(e_0) e_0'(\sigma)\sigma_L$$

$$= \left[C_I(e_I) + F + c - C_0(e_0)\right]\frac{\partial n_I}{\partial L} + S'(L) - \sigma\frac{\partial E}{\partial \sigma}\frac{\partial \sigma}{\partial L} = 0. \tag{12}$$

But now observe that $1 = \frac{\partial E}{\partial \sigma}\frac{\partial \sigma}{\partial L} - \left(e_0 - e_I\right)\frac{\partial n_I}{\partial \sigma}\frac{\partial \sigma}{\partial L}$. Hence (12) becomes

$$\sigma = S'(L) - (p - c)\frac{\partial n_I}{\partial L} > S'(L)$$

since one can show that $\frac{\partial n_I}{\partial L} < 0$.

The regulator sets the number of permits such that the permit price exceeds marginal damage. The intuition is as follows: the higher the number of permits, the lower, other things being equal, the market price for permits, and the incentive to adopt the new technology for the downstream firms. Moreover, the innovator holds down output to increase his profit. But the higher the number of permits, the less the demand for permits. Hence the regulator issues a number of permits small enough such that the market price for permits exceeds marginal social damage.

Note that the innovator always earns some rent under permits whereas he does not under taxes, at least for $F > \underline{F}$.

3.2.3 Ex Ante Commitment (Timing A)

With this timing the regulator commits herself to the level of his policy instrument before the innovator engages in R&D. This means that, in the case of a tax policy for instance, that there is one tax rate for both events, R&D success, and no R&D success. Let us first consider the case of taxes:

Taxes: The regulator's objective function is now given by the total social cost including the R&D cost

$$TSC = (1-y)SC_0 + ySC_I + R(y)$$

where SC_0 is the social cost in the case of no R&D success, and SC_I is the social cost after R&D success. Note that the behavior of both the downstream firms and the innovator is basically the same as under timing B since for both types of commitment, firms take the tax rate as given.

It is important to note, however, that in order to induce a positive level of R&D effort, the tax rate must exceed $\tau(F)$. Thus, in order to induce $y > 0$, the regulator has to accept complete adoption, i.e. $n_I = n$.

The first-order condition of the regulator's problem is now determined by:

$$0 = [SC_I - SC_0 + R'(y)]\frac{\partial y}{\partial \tau} +$$

$$y\{n[C_I'(e_I) + D'(E_I)]e_I'(\tau) - n[C_0'(e_0) - D'(E_0)]e_0'(\tau)\}.$$

Using $-C_i'(e_i) = \tau$, and solving for the tax rate we obtain:

$$\tau = \frac{\frac{\partial y}{\partial \tau}[SC_I - SC_0 + R'(y)]\frac{1}{ny} - [D'(E_0)e_0'(\tau) - D'(E_I)e_I'(\tau)]}{e_I'(\tau) - e_0'(\tau)}.$$

Observe that $e_0'(\tau) - e_I'(\tau) < 0$ holds by assumption. Hence the sign of the last term $D'(E_0)(-e'_0(\tau)) - D'(E_I)(-e'_I(\tau))$ is positive. Moreover, it is easy to show that $\partial y / \partial \tau > 0$.

If we drop the assumption $e'_0(\tau) - e'_I(\tau) < 0$, and instead assume $e'_0(\tau) = e'_I(\tau)$, then we cannot divide by $e'_0(\tau) - e'_I(\tau)$, we see that the tax term drops out, and we obtain:

$$SC_0 - SC_I - R'(y) = ny[D'(E_0) - D'(E_I)]e_0'(\tau) / \left(\frac{\partial y}{\partial \tau}\right).$$

This term is positive for convex damage functions. But this implies that the R&D-effort is less than optimal in this case. If, by contrast, the damage function is linear, we obtain optimal R&D effort. Observe further that for $e'_0(\tau) - e'_I(\tau) < 0$ and for constant marginal damage, we obtain

$$\tau = MD - \frac{SC_0 - SC_I - R'(y)}{e_I'(\tau) - e_0'(\tau)} \frac{\frac{\partial y}{\partial \tau}}{ny}.$$

From this we can only conclude that for constant marginal damage the tax exceeds marginal damage if and only if there is excessive R&D effort. We do not know, however, whether or not the second best optimal tax rate induces excessive R&D-effort.

Permits: Consider now the case where the regulator commits to a quota of permits before R&D effort. In contrast to the case of taxes the innovator can influence the market price for permits, and he can induce a number of downstream firms $n_I < n$ to adopt the new technology. In this case the second best number of permits is determined by the following rule:

$$\bar{\sigma} = D'(L) - (p-c)y\frac{\partial n_I}{\partial L} + [SC_I - SC_0 + R'(y)]\frac{\partial y}{\partial L}$$

where $\bar{\sigma} = y\sigma_I + (1 - y)\sigma_0$ is the expected price for permits, the realization of which in case of R&D success will be low, equal to σ_I, and in the case, where R&D has not been successful, will be high, namely equal to σ_0.

3.3 Comparison

A comparison yields ambiguous results in general, i.e. welfare under permits can be higher or lower than welfare under taxes in the scenarios A and B. Under scenario A permits turn out to be better if the damage function is extremely steep. This follows from Weitzman's seminal result (1974) according to which under uncertainty about the abatement cost function the expected welfare loss is smaller under permits compared to taxes if the social damage function is sufficiently steep and vice versa. With respect to R&D effort expected welfare is higher under ex ante commitment than under ex interim commitment if we compare tax policies. At least this holds true for $F > \underline{F}$, since under ex interim regulation the innovator's profit will be held down to zero by means of the second best optimal tax policy. For permits by contrast, no clear ranking is possible. The reason is that under the ex interim permit policy the innovator makes some profits, and the regulator can use different discrete policies for the cases of R&D success and no success. Under ex ante commitment, by contrast, the regulator has to choose one quantity of permits for both cases.

4 Extensions

In this section we will sketch some possible extensions of the model developed in the last section. First we will introduce asymmetry among the downstream firms, second we will show how to generalize the model for the case of upstream duopoly.

4.1 Asymmetric Downstream Firms

Asymmetry among the downstream firms can be introduced along the lines of Requate and Unold (2001) who consider a continuum of different firms adopting the new technology. Through adoption of new technology the downstream firms face a cost function $C^A(e, x)$, where x is a firm specific parameter of downstream firm $x \in [0, 1]$. Besides inducing both lower abatement and lower marginal abatement costs, the cost functions $C^A(e, x)$ satisfy the conditions $C^A_x \geq 0$ and $-C^A_{ex} \geq 0$, which means that the closer the firm specific parameter x to 0, the more suitable is the technology for that downstream firm x. Besides the variable costs $C^A(e, x)$ the downstream firms face a fixed investment cost $F(x)$ to install the new technology. Here we assume $F(0) = F$, and $F'(x) \geq 0$. This again means that the fixed cost of the new technology A is the smaller, the closer the firm specific parameter x to 0.

The asymmetry induces a downward sloping demand function for the upstream monopolist by the following relationships. A downstream firm decides to invest if

$$C^A(e^x, x) + \tau e^x + F(x) + p \leq C^0(e^0) + \tau e^0 \tag{13}$$

The firm is indifferent between investing or not if

$$C^A(e^x, x) + \tau e^x + F(x) + p = C^0(e^0) + \tau e^0 \tag{14}$$

holds. Thus (14) defines an inverse demand or willingness to pay function $p(x, \tau)$ of the downstream firms which satisfies the following properties $p_x < 0$, and, $p_\tau > 0$, i.e. it is downward sloping in x and shifting upwards through a higher tax rate τ.

Introducing asymmetry does not make much of a difference for ex post regulation. We obtain less than socially desired adoption of new technology, and too little R&D by the upstream monopolist. Asymmetry in particular turns out to be crucial for ex interim regulation. The upstream monopolist can now earn a positive rent under tax regulation. One can show that, once the new technology is available, taxes are superior to permits. The reason is that the monopolist cannot influence the price for pollution under a tax policy (which is assumed to be fixed by the regulator), whereas he can influence the price for permits by higher pricing of new technology. The general superiority of taxes, however, does not hold from an ex ante point of view. If the downstream firms are sufficiently similar, and thus the inverse demand function is sufficiently flat, the monopolist's rent becomes small if the tax rate is sufficiently high. This provides little incentives to engage in

R&D. Moreover, the monopoly profit can be higher or lower than the social value of innovation both under taxes and under a permit regime and for ex ante and ex interim regulation.

4.2 Upstream Duopoly

Finally let us study how to model upstream monopoly. For symmetric downstream firms, we obtain Bertrand competition if both upstream firms have been successful. In this case both upstream researchers do not earn any profit. It is easy to see that the R&D effort is lower than under pure monopoly since the expected profit under monopoly is given by $y_i\,(1 - y_j)\,\Pi^M - R(y_i)$ for upstream firm i, whereas it is given by $y_i\Pi^M - R(y_i)$ for a monopolist. (Here Π^M denotes the monopoly profit of the upstream firm.) Although the profit depends on the kind of regulation, this result is independent of how the downstream industry is regulated.

Matters become more interesting if we introduce asymmetry among the downstream firms because this may cause product differentiation if upstream firms offer slightly different technologies. We can model this by introducing two technologies A and B and cost functions $C^A(e^x, x)$, $F^A(x)$ and $C^B(e^x, x)$, $F^B(x)$ with $C_x^A \geq 0$, $-C_{ex}^A \geq 0$ and $F^{A\prime}(x) > 0$, but $C_x^B \leq 0$, $-C_{ex}^B \leq 0$, and $F^{B\prime}(x) < 0$. This is to be interpreted that technology A is more suitable for firms on the right hand side of the unitary interval, i.e. for those firms sufficiently close to 0, whereas technology B is more suitable for those close to 1. Depending on the adoption cost F^j and the operating cost, represented by $C^j(e^x, x)$, $j = A, B$, we may have either two "local" monopolies or competition in case that both firms are successful. In the latter case we obtain an interval of firms $[0, x_A]$ buying from upstream monopolist A, and an interval $[x_B, 1]$ buying from upstream monopolist B, whereas the firms from $[x_A, x_B]$ do not adopt any kind of new technology (see Figure 8). All results drawn from the monopoly case remain valid for this case. Total R&D effort, however, is clearly greater than in the single monopoly case and thus expected welfare increases with more upstream firms. If adoption costs are sufficiently low for those firms with a parameter close to 0.5, we have market coverage in case that both firms are successful. In this case we do not have market distortions once both upstream firms have been successful. Thus the optimal tax rate or the optimal permit price under ex interim regulation are equal to marginal damage. If downstream firms are sufficiently similar, profits of the upstream firms get close to zero, and total R&D effort may be lower than in case of pure upstream monopoly.

We have seen that if both firms were successful, there may be either two "local" monopolies or the market is covered where in the latter case no market distortion arises once the new technologies are available. This turns out to be different if we extend the downstream firms' parameter space to more than one dimension. The cost functions may, for instance, depend on two parameters. In this case it is possible to obtain real strategic interaction between the upstream firms by defining demand functions $D_A(p_A, p_B)$ and $D_B(p_A, p_B)$. In general there will now be some downstream firms which do not buy from either upstream firm (see Figure 10) and there is still a market stealing effect which was not the case in the one-dimensional case (see Figure 9). In this case, however, we obtain similar results as in the pure monopoly case. In particular the second best optimal emission tax rate will exceed marginal damage. The welfare comparisons between the different regulatory schemes and timings remain ambiguous.

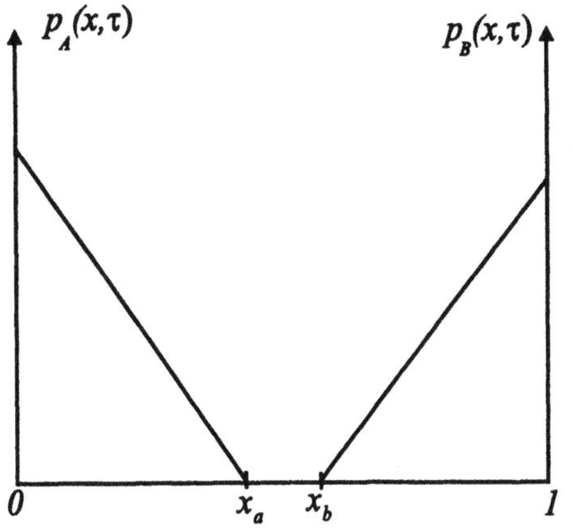

Figure 8: Two "Local" Monopolies.

5 Concluding Remarks

In this paper we have surveyed some recent findings on the incentives provided by environmental policy instruments to adopt advanced abatement technology. We have called into question the ranking of environmental policy instruments in the traditional literature, a ranking derived by comparing aggregate cost savings rather than by investigating the firms' incentives to invest in equilibrium. We have demonstrated that the comparison of environmental policy instruments leads to quite different results if the number of firms which adopt new technology is determined endogenously through equilibrium considerations.

Further we have extended the adoption model by considering an R&D process prior to the adoption phase. We have considered a non-competitive, i.e. monopolistic or duopolistic R&D industry. The focus of the analysis was on different timings, the comparison between time-consistent policies, and pre-commitment to a certain level of the policy instruments. We saw that some commitment can be welfare enhancing, but that a comparison between ex ante and ex interim commitment is ambiguous in general.

Our analysis was in the tradition of IO-literature since we considered dynamic games of a finite number of stages. In reality we have a process where research firms permanently engage in R&D, polluting firms in aggregate permanently adopt new technology, and the regulator cannot commit to the level of his policy for once and for all. Further research is necessary to account for such processes in a more complex dynamic framework.

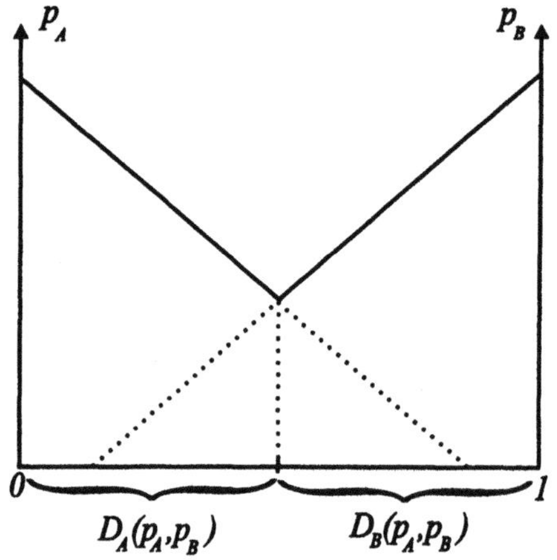

Figure 9: Market Coverage Under Upstream Duopoly.

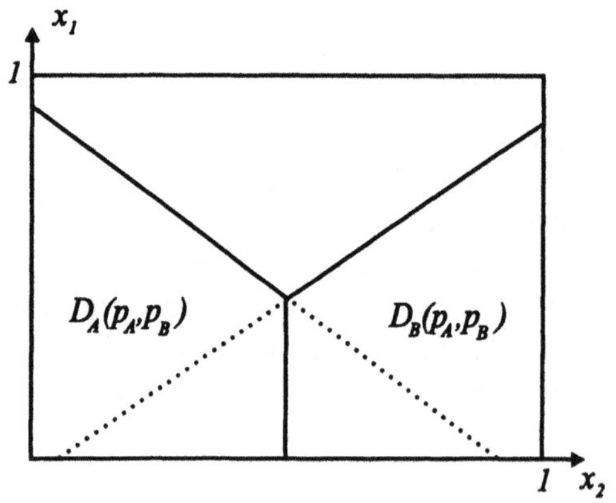

Figure 10: Duopoly in Extended Parameter Space: Some Firms Do Not Adopt.

References

Arrow, K. (1962), Economic Welfare and the Allocation of Resources for Innovations, in: Nelson, R. (Ed.), *The Rate and Direction of Inventive Activity*, Princeton.

Biglaiser, G. and J.K. Horowitz (1995), Pollution Regulation and Incentives for Pollution-Control Research, *Journal of Economics and Management Strategy*, 3 (4), 663-684.

Biglaiser, G., J.K. Horowitz, and J. Quiggin (1995), Dynamic Pollution Regulation, *Journal of Regulatory Economics* 8, 33-44.

Carraro, C. and D. Siniscalco (1994), Environmental Policy Reconsidered: The Role of Technological Innovation, *European Economic Review* 38, 545-554.

Downing, P.B. and L.J. White (1986), Innovation in Pollution Control, *Journal of Environmental Economics and Management* 13, 18-29.

Fisher, C., I. Parry, and W. Pizer (1999), *Instrument Choice for Environmental Protection When Technological Innovation is Endogenous*, Resources for the Future, Discussion Paper 99-04.

Jaffe, A., G. Newell, and R. Stavins (2000), *Technological Change and the Environment*, NBER Working Paper 7970, October.

Jung, C., K. Krutilla, and R. Boyd (1996), Incentives for Advanced Pollution Abatement Technology at the Industry Level: An Evaluation of Policy Alternatives, *Journal of Environmental Economics and Management* 30, 95-111.

Kneese, A.V. and C.L. Schulze, (1975), *Pollution, Prices and Public Policy*, Washington D.C. Brookings.

Lanjouw, J.O. and A. Mody (1995), Innovation and the International Diffusion of Environmental Responsive Technology, *Research Policy* 25, 549-571.

Malueg, D.A. (1989), Emission Credit Trading and the Incentive to Adopt New Pollution Abatement Technology, *Journal of Environmental Economics and Management* 18, 52-57.

Marin, A. (1991), Comment: Firms Incentives to Promote Technological Change in Pollution Control, *Journal of Environmental Economics and Management* 21, 297-300.

Milliman, S.R. and R. Prince (1989), Firms Incentives to Promote Technological Change in Pollution Control, *Journal of Environmental Economics and Management* 17, 247-265.

Parry, I. (1995), Optimal Pollution Taxes and Endogenous Technological Progress, *Resource and Energy Economics* 17, 69-85.

Requate, T. (1995), Incentives to Adopt New Technologies Under Different Pollution-Control Policies, *International Tax and Public Finance* 2, 295-317.

Requate, T. and W. Unold (2001), On the Incentives of Policy Instruments to Adopt Advanced Abatement Technology if Firms Are Asymmetric, *Journal of Institutional and Theoretical Economics* 157, 536-554.

Requate, T. and W. Unold (2002), Environmental Policy Incentives to Adopt Advanced Abatement Technology – Will the True Ranking Please Stand up?, forthcoming in: *European Economic Review*.

Ulph, D. (1997), Environmental Policy and Technological Innovation, in: Carraro, C. and D. Siniscalco (Eds.), *New Directions in the Economic Theory of the Environment*, Cambridge, 43-68.

Verdier, T. (1995), Environmental Pollution and Endogenous Growth: A Comparison between Emission Taxes and Technological Standards, in: Carraro, C. and J. Filar (Eds.), *Control and Game-Theoretic Models of the Environment*, Boston, 175-200.

Weitzman, M. (1974), Prices vs. Quantities, *Review of Economic Studies* 41, 477-491.

Ecological Tax Reform and Efficiency of Taxation: A Public Good Perspective*

Rüdiger Pethig

Department of Economics, University of Siegen, Hölderlinstr. 3, 57068 Siegen, Germany, *pethig@vwl.wiwi.uni-siegen.de*

Summary:
Revenue-neutral ecological tax reforms are known to yield a green dividend, but not an efficiency dividend, in general. This result is shown to be not counter-intuitive when one looks at such reforms as providing more of the costly public good 'environmental quality' financed by distortionary taxes. In that perspective the double dividend conjecture is tantamount to assuming that more environmental quality can be bought for less money, which is rather unlikely even if general equilibrium interdependencies are accounted for.

Keywords: Ecological tax reform, double dividend, environmental quality, public good.

1 Introduction

Taxing the emission of pollutants has been recommended as an 'economic' instrument of pollution control ever since economists approached environmental disruption as a problem of negative externalities in the Pigouvian tax-subsidy paradigm. Such emission taxes were shown to reduce the allocative distortions of environmental externalities caused by the market system's failure of correctly accounting for the scarcity of environmental resources. Such an efficiency enhancing externality taxation is in stark contrast to the results of modern taxation theory which shows that, in a world without lump-sum taxation and without externalities, taxes carry an excess burden and thus diminish allocative efficiency.

These two lines of argument demand for being interlinked: Suppose government is asked to collect a given amount of total tax revenue by levying both an emission tax and a distortionary labor tax. Isn't it then an obvious conclusion that changing the tax mix by raising the emission tax rate reduces both the externality

---

* Helpful comments from Daniel Weinbrenner are gratefully acknowledged. Remaining errors are the author's sole responsibility.

inefficiency and the excess burden of the labor tax? Indeed, such a taxation strategy, called 'ecological tax reform', promises a double dividend (Binswanger et al., 1983; Pearce, 1991) by improving the quality of the environment and reducing the tax system's excess burden at the same time.

In recent years the theoretical foundations of ecological tax reform and its implications have been scrutinized in a number of contributions, especially in Bovenberg and de Mooij (1994), (but see also Goulder, 1995; Parry, 1995; Schöb, 1995; Majocchi, 1996; Ruocco and Wiegard, 1997; Weinbrenner, 1999), relating to some extent the tax reform issue to the optimal tax cum externality literature (Atkinson and Stern, 1974; Sandmo, 1975). As a tendency, the theoretical investigations do not find convincing support for the tax efficiency dividend (Parry, 1995; Weinbrenner, 1999) or even reject the double dividend hypothesis like Bovenberg and de Mooij (1994: 1085) who conclude "that environmental taxes typically exacerbate, rather than alleviate, pre-existing distortions – even if revenues are employed to cut pre-existing distortionary taxes". This obvious discrepancy between informed common sense and theoretical analysis is intriguing. One may argue that the source of the discrepancy originates, perhaps, from our failure to understand the complexity of the third-best scenario of an ecological tax reform which involves two distinct inefficiencies, the externality and the distortionary taxation. Another reaction is to question the robustness of the theoretical conclusions, because carrying out conclusive comparative statics in general tax-distorted equilibrium models requires to employ very restrictive assumptions[1]. Yet another response – the one to be pursued in the present paper – is to suggest a different perspective of analyzing tax reforms in the presence of environmental externalities in an effort to obtain new insights into the expected benefits (dividends) of environmental policy in a world of distortionary taxation.

The basic focus of this paper is to look at ecological tax reform from a public good perspective rather than from a Pigouvian externality cum tax reform perspective. Our point of departure is the insight, aptly expressed by Heller and Starrett (1976: 10), e. g., that "one can think of externalities as nearly synonymous with non existence of markets". In their view an externality is "a situation in which the private economy lacks sufficient incentives to create a potential market in some good and the non existence of this market results in losses in Pareto efficiency" (ibid.). Thinking of externalities or rather public goods in terms of (non existing) markets requires, first of all, to identify the commodities traded on these fictitious markets. In our context these goods are 'environmental quality' and 'nature's assimilative services' the former being a public good and the latter a private good.

---

1 The lack of robustness of the results in the literature is a serious issue, since comparative static analysis in general equilibrium models does not yield conclusive results unless severe restrictions are introduced with respect to utility and production functions. Unfortunately, interdependence effects can bring about just every pathological result as shown in Weinbrenner (1999).

For reference purposes we build a simple model in which all disincentives of market creation are assumed to be absent: Nature's fixed endowment of assimilative services is privately owned and is sold to two different users, to the producer of environmental quality and to the consumption good industry. This fancy scenario of a complete set of competitive markets is known to constitute a first-best world with full internalization of the environmental externality. Government has no useful role to play since even the public good 'environmental quality' is privately provided in an efficient way – ignoring all the well-known problems of demand revelation and free riding.

Throughout the present paper, the first-best allocation in an economy with a full set of partly fictitious competitive markets is taken as the benchmark model of taxation – in contrast to the public finance view of ecological tax reform (e. g. Bovenberg and de Mooij, 1994) which takes as its benchmark the general competitive equilibrium in the absence of any pollution control and in the absence of harmful pollutants. To make our concept precise we assume that the government interferes with (fictitious) markets through taxes and subsidies for the purpose of fixing the market provision of environmental quality at a 'politically predetermined level'. It is shown that any (suboptimal) environmental quality standard can be tax-implemented in an efficient way. If this standard is inefficiently low, the corresponding taxation scheme includes an emission subsidy (!), whereas what is conventionally identified as the 'emission charge' turns out to be the subsidized price of the productive factor 'assimilative services' used by industry. An obvious advantage of this perspective is that we do not study the effects of distortionary taxation in an externality-distorted world but rather view taxation as the only cause of allocative inefficiency.

The paper proceeds by investigating public policies to supply environmental quality at no cost to the consumers. On the assumption that the market for assimilative services is still working and no lump-sum taxation or transfer is viable, government now has to raise a distortionary labor tax to pay for the purchase of environmental quality. It is shown that implementing a predetermined environmental quality standard by such a policy is equivalent to establishing this standard in an economy with a complete set of markets where the externality was represented as a tax distortion.

In this setting, one may want to strengthen the environmental quality standard by changing the tax mix. Rather than carrying out a full-scale comparative statics analysis (see Footnote 1) we characterize the set of possible results and consider alternative incidence hypotheses to conclude that the double dividend conjecture essentially amounts to arguing that government is able to buy more of a public good while expending less.

All considerations up to this point have been based on the obviously unrealistic assumption that assimilative services are privately owned, i.e. that the associated income from selling assimilative services (resource income, for short) is private income. If this resource income is publicly owned, as in the real world, the regulator can implement any environmental quality standard at minimum cost when she replaces the labor tax by her non-distortionary resource income. In other words, the issue of distortionary taxation vanishes unless one introduces a second public good in addition to environmental quality that is also costlessly provided to the consumers and financed by government.

With this modification of the model we obtain, in fact, the framework of analysis chosen by the studies on ecological tax reform quoted above. Hence in our perspective the ecological tax reform literature essentially aims at answering the following question:

What happens to the efficiency of the tax system

(1) when the regulator provides environmental quality along with another public good costlessly to the consumers at inefficient levels,

(2) when the purchase of these goods is financed partly by the revenue from distortionary taxes and partly by non-distortionary public resource income, and

(3) when the regulator wishes to strengthen the standard of environmental quality?

It is finally shown in Section 4.2 that paying for the increased provision of environmental quality which is one out of two publicly provided public goods with resource income and the revenue of a distortionary (labor) tax is qualitatively very similar to the simpler scenario studied in Section 4.1 of using labor tax revenue for purchasing an increment of environmental quality as the only publicly provided public good when assimilative services are privately owned. The set of possible result is again characterized along the same lines as in Section 4.1. Moreover, necessary conditions for the emergence of a double dividend turn out to be the existence of a private consumer good in addition to the 'dirty' good, if all production processes are linear, or the existence of a waste abatement technology. It is finally shown how the public good perspective of ecological tax reform is linked to the conventional analysis of ecological tax reform as exemplified in Bovenberg and de Mooij (1994). A partial equilibrium illustration shows that it is the less likely that the emission tax revenue (conventionally defined) increases as a response to an increase of the emission tax rate, the less expensive is waste abatement.

2 Tax Representations of Environmental Externality

Consider an economy producing a (private) consumer good, Y, with the help of labor, ℓ, and assimilative services, a_y, by means of the linear homogeneous production function

$$y = \underset{+\ +}{Y(\ell, a_y)} \quad \text{with domain} \quad \{(\ell, a_y) \,|\, a_y \le c\ell, c > 0 \textit{ and const.}\}. \tag{1}$$

The emission of pollutants by industry Y can be interpreted as that industry's demand for assimilative services so that a_y measures both the use of assimilative services and the emission of pollutants. The flow of these emissions affects the state of the environment represented by an index of environmental quality, q, according to the strictly declining and concave function $\tilde{Q}$ with the interpretation that $q = \tilde{Q}(a_y)$ is the environmental quality resulting from industry's use of assimilative services (alias discharge of pollutants). Nature's limited assimilative capacity is modeled in a very stylized way: It is assumed that the environment is capable of assimilating or neutralizing small amounts of pollutants (per period) without irreversible damage. But once emissions exceed some critical load, say $\bar{a}$, then serious environmental disruption occurs rendering the development unsustainable. This implied 'sustainability constraint' is formalized by $a_y + a_q = \bar{a} > 0$, $a_y \ge 0$ and $a_q \ge 0$. Hence, $\tilde{Q}$ can be transformed into a function Q defined by

$$q = \begin{cases} Q(a_q) := \tilde{Q}(\bar{a} - a_q), & \textit{if } a_q \in [0, \bar{a}]; \\ -\infty & \textit{if } a_q > \bar{a}. \end{cases} \tag{2}$$

Clearly, Q is strictly increasing and concave, and sustainability requires environmental quality to stay at or above the critical environmental quality, $Q(0)$. Introducing the convention $Q(0)=0$ equation (2) is readily interpreted as an 'ecological production function' for generating 'environmental quality' with the help of assimilative services, a_q. In this perspective the assimilative services are private factors for producing both environmental quality – which is a public good – and the private good Y. Observe, however, that the use of assimilative services is very different in both cases. Industry uses them for disposing of detrimental pollutants, whereas a_q is demanded to prevent the environment from being polluted. In this way the environmental-economic allocation problem is characterized as the conflict between competing uses of scarce assimilative services: its use for 'producing' environmental quality versus its use for absorbing industrial pollutants.

Summing up the supply side of the economy, we recall that there are two productive factors: assimilative services in fixed supply, $\bar{a}$, and labor in endogenous supply; there are two consumption goods: environmental quality as a public good and the private consumption good Y. The former is produced with assimilative services only, and the latter requires both inputs for its production.

The description of the model is completed by introducing the representative consumer's strictly quasi-concave utility function

$$u = U(y, f, q) \tag{3}$$

where $f := 1 - \ell$ represents leisure.

We now embark on the thought experiment of investigating the above model as a private ownership market economy with a complete set of competitive markets, one for each of the four commodities with market clearing prices in general competitive equilibrium. Being the owner of assimilative services, the representative consumer sells these services to the producers of good Y and of environmental quality. Such a market for assimilative services differs from markets in tradable emission permits as usually discussed in the literature, because in the latter permits are not (designed to be also) purchased for the purpose of preventing the emission of pollutants which is precisely the reason why the producer of environmental quality demands assimilative services in the model of the present paper. This producer runs a profit-maximizing and price-taking private firm and thus establishes the link between the markets for assimilative services and environmental quality.

Denote by p_y the price of good Y, by p_q the price of environmental quality, by p_a the price of assimilative services, and by w the wage rate. Let t be the labor tax rate, θ the tax rate on the consumption of environmental quality, and τ the tax rate on the use of assimilative services. The tax rate t is assumed to be non-negative, but θ and τ are unrestricted in sign[2]. Since we want to consider the possibility of applying different tax rates to different uses of assimilative services, we allow for splitting τ into a tax rate τ_y levied on the use of assimilative services by the industry and a tax rate τ_q levied on the use of these services by the producer of environmental quality.

At equilibrium prices, maximum profit is zero in both sectors provided that both production functions Y and Q are linear homogeneous[3]. With this simplifying assumption the consumer's budget constraint is

$$(w - t)\ell + p_a \bar{a} + \sigma = p_y y + (p_q + \theta) q \tag{4}$$

[2] Unless it is particularly relevant to refer to $\theta < 0$ or $\tau < 0$ as rates of subsidy we usually call θ and τ tax rates independent of their sign.

[3] Strict concavity of Q appears to be realistic in some empirical applications (Fiedler 1994). But here we stick to linear Q for expository convenience.

In (4), σ is a lump-sum transfer (or tax, if negative) from the government to the consumer which will be explicitly set zero when appropriate. The pertaining government budget constraint is

$$\theta q+\tau_q a_q+\tau_y a_y+t\ell-\sigma=0. \tag{5}$$

Throughout the present paper, corner solutions will be disregarded. As a consequence, for given θ, τ_q, τ_y and t the equilibrium allocation is characterized by the equations (1)-(5) combined with the following first-order conditions of profit and utility maximization, respectively:

$$p_y Y_\ell(\ell,\ a_y)=w \tag{6a}$$

$$p_y Y_a(\ell,\ a_y)=p_a+\tau_y \tag{6b}$$

$$p_q Q_a(a_q)=p_a+\tau_q \tag{6c}$$

$$p_y U_f(y,\ f,\ q)=(w-t)\cdot U_y(y,\ f,\ q) \tag{6d}$$

$$p_y U_q(y,\ f,\ q)=(p_q+\theta)\cdot U_y(y,\ f,q) \tag{6e}$$

Let us first recall the conditions under which a full market equilibrium is Pareto efficient:

Proposition 1:

(a) A general competitive equilibrium allocation is Pareto efficient, if and only if $t=\theta=0$ and $\tau_q=\tau_y\geq 0$.

(b) A Pareto efficient competitive equilibrium exhibits $p_a=p_q\,Q_a\,(a_q)>0$ if $U_q>0$, $U_y>0$, $Q_a>0$ over the set of feasible allocations and if $Y_a(\ell,\ c\ell)=0$ for all $\ell\in[0,1]$.

Since corner solutions are suppressed, Proposition 1a follows immediately from the Lagrangean

$$U(y,\ f,\ q)+\lambda_y\left[Y(\ell,\ a_y)-y\right]+\lambda_q\left[Q(a_q)-q\right]$$

$$+\lambda_a\left(\bar{a}-a_y-a_q\right)+\lambda_\ell\left(1-f-\ell\right) \tag{7}$$

by writing down the first-order conditions as equalities and assigning $\lambda_\ell=w$, $\lambda_y=p_y$, $\lambda_q=p_q$ and $\lambda_a=p_a+\tau$. The resultant equations coincide with the equations (6a)-(6e), if and only if $t=\theta=0$ and $\tau_q=\tau_y=\tau\geq 0$.

Proposition 1a states the obvious: In a world with a full set of competitive markets the best tax-subsidy policy is no policy. The government has no meaningful role to play. With all tax rates being zero the equations (6a)-(6e) readily imply the (Samuelson) condition of Pareto optimal public good provision[4]

$$\frac{U_q}{U_y}=\frac{Y_a}{Q_a}. \tag{8}$$

In (8), U_q/U_y is the marginal willingness-to-pay for environmental quality of the representative consumer (or, equivalently, the marginal pollution damage). Y_a/Q_a is the marginal production cost of environmental quality in terms of good Q (marginal abatement cost), and Y_a is the marginal abatement cost in terms of good Y.

The conditions listed in Proposition 1b secure that environmental quality and assimilative services are not free goods. Since Pareto efficient situations where $p_q=p_a=0$ do not involve serious pollution problems, they will be ignored in what follows. It is worth emphasizing that the scarcity price of assimilative services ($p_a>0$) required for efficiency is by no means a distortionary input tax. On the contrary, in the present model environmental disruption rather occurs, if and only if government interferes with market prices through taxes or subsidies. The reason why uniform taxation of assimilative services ($\tau_q=\tau_y>0$) does not cause allocative distortions is, of course, that assimilative services are in fixed supply ($\bar{a}$).

The allocative effects of differential taxation of assimilative services are characterized by

Proposition 2: The allocative impact of τ_q and τ_y exclusively depends on the difference $\tau_q-\tau_y$.

To see this, suppose that the quantity vector $h(\delta):=(a_q, a_y, \ell, q, y)$ and the price vector $p(\delta):=(p_a, p_q, p_y, w)$ constitute a competitive equilibrium for given tax rates $\delta:=(\tau_q, \tau_y, \theta, t)$. Consider now $\Delta\tau$ positive or negative (but sufficiently small in absolute value as to avoid corner solutions) and define

$$\delta':=\left(\tau_q'=\tau_q+\Delta\tau,\ \tau_y'=\tau_y+\Delta\tau,\ \theta'=\theta,\ t'=t\right) \quad \text{and}$$
$$p(\delta'):=\left(p_a'=p_a-\Delta\tau,\ p_q'=p_q,\ p_y'=p_y,\ w'=w\right).$$

It follows from the equations (1)-(6e) that $p(\delta')$ and $h(\delta')=h(\delta)$ constitute an equilibrium for the tax rates δ'. It is an immediate corollary of Proposition 2 that if an allocation $h(\delta)$ is supported as an equilibrium by prices $p(\delta)$ and tax rates δ

4 The property of the Samuelson condition of summing over the marginal willingness-to-pay of all individuals is missing here, because by assumption there is only a single (representative) consumer in the present model.

exhibiting $\tau_q \neq 0$, $\tau_y \neq 0$ and $\tau_q \neq \tau_y$ then there are prices $p(\delta'')$ and tax rates δ'' with either ($\tau''_q = 0$ and $\tau''_y \neq 0$) or ($\tau''_q \neq 0$ and $\tau''_y = 0$) which also support $h(\delta)$ in a competitive equilibrium. This can be easily checked by reconsidering δ' and $p(\delta')$ as defined above and setting, alternatively, $\Delta\tau = -\tau_q$ or $\Delta\tau = -\tau_y$.

Proposition 3: In competitive equilibrium with t ≥ 0 environmental quality is (locally)

$$\begin{Bmatrix} \textit{underprovided} \\ \textit{efficiently provided} \\ \textit{overprovided} \end{Bmatrix} \quad \textit{if and only if} \quad \theta\, Q_a + \tau_q - \tau_y \begin{Bmatrix} > \\ = \\ < \end{Bmatrix} 0.$$

Proposition 3 is straightforward from comparing (8) to

$$\frac{U_q}{U_y} - \frac{Y_a}{Q_a} = \frac{\theta Q_a + \tau_q - \tau_y}{p_y Q_a} \tag{8'}$$

which is implied by (6b), (6c) and (6e). In contrast, (8) is required for Pareto efficiency. The LHS of (8'), here referred to as the internalization gap, measures the extent of allocative distortion caused by an inappropriate level of environmental quality. It is important to observe that this internalization gap is uniquely determined by θ, τ_q and τ_y, but does not depend on $t \geq 0$. Note also that equal absolute changes of τ_y and τ_q ($d\tau_q = d\tau_y$) do not affect the allocation thus confirming Proposition 2.

The welfare impact of small changes in tax rates is readily spelled out in

Proposition 4: Suppose the tax rate $\delta = \theta, \tau_y, \tau_q, t$ is marginally changed. The resultant change in utility is given by

$$\frac{du}{\lambda d\delta} = t\frac{d\ell}{d\delta} + \tau_y\frac{da_y}{d\delta} + \tau_q\frac{da_q}{d\delta} + \theta\frac{dq}{d\delta} = t\frac{d\ell}{d\delta} + \left(\tau_y - \tau_q\right)\frac{da_y}{d\delta} + \theta\frac{dq}{d\delta} \tag{9}$$

where λ is the Lagrange multiplier (marginal utility of income) associated to the problem of maximizing (3) subject to (4).

To derive (9), consider the total differential of (3) and then substitute the first-order conditions $U_f = -\lambda(w-t)$, $U_y = \lambda p_y$ and $U_q = \lambda p_q + \lambda\theta$ implied by maximizing utility subject to (4). One obtains

$$du = -\lambda\left(w-t\right)d\ell + \lambda\, p_y\, dy + \lambda\, p_q\, dq + \lambda\theta\, dq\,, \tag{10}$$

and this equation is turned into (9) by totally differentiating (1) and (2) and by making use of (6a)-(6c) and $da_q = -da_y$.

Proposition 4 tells us that whenever one of the tax rates is positive in the initial situation, its increase ($d\delta > 0$) exacerbates the tax distortion whenever it erodes the base of any of the prevailing taxes. To determine the sign of the terms $d\ell / d\delta$, $da_y/d\delta$ and $dq/d\delta$ in (9) it would be necessary to carry out a full-scale comparative statics analysis which is not the focus of the present paper, however.

3 Tax Implementations of Environmental Quality Standards in an Economy with a Full Set of Competitive Markets

The observation that θ, τ_q and τ_y determine the internalization gap (Proposition 3) suggests to turn the argument around as follows: Suppose the government aims at securing a certain – somehow politically determined – level of environmental quality, called environmental quality standard, q_s, hereafter. Are there tax rates θ, τ_q and/or τ_y to support any given standard q_s which the government may wish to implement? The affirmative answer is formalized as follows:

Proposition 5: Let $q_s \in [0, Q(\bar{a})]$ and let $\alpha(q_s) := (a_{qo}, a_{yo}, \ell_o, q_s, y_o)$ be the allocation that is Pareto efficient relative to q_s.

(a) $\alpha(q_s)$ is implemented by any tax policy (τ_q, τ_y, θ, t, σ) satisfying (5), t=0 and (τ_q, τ_y, θ) $\in M(q_s)$, where

$$M(q_s) :=$$

$$\left\{ (\tau_q, \tau_y, \theta) \,\middle|\, \frac{U_q(\ell_o, q_s, y_o)}{U_y(\ell_o, q_s, y_o)} = \frac{Y_a(a_{yo}, \ell_o)}{Q_a(a_{qo})} + \frac{\theta Q_a(a_{qo}) + \tau_q - \tau_y}{p_{yo} Q_a(a_{qo})} \right\} \tag{11}$$

(b) Consider the tax policies characterized by t=0 and, respectively, $\tau_q = \tau_y = 0$, θ free (policy A) or $\tau_q = \theta = 0$, τ_y free (policy B). Suppose $q_s \in [0, Q(\bar{a})]$ is implemented by policy A with (θ_A, $\sigma_A = \theta_A q_s$). Then there is policy B with ($\theta_B = 0$, $\tau_{yB} = -\theta_A Q_{aA}$, $\sigma_B = -\tau_{yB} a_{yA}$) which is equivalent to policy A in the sense that it also implements q_s and exhibits the same equilibrium allocation.

To validate the efficiency claim of Proposition 5a we add the terms "$+\lambda_{so}(q - q_s) + \lambda_{su}(q_s - q)$" to the Lagrangean (7) and then determine the first-order conditions of an interior solution. The assignment of terms

$$\lambda_\ell = w,\ \lambda_y = p_y,\ \lambda_q = p_q,\ \lambda_a = p_a \text{ and } \theta = \lambda_{su} - \lambda_{so}$$

(implying $\tau_q = \tau_y = t = 0$) secures an efficient implementation of the quality standard q_s. Obviously, any $(\tau_q, \tau_y, \theta) \in M(q_s)$, $(\tau_q, \tau_y, \theta) \neq (0, 0, \lambda_{su}-\lambda_{so})$ also implements q_s efficiently.

To prove Proposition 5b we recall that, by definition, policy A is characterized by $w_A = p_{yA} Y_{\ell A}$, $p_{aA} = p_{qA} Q_{aA}$, $p_{aA} = p_{yA} Y_{aA}$ and $p_{yA} U_{qA} = (p_{qA} + \theta_A) U_{yA}$. We now define $p_{qB} := p_{qA} + \theta_A$ so that $p_{aA} = p_{qA} Q_{aA}$ can be rewritten as

$$p_{aA} = \left(p_{qB} - \theta_A\right) Q_{aA} \text{ or } p_{aA} + \theta_A Q_{aA} = p_{qB} Q_{aA} .$$

Define also $p_{aB} := p_{aA} + \theta_A Q_{aA}$ and $\tau_{yB} := -\theta_A Q_{aA}$ to obtain $p_{aA} = p_{aB} + \tau_{yB}$. The claim is that with $w_A = w_B$, $p_{yB} = p_{yA}$ and θ_B=0 the equilibrium allocation under policy A is also supported by w_B, p_{yB}, p_{aB}, p_{qB} when the producers' price of assimilative services is p_{aB} in the production of environmental quality and $p_{aB} + \tau_{yB}$ in sector Y. All tax parameters with subscript B clearly constitute a policy B as defined above. The equilibrium quantities a_{qA}, a_{yA}, ℓ_A and y_A are compatible with (6a)-(6e) if all prices and tax rates with subscript A are appropriately substituted by those with subscript B. It remains to be shown that the consumer's budget equation

$$w_A \ell_A + p_{aA} \bar{a} + \sigma_A = p_{yA} y_A + \left(p_{qA} + \theta_A\right) q_s$$

is also satisfied for ℓ_A and y_A when w_A, p_{aA}, p_{yA}, p_{qA}, θ_A is replaced by w_B, $p_{aB} + \tau_y$, p_{yB}, $p_{qB} - \theta_A$, θ_B = 0, respectively. This substitution turns the budget equation into

$$w_B \ell_A + \left(p_{aB} + \tau_y\right)\bar{a} + \theta_A q_s = p_{yB} y_A + p_{qB} q_s ,$$

which coincides with the budget equation under policy B (with $q=Q_a a_q$), indeed, since

$$\tau_y \bar{a} + \theta_A q_s = \tau_y \bar{a} - \frac{\tau_y}{Q_a} q_s = \tau_y \left(\bar{a} - a_q\right) = \tau_y a_y = \sigma_B .$$

Therefore, under policy B the consumer chooses $\ell_B = \ell_A$ and $y_B = y_A$. In an analogous and straightforward way one can show that an equilibrium allocation under policy B can be supported by policy A with appropriate changes in prices and tax rates.[5]

---

5 The proof of Proposition 5b relies on the assumption that Q is a linear function (see Footnote 3). It would be possible but tedious to construct a scheme for subsidizing emissions that extends Proposition 5 to cases where the function Q is strictly concave.

The message of Proposition 5a is that if, for whatever reason, the provision of environmental quality is to deviate from its Pareto efficient level this can be achieved in a cost minimizing way by choosing an appropriate tax mix $(\tau_q, \tau_y, \theta) \in M(q_s)$ – provided that the total tax revenue is transferred back to the consumer in a lump-sum way.[6]

While Proposition 5a is a general equivalence result, there are also interesting special cases, defined as policies A and B in Proposition 5b which will be the main focus of the subsequent analysis. The special equivalence result of Proposition 5b is striking: In order to implement in a cost minimizing way a predetermined environmental quality standard falling short of its Pareto efficient level, one can either tax consumer demand for environmental quality (policy A) or subsidize (!) the emission of pollutants (policy B). It is also important to observe that there is a special case of policy B, namely $\tau_y = -p_a$, which, in conventional perspective, characterizes the absence of any pollution control (laissez-faire). The state of 'policy inactivity' emerges here as the result of a particular form of policy B which, by the way, implements the corresponding (low) environmental quality efficiently!

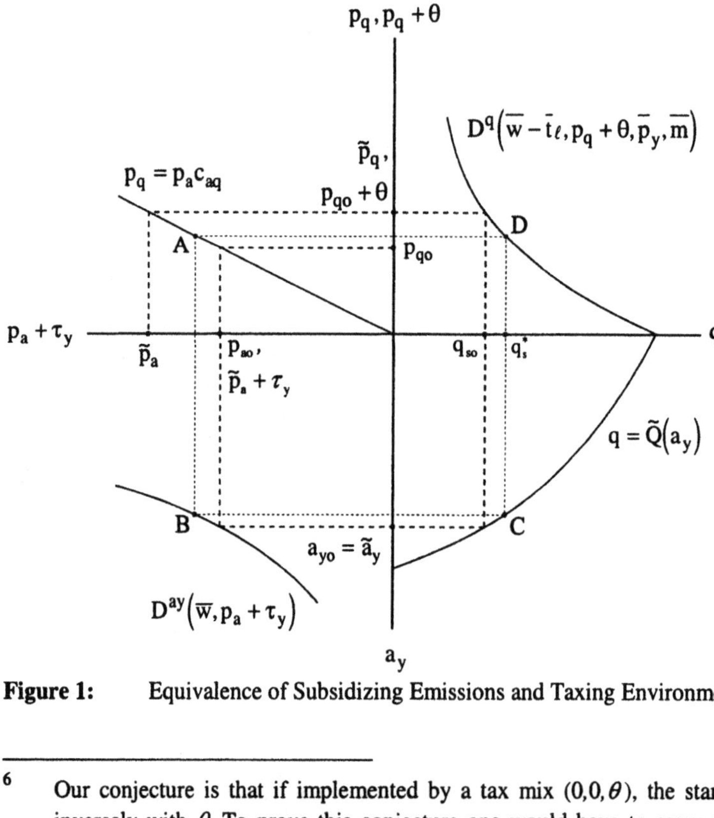

Figure 1: Equivalence of Subsidizing Emissions and Taxing Environmental Quality.

6 Our conjecture is that if implemented by a tax mix $(0,0,\theta)$, the standard q_s varies inversely with θ. To prove this conjecture one would have to carry out a full-scale comparative static analysis, however.

Figure 1 provides a partial equilibrium illustration of implementing the standard, q_{so} alternatively, by policy A (as indicated by q_{so}, a_{yo}, p_{ao}, p_{qo} or $p_{qo}+\theta$) or policy B (as indicated by $q_{so}, \tilde{a}_y, \tilde{p}_a + \tau_y$ or $\tilde{p}_a, \tilde{p}_q$). To be supported from the supply side, (lower parts of Figure 1) the standard q_{so} requires the demand $a_{yo} = \tilde{a}_y$ for assimilative services by industry Y which is generated, in turn, when the price for assimilative services is $p_{ao} = \tilde{p}_a + \tau_y$. On the other hand, the demand-side price of the environmental quality standard q_{so} is $\tilde{p}_q = p_{qo} + \theta$ (upper right part of Figure 1). Since profit maximizing production of environmental quality (with production function $q c_{aq} = a_q$) requires $p_q = p_a c_{aq}$, the tax rate θ drives a wedge between the consumer's price, $p_{qo} + \theta$, and the producer's price, p_{qo}, of environmental quality. Correspondingly, τ_y drives a wedge between the price for assimilative services faced by producers of good Y, $\tilde{p}_a + \tau_y$, and the price faced by the producer of environmental quality, $\tilde{p}_a$. Observe also that if in Figure 1 q_s^* happened to be the standard to be implemented, then one would have θ=0 in case of policy A and τ_y =0 in case of policy B implying that q_s^* is Pareto efficient.

The policies A and B characterized above are taken as the basis of our further investigations of a gallery of tax policies designed to implement predetermined levels of environmental quality. Our aim is to proceed from the full-scale market model to the conventional price-and-standard approach in an effort to improve the understanding of the latter by studying the former. Table 1 provides an overview of all tax policies to be investigated. Since $\tau_q \equiv 0$ in the remainder of this paper we write $\tau_y = \tau$ in what follows for notational convenience. The policies C_A and C_B still presuppose the working of all competitive markets, as do the policies A and B, while the policies D to G have in common that the environmental quality is costlessly provided to the consumers. Another distinguishing criterion is that policies A to D assume private ownership of assimilative services, while these services are publicly owned in case of the policies E to G.

Before concluding Subsection 3 with its scenario of a complete set of competitive markets we briefly consider the policies C_A and C_B in Table 1 that are extended versions of the policies A and B, respectively, in that they include labor taxation and place, in addition, a constraint on total tax proceeds (to be recycled to the consumer) as specified in Table 1.
Observe, that the tax rate t cannot be freely chosen under the policies C_A and C_B because, essentially, these policies seek to reach two targets with the tax instruments (t, θ) or (t, τ), respectively: the implementation of q_s (guided by θ or τ) and the collection of a specified amount of total tax revenue $p_q q_s$.[7] Following the argument in the proof of Proposition 5b, the policies C_A and C_B can be shown

---

7 Quite obviously, the policies C_A and C_B are difficult to handle for the regulator, because in order to collect the prescribed total tax revenue and to make the supply of environmental quality equal to q_s she must engage in a trial and error process of adjusting, respectively, (t, θ) and (t, τ).

to be equivalent, and if there are two equivalent policies C_A and C_B then $t_A < t_B$. To see this, recall from Table 1 that $t_A \gtrless t_B$ is equivalent to $p_{qA}\, q_s \gtrless p_{qB}\, q_s - \tau a_y$.

Making use of the information in the proof of Proposition 5b that $p_{qB} = p_{qA} + \theta_A$ and, $\tau = -\theta_A Q_a$ one clearly obtains

$$p_{qA}\, q_s \gtrless p_{qB}\, q_s - \tau a_y = p_{qA}\, q_s + \theta_A\, q_s + \theta_A\, Q_a\, a_y \text{ or } \theta_A\left(q_s + Q_a\, a_y\right) \lesseqgtr 0$$

which is satisfied for the 'greater' sign only since all constituent factors on the left hand side are positive.

The taxing rules defining the policies C_A and C_B do not seem to be very meaningful at first glance, since they cause allocative inefficiency ($t > 0$) without assigning a useful role to government. The reason why these policies are of some analytical interest will be clarified in the subsequent section.

Table 1: A Gallery of Environmental Tax Policies.

Type of Policy*)	Tax Rates Set Zero	Public Transfer to Consumers	Government Budget	Implications
A	t, τ	θq_s	$\sigma = \theta q_s$	Efficient implementation of q_s
B	t, θ	τa_y	$\sigma = \tau a_y$	Efficient implementation of q_s
C_A	τ	$\theta q_s + t\ell$	$(p_q + \theta) q_s = \sigma$ $= \theta q_s + t\ell$	$p_q q_s = t\ell$
C_B	θ	$\tau a_y + t\ell$	$p_q q_s = \sigma = \tau a_y + t\ell$	$p_q q_s - \tau a_y = t\ell$
D_A	τ	0	$(\pi + \theta) q_s = \theta q_s + t\ell$	$\pi q_s = t\ell$
D_B	θ	0	$\pi q_s = \tau a_y + t\ell$	$\pi q_s - \tau a_y = t\ell$
E_A	t, τ	$p_a a_y$	$(\pi + \theta) q_s + \sigma$ $= \theta q_s + p_a \bar{a}$	Efficient implementation of q_s
E_B	t, θ	$(p_a + \tau) a_y$	$\pi q_s + \sigma$ $= \tau a_y + p_a \bar{a}$	Efficient implementation of q_s
F	t, τ, θ	$p_e a_y$	$\sigma = p_e a_y$	Conventional efficient price-and-standard policy
G_A	τ	0 for $p_z z > p_a a_y$	$(\pi + \theta) q_s + p_z z$ $= \theta q_s + t\ell + p_a \bar{a}$	$p_z z = t\ell + p_a a_y$
G_B	θ	0 for $p_z z > (p_a + \tau) a$	$\pi q_s + p_z z =$ $\tau a_y + t\ell + p_a \bar{a}$	$p_z z = t\ell + (p_a + \tau) a_y$

*) $\bar{a}$ is private property in policies A to D and public property in policies E to G.

4 Costless Provision of Environmental Quality

4.1 Private Ownership of Assimilative Services

Suppose now that there is no market for environmental quality any more and the environmental quality is, instead, costlessly provided by the government at some level $q=q_s$. The consumer contends herself with the prevailing environmental quality at zero price. Since the producer of environmental quality cannot sell her output at positive prices she will be reluctant to demand positively priced assimilative services. In other words, the break-down of the market for environmental quality causes the market for assimilative services (as established in the previous model) to also collapse even though the assumption of private ownership of assimilative services is maintained. But rather than closing down both markets, we aim at designing an allocation procedure, as an intermediate step, that makes as much use as possible of allocative guidance by prices under the constraint, however, that environmental quality is provided to the consumer at no cost.

Indeed, the breakdown of the market for assimilative services can be avoided (in an admittedly somewhat artificial way) by adopting an institutional arrangement which we refer to as policy D (Table 1). More specifically, an environmental quality standard, q_s, is said to be implemented by policy[8] D,

(1) if the regulator provides a correct estimate[9], π, of the consumer's marginal willingness-to-pay for q_s,

(2) if she offers the producer of environmental quality to pay πq for the provision of quantity q; and

(3) if she then chooses her tax instruments τ and t so that

$$q = q_s, \quad \sigma = 0 \quad \text{and} \quad \tau a_y + t\ell = \pi q_s . \tag{10}$$

---

8 To promote our argument it suffices to restrict our attention to policy D_B. For notational simplicity we refer to the policies C_B and D_B as policies C and D, respectively, unless it is of interest to explicitly distinguish their types A and B.

9 The concept of π in the above description of policy D simply means that the regulator is asked to estimate the fictitious market price p_q. If we denote by $D^q(w\text{-}t, p_q, p_y, m)$ the consumer's demand for environmental quality in the full-fledged market model (with m for non-labour income and transfers), then the regulator's task is to elicit π as implicitly determined by the equation $q_s = D^q(w\text{-}t, \pi, p_y, m)$. The informational requirement of obtaining a good estimate π is very demanding, thus rendering this kind of pollution control fairly stylized. But on the other hand, if available valuation techniques should ever be applied systematically, π is the kind of information one would like to obtain and use.

The common feature of the policies C and D is that both make use of $\tau \neq 0$ and $t \neq 0$. Profit maximizing behavior in both sectors is the same under these policies except that under policy D the output price for the producer of environmental quality is not the market price but rather the regulator's correct estimate. The principal difference between both policies is, however, that lump-sum taxation or transfer is denied in policy D but not in policy C. The model of policy D implies that the consumer chooses ℓ and y as to maximize

$$U\left(1-\tilde{\ell},\, q_s,\, \tilde{y}\right) \text{ subject to}$$

$$(w-t)\tilde{\ell} + p_a \bar{a} = p_y \tilde{y}, \tag{11}$$

whereas in the model of policy C the consumer chooses ℓ, y and q to maximize

$$U\left(1-\tilde{\ell},\, \tilde{q},\, \tilde{y}\right) \text{ subject to}$$

$$(w-t)\tilde{\ell} + p_a \bar{a} + \sigma = p_y \tilde{y} + p_q \tilde{q} \quad with \quad \sigma = p_q q_s\,. \tag{11'}$$

Irrespective of these important differences, closer inspection of policies C and D reveals their equivalence:

Proposition 6: The policies C and D are equivalent in the sense that the quantities and prices of the pertaining general competitive equilibria are the same, if the same environmental quality standard is implemented.

To put Proposition 6 into more formal terms, suppose that the environmental quality standard q_s is implemented, alternatively, by policy C or policy D and characterize the respective results by the vector

$$h^* := \left(p_a^*,\, p_q^*,\, p_y^*,\, w^*;\, \tau^*,\, t^*;\, a_q^*,\, a_y^*,\, \ell^*,\, y^*\right)$$

for policy C and by the vector

$$h^\circ := \left(p_a^\circ,\, \pi^\circ,\, p_y^\circ,\, w^\circ;\, \tau^\circ,\, t^\circ;\, a_q^\circ,\, a_y^\circ,\, \ell^\circ,\, y^\circ\right)$$

for policy D. Then the claim of Proposition 6 is that $h^* = h^\circ$. To prove this, consider first the case that q_s is implemented by policy C so that h^* prevails and one has $\sigma^* = p_q\, q_s = \overset{\circ}{\tau} a_y^* + \overset{\circ}{t} \ell^*$, in addition. We want to show that policy D implements q_s at prices (p_a^*, $\overset{\circ}{\pi} = p_q^*$, p_y^*, w^*) and tax rates (τ^*, t^*, $\overset{\circ}{\sigma} = 0$). The pertaining profit maximizing productions are obviously a_q^* and (ℓ^*, a_y^*), respectively. Under policy C the budget constraint satisfies

$$\left(w^* - t^*\right)\ell^* + p_a^* \bar{a} + \sigma^* = p_y^* y^* + p_q^* q_s \tag{12}$$

with $\sigma^* = p_q^* q_s = \tau^* a_y^* + t^* \ell^*$.

Let $(\tilde{\ell},\ \tilde{y})$ maximize $U\left(1-\ell,\ q_s,\ y\right)$ subject to (11) for $\left(p_a^*,\ \pi^\circ,\ p_y^*, w^*,\ t^*\right)$.

Then $(\tilde{\ell},\ \tilde{y})$ clearly satisfies

$$\left(w^* - t^*\right)\tilde{\ell} + p_a^* \bar{a} = p_y^* \tilde{y}. \tag{13}$$

If we add the term '$+\overset{\circ}{\pi}\, q_s$' to both sides of (13), this modified equation coincides with equation (12). Hence $(\tilde{\ell},\ \tilde{y}) = (\ell^*,\ y^*)$ for $\overset{\circ}{\pi} = p_q^*$. The second part of the proof starts from $\overset{\circ}{h}$ (policy D) to demonstrate that $\overset{\circ}{h}$ with $\sigma = \overset{\circ}{\pi}\, q_s$ is also an implementation of q_s by policy C. This can be shown along similar lines as above and is therefore omitted here.

The important message of Proposition 6 is that implementations of quality standards by policy D can be looked at as if these standards were implemented by policy C. The analytical advantage being that all relevant scarcity indicating prices are available (to the analyst) and the allocative displacement effects originating either from the environmental externality or from distortionary labor taxes are still uniformly treated as tax-induced distortions.

The fact that the policy C as well as policy D employs two different taxes provides the regulator with the opportunity of changing the tax mix subject to the respective budget and transfer constraints. More specifically, suppose, an inefficiently low environmental quality standard q_s is implemented by policy C_B and assume the regulator aims at improving environmental quality. To this end she may want to consider τ as a discretionary policy instrument and raise the rate τ by $d\tau>0$ leaving the change of environmental quality to the market forces which are supported by appropriate adjustments of t for the purpose of keeping the government budget balanced. If $\tau < 0$ in the initial situation, the conjecture of 'normal reaction' is that the autonomous change $d\tau > 0$ (reduction of the emission subsidy) leads to a new equilibrium differing from the initial one by $dq>0$ and $dt \neq 0$. Similarly, if an environmental quality standard q_s had been implemented by policy C_A, the corresponding revenue neutral tax policy would be to set $d\theta < 0$ with the expectation that $dq>0$.

These tax policies bear some similarity to ecological tax reforms as discussed in the literature because both approaches deal with tax induced pollution control when non-distortionary taxes are not available. But the differences are significant, since in the present approach assimilative services are privately owned and the tax rates τ and θ, respectively, are policy instruments. Observe, moreover, that the tax policies described above are no tax reforms in the narrow sense because they violate revenue neutrality, in general. It is clearly viable, to carry out an exogenous tax shift ($d\tau > 0$ or $d\theta < 0$) obeying the government balanced-budget constraint

$$\frac{d(\pi q)}{d\theta} = \frac{d(t\ell)}{d\theta} \text{ in case of policy } D_A \text{ or}$$

$$\frac{d(\pi q)}{d\tau} = \frac{d(\tau a_y)}{d\tau} + \frac{d(t\ell)}{d\tau} \text{ in case of policy } D_B.$$

But revenue neutrality (being a constitutive property of a conventional tax reform) would require, in addition, $d\sigma = 0$ implying

$$\frac{d[(\pi_A + \theta)q]}{d\theta} = \frac{d(\pi_B q)}{d\tau} = 0.$$

Clearly, it is not feasible, in general, to simultaneously satisfy both the balanced government budget and revenue neutrality.

To gain additional insights into the nature of ecological tax reform we now investigate the impact of the above policies without the revenue neutrality constraint but with the requirement of balancing the government budget. In addition, motivated by the observation that the target of public policy is not a particular level of the subsidy rate τ or tax rate θ but rather the level of environmental quality provided, we consider, in what follows, an initial quality standard q_s implemented by policy D_A or D_B and use this policy again for implementing the new standard $q_s + dq > q_s$. If q is increased by $dq > 0$ balancing the government budget requires, in view of Table 1,

$$\frac{d\alpha}{dq} = \frac{d(t\ell)}{dq} \tag{14}$$

where $\alpha = \pi q_s$ in case of policy D_A and $\alpha = \pi q_s = \tau a_y$ in case of policy D_B. Observe that moving from q_s to $q_s + dq$ induces endogenous changes $d\tau$ and $d\theta$, respectively, as well as dt.[10] In view of (9) the welfare implications of such a policy are

$$\frac{du}{\lambda dq} = \theta + \tau \frac{da_y}{dq} + t \frac{d\ell}{dq} \tag{15}$$

---

10 Such a policy of standard tightening by applying either policy D_A or D_B has a very broad interpretation beyond environmental policy: Essentially, we investigate the impact of a public policy of increasing the public (and costless) provision of any arbitrary public good when the public good has to be paid for by the revenues from distortionary taxes.

where either $t>0$, $\theta \neq 0$ and $\tau = 0$ (policy D_A) or $t > 0$, $\tau \neq 0$ and $\theta = 0$ (policy D_B). To fix our ideas we assume that in the initial situation an inefficiently low environmental quality standard q_s is implemented. Then the ecological dividend of tightening the standard ($dq > 0$) is $\theta > 0$ in case of policy D_A and $\tau(da_y/dq) > 0$ in case of policy D_B. The decisive question is, however, which sign the term $t(d\ell/dq)$ has in (15). This partial welfare effect is clearly associated to the labor tax distortion (since it would vanish in case of lump-sum taxation ($t = 0$)) and is therefore interpreted as the tax efficiency impact of enhancing environmental quality. If $t(d\ell/dq) > 0$, the policy of environmental quality improvement is said to exhibit a double dividend: the ecological dividend (which is secured by definition of the policy performed) and the labor tax efficiency dividend.

It is worth mentioning that the concept of dividends introduced above here differs from that employed in Bovenberg and de Mooij (1994). To see this observe that in case of policy D_A (8') and (15) yield the welfare change

$$\frac{du}{\lambda dq} = \underbrace{\theta}_{(a)} + \underbrace{t\frac{d\ell}{dq}}_{(b)} = \underbrace{\frac{p_y U_q}{U_y}}_{(c)} + \underbrace{\left[t\frac{d\ell}{dq} - \frac{t_e}{Q_a}\right]}_{(d)}, \tag{15'}$$

where $t_e := p_y Y_a$ denotes the 'emission tax rate' in the conventional perspective (see also Proposition 9 below) which is set equal to the marginal abatement cost by profit maximizing firms. Bovenberg and de Mooij (1994) look at the tax rates (t, t_e) as representing the tax system. Consequently they consider (d) in (15') as the efficiency-of-taxation dividend whereas (c) is the ecological dividend. In contrast, our market perspective suggests to look at the tax system as consisting of tax rates (θ, t) so that (a) and (b) in (15') gives us the overall efficiency-of-taxation effect with respect to (θ, t). Note that θ is clearly the ecological dividend which would still exist if the labor tax would be substituted by lump-sum taxation. Correspondingly, the term (b) in (15') is the change in efficiency of the labor tax, the labor tax efficiency dividend. Observe finally, that while θ becomes negative in case of 'over-internalization', the ecological dividend (c) is always positive.

As a next step of the analysis it would be natural to provide the information about the sign of $d\ell/dq$ from a full-scale comparative static analysis. However, a reasonably general comparative static analysis which avoids the preoccupation with special cases is beyond the scope of the present paper. For more details see Weinbrenner (1999) and also Footnote 1, above. We therefore contend ourselves with offering a classification of feasible results. Recall that (14) holds for both policy D_A or D_B. Rewriting (14) yields

$$\frac{d\alpha}{dq} = \frac{d(t\ell)}{dq} = \frac{d(t\ell)}{dt} \cdot \frac{dt}{dq}. \tag{14'}$$

If $\frac{dt}{dq}$ from (14') is substituted in $\frac{d\ell}{dq} = \frac{d\ell}{dt} \cdot \frac{dt}{dq}$ we obtain $\frac{d\ell}{dq} = \frac{\frac{d\ell}{dt} \cdot \frac{d\alpha}{dq}}{\frac{d(t\ell)}{dt}}$.

Obviously, the sign of $d\ell/dq$ which determines the sign of the tax efficiency dividend depends on the sign of the terms on the RHS of the above equation. The following hypotheses seem to be supported on plausibility grounds:

(H1) If the rate δ of a tax with base b is increased then its base erodes: $db/d\delta < 0$.

(H2) The revenue δb of a tax with rate δ rises when the tax rate is increased: $d(\delta b) / d\delta > 0$ (i.e. the 'Laffer curve' is positively sloped in the relevant domain).

(H3) Providing a greater amount of environmental quality requires an increase in $d\alpha / dq > 0$ with $\alpha = \pi q_s$ in case of policy D_A and $\alpha = \pi q_s - \tau a_y$ in case of policy D_B.

In our view, the hypothesis H3 is not artificial since even if general equilibrium effects are accounted for, common sense tells us that buying more of a good requires spending more money. To see the rational of spending more money in case of policy D_B suppose environmental quality is underprovided in the initial situation ($\tau < 0$). Then

$$\frac{d(\tau a_y)}{dq} = \frac{d(\tau a_y)}{d\tau} \cdot \frac{d\tau}{dq} > 0$$

if H2 holds (implying $d(\tau a_y) / d\tau > 0$ and if firms are induced to pollute less by reducing the rate of emission subsidy in absolute terms ($d\tau / dq > 0$)).

Proposition 7: Suppose an inefficiently low environmental quality standard q_s is implemented either by policy D_A or D_B and the environmental quality standard q_s is to be strengthened ($dq > 0$).

- There are various constellations, as listed in Table 2, in which the efficiency of taxation is improved or worsened;
- No double dividend accrues, if H1, H2 and H3 hold;
- A double dividend accrues, if H1 and H2 hold, but H3 does not hold.

Table 2: Feasible Impacts of Improving Environmental Quality*.

Case	1	2	3	4	5	6	7	8
$\frac{d\ell}{dt}$	-	-	-	-	+	+	+	+
$\frac{d(t\ell)}{dt}$	+	-	+	-	+	-	+	-
$\frac{d\alpha}{dq}$	+	+	-	-	+	+	-	-
$\frac{d\ell}{dq}$	<0 no TD*)	>0 TD	>0 TD	<0 no TD	>0 TD	<0 no TD	<0 no TD	>0 TD

*) TD = Tax efficiency dividend.

To prove Proposition 7b and 7c set $\alpha=\pi q_s$ in case of policy D_A and $\alpha=\pi q_s - \tau a_y$ in case of policy D_B and rewrite (14) as

$$\frac{d\alpha}{dq} = \frac{d(t\ell)}{dq} \equiv \frac{d(t\ell)}{dt} \cdot \frac{dt}{dq} \equiv \left(\ell + t\frac{d\ell}{dt}\right)\frac{dt}{dq} \qquad (14")$$

Combining H1, H2 and H3 with (14") clearly takes us to column 1 of Table 2 while (14") combined with H1, H2 and $d\alpha / dq < 0$ corresponds to column 3 of Table 2.

In the analysis up to now, income from assimilative services, $p_a \bar{a}$, remained untaxed. Let us briefly consider the case that, in addition to τ and t, the regulator has at her disposal a resource income tax with rate $t_a > 0$ and chooses her tax instruments so that

$$\sigma = 0 \quad \text{and} \quad t_a \bar{a} + \tau a_y + t\ell = \pi_q q_s. \qquad (10')$$

Accordingly, the consumer's budget constraint (11) is modified to read

$$(w-t)\ell + (p_a - t_a)\bar{a} = p_y y. \qquad (11")$$

In this scenario different combinations of (τ, t_a, t) can be chosen to implement one and the same standard q_s. But since the resource is in fix supply, $\bar{a}$, any increase of t_a causes income effects, in general, but no distortionary substitution effects. It is therefore very convenient for the regulator to use the revenue $t_a \bar{a}$ of the resource income tax to pay for the provision of environmental quality in (partial) substitution for the distortionary labor tax. The straightforward consequence of this strategy is that if the regulator chooses t=0 and t_a = $(p_a a_q - \tau a_y)/\bar{a}$ the resultant allocation is efficient.

4.2 Public Ownership of Assimilative Services

Clearly, owing to fixed resource endowment resource income taxation amounts to (partially) expropriating the private resource owners. In this perspective, $t_a \in]0, p_a[$ is a property rights regime of shared ownership, where the private owner's share of resource income is $(p_a t_a) / p_a$ and the government's share is t_a / p_a. The polar case is the confiscatory tax rate $t_a = p_a$ which essentially means that the property rights for assimilative services are shifted completely from the consumer to the government. In the remainder of this paper we assume that the government owns the resource endowment, $\bar{a}$. Thus we now consider the institutional setting of policy E (Table 1) in which the regulator chooses her tax instruments so that

$$p_a \bar{a} + \tau a_y + t\ell = \pi q_s + \sigma \quad \text{with} \quad \sigma = \min\left[0, (p_a + \tau) a_y\right]. \qquad (10'')$$

The consumer's budget constraint (11") is modified to read

$$\sigma + (w - t)\ell = p_y y. \qquad (11''')$$

Observe that in contrast to policy D policy E allows for positive transfers to the consumer, in order to avoid the fancy implication of negative tax rates on labor income. The allocative consequences of this policy are straightforward:

Proposition 8: If the environmental quality standard q_s is implemented by policy E_B and $p_a + \tau \geq 0$, then t =0 and the resultant allocation is efficient (as under policy B).

To see this, notice that the government budget constraint (10") can be rewritten as $\sigma = (p_a + \tau)\, a_y + t\ell$. The inequality $p_a + \tau \geq 0$ implies t=0 and this policy implements q_s in an efficient way. The condition $p_a + \tau \geq 0$ in Proposition 8 is by no means restrictive, because $p_a + \tau = 0$ is the no-policy or laissez-faire state (see above). Clearly, $\sigma = (p_a + \tau)\, a_y \geq 0$ is the government's residual resource income after having paid for the standard q_s and the emission subsidy.

Recall that in all policies A to E discussed above we used the concept of politically determined and 'price implemented' environmental quality standard which plays an important role in environmental economics since Kneese and Bower (1968), Baumol and Oates (1971) and Dales (1972). But in contrast to those contributions we introduced a fictitious market for assimilative services in which these services are also demanded for the protection of the environment. To see how the approach of the present paper relates to conventional price and standard approaches consider policy F (Table 1) that envisages a regulator issuing the amount $a_{ys} = Q^{-1}(q_s)$ of tradable emission permits, selling them to industry Y at the market clearing price, p_e, and setting $\sigma = p_e a_{ys}$. Assimilative services used for the generation of environmental quality are not marketed anymore in the setting of policy F.

Obviously, this policy works through establishing a (competitive) market for emission permits as discussed in the literature since Dales (1972). Such a market is, in fact, a partial substitute for the market of assimilative services as employed above. The producer of environmental quality is indifferent between any positive or zero production because she has neither costs (p_a=0) nor positive revenue (p_q=π=0). An alternative institutional arrangement equivalent to policy F at the level of abstraction applied here is, of course, the charge and standard scheme as suggested by Baumol and Oates (1971). In this scheme the regulator does not issue permits but she levies, instead, a charge on each unit of pollutants emitted. In a trial and error procedure this charge has to be adjusted until the standard q_s is met. The associated emission charge, t_e clearly equals the market clearing price, p_e, on the market of tradable emission permits.

Proposition 9:

(a) The implementation of any environmental quality standard q_s by policy F is efficient.

(b) Let the standard q_s be efficiently implemented, alternatively, either by policy E or policy F and denote the associated prices by (p_{aE}, π_E, p_{yE}, w_E, τ_E) in case of policy E and by (p_{aF}=0, π_F=0, p_{yF}, w_F, τ_F=0, p_e) in case of policy F. Then the equilibrium allocations are identical with (p_{yF}, w_F) = (p_{yE}, w_E) and p_e = $p_{aE} + \tau_E$.

Proposition 9 is straightforward. It provides the last 'missing' link between conventional price and standard environmental policy schemes and the world of fictitious markets discussed in the previous sections of this paper. Observe that with policy F the environmental quality standard q_s is not only provided to the consumer at no cost but is also (seemingly) costless for the government. However, policy E with the pertaining public budget (10") reveals the costs of providing environmental quality. Looking at policy F from the 'market perspective' of policy E makes also precise that in Pigouvian language $p_a+\tau$ is interpreted as an 'emission charge' and σ=(p_a+τ) a_y as the proceeds from this emission charge. This terminology amounts to misinterpreting a factor price p_q net of subsidy τ as a tax rate!

After having introduced public property of assimilative services, we saw that the public resource income could be used to completely substitute the proceeds from the distortionary labor tax to pay for the provision of environmental quality. As a consequence, the problem of inefficient taxation being at the heart of the debate on ecological tax reform simply disappeared. To reintroduce it while maintaining the (realistic) assumption of public ownership, a second public good must be considered in the model (in addition to environmental quality) that is also costlessly provided to the consumers in fixed supply and whose cost exceeds the excess resource income $(p_a+\tau)\ a_y$. In Table 1 we denoted by policy G the implementation of an environmental quality standard when the amount z_s of an additional public good Z is publicly provided. If policy G applies and the environmental quality standard is to be exogenously tightened then balancing the public budget requires to satisfy the equation (14) with $\alpha = \pi q_s + p_z z_s - p_a \bar{a}$ in case of policy G_A and with $\alpha = \pi q_s - \tau a_y + p_z z_s - p_a \bar{a}$ in case of policy G_B. With these specifications of α in H3, equation (14') and Table 2 can be immediately applied to show that Proposition 7 carries over for the policies G_A and G_B. Observe that the only difference is the contents of hypothesis H3 through the different specifications of α. More specifically the difference between H3 under policies D and G is the term

$$\frac{d\left(p_z z_s - p_a \bar{a}\right)}{dq}. \tag{16}$$

Since z_s and $\bar{a}$ are constant, (16) differs from zero only as a consequence of general equilibrium price effects ($dp_z \neq 0,\ dp_a \neq 0$). In fact, if environmental quality is taken as numéraire ($\pi = 1$) and the technology of producing environmental quality is linear (p_a = *const.*) then (16) reduces to the term $z_s \cdot dp_z / dq$. If p_z remains unchanged (because good z is the numéraire or is linked to the numéraire by a linear technology) the basic difference is the impact of changing resource price p_a. A plausible sign of (16) is difficult to establish. In other words, except for some ambiguous general equilibrium price effects, the welfare impact of improving environmental quality in a model with two public goods and public ownership of assimilative services is very similar to that in case of private ownership of assimilative services and environmental quality as the only public good (Section 4.1).

Proposition 10: Suppose, the public goods Z and Q are publicly provided in fixed supply via policy G_A or G_B and the level of environmental quality is exogenously raised (dq>0).

(a) Propositions 7a and 7b hold (with an appropriate modification of α in H3).

(b) Necessary conditions for a double dividend are the existence of private goods in addition to good Y if all production processes are linear or the existence of waste abatement technologies (in the absence of additional private goods).

Proposition 10a summarizes the arguments given above. To prove Proposition 10b consider an economy producing environmental quality Q, the public good Z, the private good Y and additional private goods $X_1 ... X_n$ ($n \geq 0$). Denote by c_{vw} the input of factor V per unit of good W produced. Then we have $c_{ay}\, dy = da_y$, $c_{aq}\, dq = da_q$, $c_{\ell z}\, dz = d\ell_z$, $c_{\ell y}\, dy = d\ell_y$ and $c_{\ell xi}\, dx_i = d\ell_{xi}$ for $i = 1, ..., n$. Observe that $da_y = -da_q$ and $d\ell = d\ell_y + \Sigma_i\, d\ell_{xi}$ (owing to $dz = d\ell_z = 0$). This yields

$$\frac{d\ell}{dq}-\sum_{i=1}^{n}\frac{d\ell_{xi}}{dq}=c_{\ell y}\frac{dy}{dq}=-\frac{c_{\ell y}\,c_{aq}}{c_{ay}}<0\,. \tag{17}$$

Hence in this model the policy of improving environmental quality does not yield a double dividend if $\sum_{i=1}^{n}\frac{d\ell_{xi}}{dq}\leq 0$.

The last inequality trivially holds if there are no goods $X_1, ..., X_n$ $(n=0)$ which proves the first part of Proposition 10b. If $n > 0$, however, a double dividend ($d\ell/dq > 0$) may emerge if the production of some of the goods X_1 to X_n expands significantly which is conceivable in case of special relationships of substitutability or complementary between the goods Q, Y and $X_1,..., X_n$ (see e. g. Parry 1995). Another possibility of obtaining $d\ell/dq > 0$ as a result of strengthening the environmental quality standard is to assume substitutional technology for good Y as allowed for in the general formulation of the production function (1) implying intra-industrial waste abatement.

Considering (1) and setting $n = 1$ turns (17) into

$$\frac{d\ell}{dq}-Y_{\ell}\frac{d\ell_{y}}{dq}=-c_{aq}\,Y_{a}<0\,. \tag{17'}$$

This equation shows immediately that $d\ell_y/dq > 0$ is necessary for $d\ell/dq > 0$.

If substitution between labor input and the use of assimilative services is feasible, industrial emissions a_y are likely to be reduced with less than proportional reductions in output y and labor input ℓ_y. Using labor for waste abatement in addition to labor input in the production process proper may, in fact, increase total labor demand in industry Y even in the likely case that its output shrinks owing to an increase in its relative price.

In concluding our tour d'horizon we return to the conventional analysis of ecological tax reform by demonstrating that the public goods perspective adopted here fits well into – and smoothly extends – the view of the prevailing literature.

Proposition 11: Suppose the technology of producing environmental quality is linear ($Q_{aa} = 0$) and the expenditure for the public good Z is constant [$d(p_z z_s)/dq = 0$]. Under the regime of policy G_B (where $\alpha=\pi q_s+p_z z_s-\tau a_y+p_a\,\bar{a}$) a policy change $dq > 0$ implies that the change in labor tax revenue is equal to but opposite in sign to the change in 'emission tax revenue', conventionally defined. In formal terms:

$$\frac{d\alpha}{dq}=-\frac{d\left[(p_a+\tau)a_y\right]}{dq}=\frac{d(t\ell)}{dq}\,. \tag{14'''}$$

This result is straightforward from observing that the government budget equation $p_z z_s + \pi q_s = \tau a_y + p_a \bar{a} + t\ell$ is equal to

$$p_z z_s = (p_a + \tau) a_y + t\ell \tag{18}$$

under the assumption that the technology of producing environmental quality is linear. When combined with the assumption $d(p_z z_s) / dq = 0$ this observation yields (14'''). Interpreting $p_a + \tau$ as an 'emission tax rate' and $(p_a + \tau) a_y$ as the associated emission tax revenue, equation (18) shows, indeed, that the problem of increasing the supply of one out of two public goods (here: good *Q*) paid for by public resource income and labor tax revenue has been transformed into the problem of costlessly supplying a fixed amount of a single public good (good *Z*) which is financed by two 'taxes': an 'emission tax' with rate $p_a + \tau$ and a labor tax.

The very least equation reflects, indeed, the conventional revenue-neutral ecological tax reform as investigated, e. g., by Bovenberg and de Mooij (1994). Their calculations yield $dt < 0$, $d\ell < 0$ as well as $d[(p_a + \tau) a_y] > 0$ and $d(t\ell) < 0$. Observe that these results contradict both H1 and H2 while H3 is satisfied. For $\alpha = p_z z_s - (p_a + \tau) a_y$ (and $d(p_z z_s) / dq = 0$) the conclusions of Bovenberg and de Mooij correspond to column 6 in Table 2.

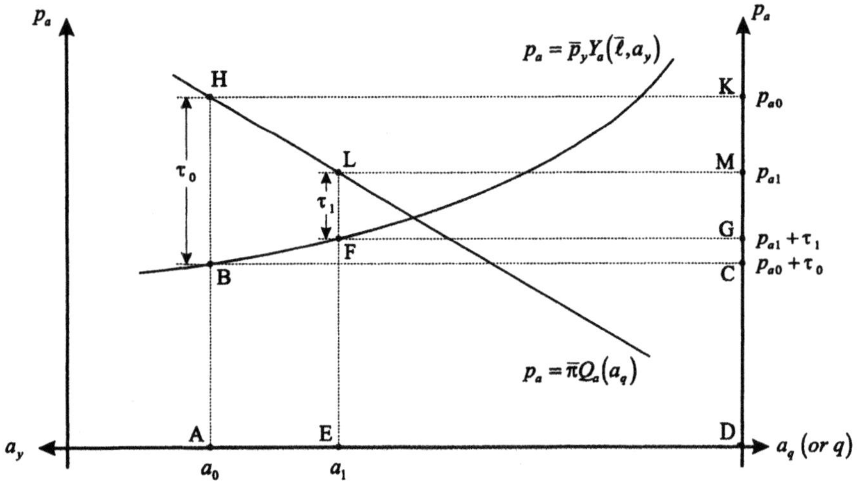

Figure 2: The Market for Assimilative Services (Policy G_B, Partial Equilibrium).

According to the conventional view, $d[(p_a + \tau) a_y] / dq < 0$ (or $d\alpha < 0$) means that the 'emission tax revenue' increases with $dq > 0$. Figure 2 presents a partial equilibrium example for $d[(p_a + \tau) a_y] / dq < 0$ presupposing the possibility of waste abatement in the course of producing good *Y*. More specifically, suppose $q_0 = \tilde{Q}(a_0)$ is changed to $q_1 = \tilde{Q}(a_1) > q_0$. Then the 'emission tax revenue' decreases from $(p_{a0} + \tau_0) a_0$ (or ABCD) to $(p_{a1} + \tau_1) a_1$ (or EFGD). At the same time the amount of the emission subsidy is reduced from $|\tau_0 a_0|$ (or BHKC) to $|\tau_1 a_1|$ (or FLMG). This result requires low abatement costs or a highly elastic demand of industry Y for assimilative services. It cannot be excluded, of course, that general equilibrium repercussions in p_y, π and ℓ override this partial equilibrium effect.

5 Concluding Remarks

This paper focused on the public provision of the public good 'environmental quality' using a benchmark model with a complete set of competitive markets and investigated various ways to finance that public good including subsidies, (distortionary) taxes and resource income from selling assimilative services. Raising the level of environmental quality in a full-scale market economy was related to the concept of ecological tax reform as discussed in the literature. In the first part of the paper, assimilative services had been assumed to be privately owned, but since public ownership is the empirically relevant case, a second public good (in addition to environmental quality) has finally been introduced that is also costlessly provided and whose cost exceeds the (excess) resource income – following the theoretical literature on ecological tax reform.

In this scenario, government essentially provides fixed amounts of two different public goods and has two different sources to pay for them: resource income from assimilative services (in fixed supply) and labor tax revenue. To increase the supply of one of the public goods requires, under plausible conditions, to increase the labor tax proceeds and thus the pertaining excess burden, unless general equilibrium price effects override this 'normal' reaction. The principal analytical difficulty is that one deals with multiple inefficiencies associated with labor taxation and with the non-optimal predetermined quantities of two public goods. A full characterization of conditions under which labor supply will increase when the environmental quality standard is strengthened – which is the necessary and sufficient condition for an ecological tax reform to yield a second positive dividend – will hardly be possible except for very restrictive assumptions on preferences and technologies because the comparative statics become very messy even in simple models. But our analysis suggests that the double dividend conjecture (Binswanger et al., 1983; Pearce, 1991) which was considered an 'informed common sense' argument is not sound on the basis of another economic common sense arguments: Our public good perspective of ecological tax reform suggests that buying more environmental quality typically requires to raise additional distortionary tax revenue which, in turn, tends to increase the inefficiency of taxation and hence prevents a double dividend.

It should be emphasized that the analysis of the present paper does not challenge the validity of the results obtained in the conventional ecological tax reform studies. But we conclude that applying the concept of tax reform from public finance to deal with 'Pigouvian taxes' on environmental externalities does not handle the issue in a satisfactory way. We demonstrated that important insights into ecological tax reform are gained if it is studied in models that allow for more specific structure with regard to assimilative services as productive factors along with labor or other 'conventional' inputs and intra-industrial waste abatement. The concept of tax representation of public goods that are publicly and costlessly provided (at inefficient levels) turns out to be useful. This concept allows for looking at allocative distortions from externalities or public goods as being tax induced distortions and thus offers a uniform approach to tax reforms in second best worlds.

A final message of this paper is that our market approach to ecological tax reform yielded new insights in scarcity prices, tax rates and distortionary tax wedges. In our view the previous discussion on ecological tax reform suffered from mixing up resource prices, tax revenues, resource incomes and even emission subsidies and emission taxes. One may continue to use conventional public finance notation, but it should be acknowledged that tax reforms in a world with environmental externalities have special features as compared to tax reforms in a world where zero taxation is Pareto efficient.

References

Atkinson, A.B. and N.H. Stern (1974), Pigou, Taxation and Public Goods, *Review of Economic Studies* 41, 119-128.

Baumol, W.J. and W.E. Oates (1971), The Use of Standards and Prices for the Protection of the Environment, *Swedish Journal of Economics* 73, 42-54.

Binswanger et al. (1983), *Arbeit ohne Umweltzerstörung. Strategien für eine neue Wirtschaftspolitik*, Frankfurt/M.

Bovenberg, A.L. and R.A. de Mooij (1994), Environmental Levies and Distortionary Taxation, *American Economic Review* 94, 1085-1089.

Dales, J.H. (1968), *Pollution, Property and Prices*, Toronto.

Fiedler, K. (1994), Naturwissenschaftliche Grundlagen natürlicher Selbstreinigungsprozesse in Wasserressourcen, *Zeitschrift für Umweltpolitik und Umweltrecht* 17, 323-355.

Goulder, L.H. (1995), Environmental Taxation and the 'Double Dividend': A Reader's Guide, *International Tax and Public Finance* 2, 157-184.

Heller, W.P. and D.A. Starrett (1976), On the Nature of Externalities, in: Lin, St.A.Y. (Ed.), *Theory and Measurement of Economic Externalities*, New York, 9-22.

Majocchi, A. (1996), Green Fiscal Reform and Employment: A Survey, *Environmental and Resource Economics* 8, 375-397.

Parry, I.W.H. (1995), Pollution Taxes and Revenue Recycling, *Journal of Environmental Economics and Management* 29, 64-77.

Pearce, D.W. (1991), The Role of Carbon Taxes in Adjusting to Global Warming, *Economic Journal* 101, 938-948.

Ruocco, A. and W. Wiegard (1997), Green Tax Reforms. Understanding the Double Dividend Hypothesis, *Zeitschrift für Umweltpolitik und Umweltrecht*, 171-198.

Sandmo, A. (1975), Optimal Taxation in the Presence of Externalities, *Swedish Journal of Economics* 77, 86-98.

Schöb, R. (1995), *Ökologische Steuersysteme. Umweltökonomie und optimale Besteuerung*, Frankfurt a. M.

Weinbrenner, D. (1999), *Ökologische Steuerreform. Wirkungszusammenhänge zwischen Emissions- und Fiskalsteuern*, Wiesbaden.

Optimal Intertemporal Pricing of Resource Stocks: The Case of Fossil Fuel Extraction and Atmospheric CO_2 Deposits

Klaus Conrad

Department of Economics, Mannheim University, L7, 3-5, 68131 Mannheim, Germany, *kconrad@rumms.uni-mannheim.de*

Summary:
The purpose of this paper is to extend the dynamic resource allocation problem by including stock externalities like accumulated CO_2 and SO_2 emissions as well as flow externalities like pollutants which can be abated (SO_2). The objective is to examine how the evolution of energy-, CO_2- or SO_2-tax rates can address these problems in an optimal way. The concern about the time profile of an energy tax arises from the fact that fossil fuels are an exhaustible resource and that global warming, being a consequence of carbon accumulation in the atmosphere, is a stock externality problem. We use a micro model of a firm, which maximizes profits, uses energy as one of its inputs and is confronted with a varying energy tax. It reacts by substitution, by changing its output level or by purchasing abatement equipment. The government is well aware about firms' reaction on price signals. It maximizes a stream of social welfare by choosing an optimal path of its instrument – an energy tax. Our analyses supports the idea of a tax first rising and later falling over time.

Keywords: Intertemporal carbon tax, fossil fuel taxation, stock accumulation (CO_2, SO_2), fossil fuel extraction.

1 Introduction

There is a substantial literature on resource use and economic activity with accumulative pollution.[1] In the prototype of pollution control models economic benefits depend positively on emissions (via production and consumption) and negatively on the pollution stock level. Of interest are time paths that maximize the present value of welfare taking into account the effects of accumulative

1 The seminal article on the dynamics of pollution control is by Keeler et al. (1972).

pollution on production possibilities and utility from consumption.[2] Some papers look at pollution and optimal capital accumulation and others at pollution and optimal investment in cleaning-up activities. Since pollution is essentially a problem of missing markets, the emission fee corresponds to the value of the co-state variable, i.e. to the social price of an additional unit of pollution. Most studies address only the stock externality (CO_2) and not the exhaustibility of a resource (fossil fuel). In those studies economic activity and pollution converge to a steady state in which the (constant) flow of additions to the pollution stock is offset by the environment's assimilative capacity.[3] In all studies the shadow price of the resource is interpreted as a pollution tax. The government controls quantities but not the path of a tax to achieve optimal allocation. The price mechanism is not at work to influence producer and consumer behavior.

Our approach differs from the standard approach which can be found in the literature. First, it is not the kind of macro-approach, where the utility of energy enters into a government's objective function and where the country incurs cost of exploration. We consider a profit maximizing firm which uses energy as one of its inputs to produce sectoral output. The firm is therefore not a mining company nor does the government own the mine. The firm has to buy energy from a resource country and has to pay taxes on it to its domestic government. The government maximizes a welfare function where consumer and producer surplus, environmental damage and tax revenues are part of the welfare function. Second, its control variable is not a quantity like consumption or energy consumption but an energy tax. The components of the energy tax and its path depend on the constraints the economy faces. One constraint is that energy use contributes to exhausting the CO_2-fill capacity of the atmosphere. Although this is a global problem, the government does not take a free-rider position but favors unilateral actions. If absorption of CO_2 is of a significant order, the problem can be addressed in the context of a renewable resource. Then it is possible to study the optimal path of the tax towards a steady state, to explore the property of this steady state and to do comparative statics.

Third, we add the constraint to our model that fossil energy is also a non-renewable resource. The linkage of energy use to two stock variables is not new,

---

2 See Plourde (1972), Forster (1977) or Conrad and Olson (1992).

3 There are many extensions of the prototype model. In Tahvonen (1995), damage depends also on the rate of the stock accumulation. Tahvonen (1997) extends the stock pollution problem by including a non-polluting backstop technology. The topic in Toman and Palmer (1997) is how an accumulative pollutant should be phased out over time (e.g. toxic waste). Tarcin and Tahvonen (1996) replace the assumption of a constant rate of decay of atmospheric carbon by two different models of decay and show that this extension changes the basic dynamic properties of the optimal carbon tax. In Toman and Withagen (2000) the exhaustion of the assimilation capacity of the environment by pollution accumulation results in a non-convexity of the problem that gives rise to multiple potential optima.

of course (e.g. Ulph and Ulph, 1994), but our approach is it because we assume that although the government is neither the owner of the stock of fossil fuel nor is it able to solve the CO_2 problem as well as the energy exhaustion problem by a unilateral action, it wishes to contribute to the solution of these problems by introducing an energy tax. It is benevolent in the sense that it is, in addition to the unilateral action with respect to global warming, concerned about the limited availability of fossil fuel resources in the near future. It is an interesting result of our approach that a government of a country without resources of coal or crude oil, knowing only about the profit-maximizing behavior of its firms, will choose the same tax structure as in Ulph and Ulph (1994) where a resource owning government chooses energy consumption in a macro-economic framework.

Instead of carrying out a detailed technical analysis of the properties of an optimal path of the tax, we fourthly take into account that fossil fuel is also the source of another air pollutant, SO_2. The impact of SO_2 on the environment (e.g. acid rain) can be mitigated by investing in abatement measures. We therefore introduce in addition to the energy tax a SO_2 tax to cope with the damage from the accumulation of two stock pollutants. Although one could expect a complicated mix of tax interactions, the tax rules turn out to be very simple.

2 CO_2 Emissions as an Intertemporal Resource Allocation Problem

2.1 The Capacity of the Atmosphere for CO_2 as a Resource

We treat CO_2 emissions as waste disposal which decumulates a resource for CO_2 deposits in the atmosphere. With $\overline{DE}$ we denote the critical deposit capacity of the atmosphere for CO_2. If $\overline{DE}$ is reached, this is equivalent to waste disposal when the landfill is full. CO_2 emissions which are right now in the atmosphere are $P(t)$, and the resource $R(t)$ we want to preserve is

$$R(t) = \overline{DE} - P(t) \tag{1}$$

that is, $R(t)$ is the capacity left for future CO_2 emissions. The change in pollution over time is

$$\dot{P} = e \cdot E - G(P) = e \cdot E - G\left(\overline{DE} - R(t)\right) \tag{2}$$

where E is energy (fossil fuel) and e is a CO_2 emission coefficient. $G(P)$ is the natural growth of the CO_2 fill capacity due to the absorption from the atmosphere to other carbon sinks (mainly the ocean). It is $GP(P)>0$, i.e. the more CO_2

emissions are in the atmosphere, the more can be absorbed. In terms of $G(\overline{DE}-R(t))$, it is $G_R<0$ which means that the higher the CO_2 storage capacity, the less CO_2 exists for pollution decay. Therefore, the pollution variable $P(t)$ causes a permanent reduction of CO_2 fill capacity in the atmosphere:

$$P(t) = \int_{-120+t}^{t} \left(e \cdot E(s) - G\left(\overline{DE} - R(s)\right)\right) ds \tag{3}$$

where t_0=-120 is the approximate duration time of a CO_2-unit in the atmosphere. Because of (1) and (2), $\dot{R}(t)$ is

$$\dot{R} = -\dot{P} = G\left(\overline{DE} - R\right) - e \cdot E . \tag{4}$$

Since we cannot control what has happened in the past, CO_2 emissions at t=0 are given,

$$\overline{P}(0) = \int_{-120}^{0} -\dot{R}(t)\, dt \tag{5}$$

and so is the resource stock $R(0)=\overline{DE}-\overline{P}(0)$.[4] The objective of environmental policy is to extend the time span of fossil fuel usage until a final time T where a backstop technology is available. At time T the inequality

$$R(T) = R(0) - P(T) \geq 0 \tag{6}$$

should hold. When the capacity for CO_2 fill is under control, either the capacity is exhausted ($R(T)=0$) or some capacity is left ($R(T)>0$), but not used anymore for CO_2 emissions. For the years after T either a steady state situation must hold in which CO_2 emissions are limited by natural absorption, that is

$$\dot{R}(t) = 0 \;\text{ or }\; G\left(\overline{DE} - R(t)\right) = e \cdot E(t) \;\text{ for }\; t > T , \tag{7}$$

or a backstop technology is available. In principle we could think of two state variables, the capacity still available in the atmosphere, $R(t)$, and the pollution stock $P(t)$ that has accumulated over the centuries. However, assuming that extracting capacity equals the net accumulation of CO_2 emissions, that is (4), there is thus only one stock variable and $E(t)$ is our control variable to preserve the CO_2 fill capacity $R(t)$. Since all countries contribute to the CO_2 problem, E should be the aggregated energy consumption. Since we wish to analyze the CO_2 tax policy of a single country, we assume that energy consumption of the rest of the world is

4 We do not intend to model a policy that increase the capacity $R(0)$, that is $E<G(\overline{DE}-R)$.

constant as are the yearly CO_2 emissions. To simplify the analysis we assume that this aspect is taken into account by an appropriate critical limit $\overline{DE}$.

2.2 Maximizing the Performance of the Economy by Choosing the Path of a CO_2 Tax

In order to find the time profile of the resource price, the government maximizes the present value of the sum of consumers' and producers' surplus and of the "ecological surplus" which consists of CO_2 tax revenues minus damage from global warming:

$$\max_{\{\tau_E(t)\}} \int_0^T e^{-rt} \left\{ \int_0^{x(\tau_E)} p(\varsigma)d\varsigma - C\left[x(\tau_E), \overline{DE} - R, PE(\tau_E)\right] + \tau_E \cdot E - D\left(\overline{DE} - R\right) \right\} dt. \quad (8)$$

It is τ_E the energy tax rate, $x(\tau_E)$ is output of the energy intensive industry, $p(x)$ is the market demand function for energy intensive goods, $D(\overline{DE}-R)$ is the damage in money terms caused from global warming ($D_R<0$),[5] r is the discount factor, and $C(\cdot)$ is a cost function with PE as the price for energy which is the only variable input price. The resource stock variable enters the cost function and represents the aspect that more pollution, that is a reduction in the CO_2 fill capacity R, will raise the cost of production; hence $C_R<0$. Possible negative productivity effects follow from the state of health, flooding or hurricanes. If we extend the model by including more industries, than R would also affect the productivity in agriculture.

The government is a Stackelberg leader and has perfect information on producers' behavior, that is, supply x is derived from profit maximizing behavior under price taking behavior:

$$\max_x \quad \pi = p \cdot x - C\left[x, \overline{DE} - R, PE(\tau_E)\right] \quad (9)$$

and cost minimizing energy demand E is derived from $C_{PE}=E[\, x, \overline{DE} - R, PE(\tau_E)\,]$ (Shephard's lemma). From the FOC of (9), i.e. $p=C_x$, follows that supply x depends also on R and $PE(\tau_E)$, hence $x=x(p, \overline{DE}-R, PE(\tau_E))$. The government wishes to control the energy demand by choosing a time path of the control variable $\tau_E(t)$. This tax affects not only E, but also substitution of other inputs for E, marginal costs, and supply and demand. The current value of the Hamiltonian for the problem (8) subject to (4) is:

---

5 The case of a catastrophe if $R\rightarrow 0$ could be considered by a damage function $D\left(\frac{\overline{DE}-R}{R}\right)$ where damage and marginal damage become extremely high when $R\rightarrow 0$.

$$H = \int_0^{x(\tau_E)} p(\varsigma)\, d\varsigma - C\left[x(\tau_E), \overline{DE} - R, PE(\tau_E)\right] + \tau_E \cdot E - D\left(\overline{DE} - R\right) + q_R \cdot \dot{R} \qquad (10)$$

where q_R is the current value of the resource price or user cost.[6] After substitution of the dynamic constraint (4) into (10), the FOC are,[7] assuming an interior solution,

$$H_{\tau_E} = 0 \qquad \text{and} \qquad \dot{q}_R = r\, q_R - H_R \qquad (11)$$

and the dynamic constraint, recovered from the Hamiltonian,[8]

$$\dot{R} = H_{q_R} = G\left(\overline{DE} - R\right) - e \cdot E\,. \qquad (12)$$

The maximum principle yields[9]

$$H_{\tau_E} = (p - C_x)\frac{\partial x}{\partial \tau_E} - E\frac{d\, PE}{d\, \tau_E} + E + \tau_E \frac{d\, E}{d\, \tau_E} - q_R \cdot e\, \frac{d\, E}{d\, \tau_E} = 0$$

$$\text{where } \frac{d\, E}{d\, \tau_E} = \frac{\partial E}{\partial \tau_E} + \frac{\partial E}{\partial x}\frac{\partial x}{\partial \tau_E} < 0\,.$$

Since $p = C_x$ and $PE = PE_0 + \tau_E$, we obtain $\tau_E = e \cdot q_R$; the energy tax is a CO_2 tax and proportional to the resource price q_R of the remaining capacity for CO_2 storage. Thus q_R represents the shadow cost of CO_2 or, equivalently, the benefit

6 Since energy E is derived from Shephard's lemma, it therefore depends on the same variables as the cost function.

7 We suppress the time subscript t in H_t, $q_r(t)$ and $\tau_E(t)$ although all variables depend on t.

8 We could also postulate by an additional restriction that the critical concentration $\overline{DE}$ must not be exceeded by time T and maintained after that point in time by adding the inequality restriction $R(t) \geq 0 \ \forall t \geq T$. Then for $t \geq T$ we must form the Lagrangeian $L = H + \eta \cdot R$. The term η complicates the matter and we neglect this aspect (see Goulder and Mathai, 2000).

9 It is $x(\tau_E)$ in (8) a short-cut of $x\,[\,p(x), \overline{DE}\text{-}R, PE(\tau_E)]$ i.e. the government knows via the demand function the effect on the price of controlling the quantity. The first expression in H_{τ_E} should be

$$H_{\tau_E} = (p - C_x)\frac{d\, x}{d\, \tau_E} \quad \text{where} \quad \frac{d\, x}{d\, \tau_E} = \frac{\partial x}{\partial p}p' + \frac{\partial x}{\partial PE}\frac{d\, PE}{d\, \tau_E}\,.$$

from an incremental amount of avoided CO_2 (a small increase of the CO_2 storage capacity).[10] The portfolio balance condition in (11) yields:[11]

$$\dot{q}_R = (r - G_R)\, q_R + C_R + D_R \,. \tag{13}$$

There is a tendency of the tax rate to grow at the rate $r + |G_R|$, but there is also a tendency for it to decline over time because an extra amount of CO_2 later on would inflict marginal damages over a shorter time horizon.

2.3 Basic Characteristics of the CO_2 Tax

For the interpretation of the slope of the optimal time path of q_R we consider four cases:

(I) $r = 0$, $G_R = 0$: The CO_2 tax is falling since a unit of CO_2 emitted today reduces the CO_2 fill capacity at all future dates, and so the increase in the cost of production $(C_R < 0)$ as well as the increase in damage from global warming $(D_R < 0)$ is greater than a unit of CO_2 emitted at any later date. Therefore, a unit today should be taxed more heavily than a unit tomorrow. The required intertemporal adjustment is brought about if the tax falls at exactly the rate of marginal damage from global warming and from its negative impact on production.

(II) $r = 0$, $G_R < 0$: If a fraction of CO_2 emitted today is absorbed, this reduces future marginal damage. Therefore, the aspect described above is somewhat mitigated, and the tax could fall at a lower rate.

(III) $r > 0$, $G_R < 0$, C_R and D_R small: This is the situation at the beginning of the planning period. The tax is low and rising. The CO_2 emissions contribute to social welfare due to the use of energy in production. If the value q_R of the CO_2 fill capacity increases less than postulated in (13), then more CO_2 should be emitted to raise today's social welfare from material well-being.

(IV) $r > 0$, $G_R < 0$, C_R and D_R high: If a steady state is out of reach then the damage and cost aspects from a low CO_2 fill capacity might finally dominate and it is thus possible for the optimal carbon tax to decline over time. As shown by Goulder and Mathai (2000), under standard assumptions on $G(\cdot)$ and $D(\cdot)$ the optimal carbon tax will rise forever, if the optimized path of CO_2 slopes upward.

---

10 In most of the literature the shadow value of a unit of CO_2 is negative since damage is a function of the stock of accumulated CO_2; i.e. $D(P)$, so that CO_2 is a "bad" from the social planner's perspective. In this paper, the CO_2 capacity of the atmosphere is a "good" and the shadow value q_R is therefore positive.

11 We omit the term $-(\tau_E - q_R \cdot e)E_R$ which is zero because of $\tau_E = q_R \cdot e$ and the term $(p - C_x)x_R$ because of $p = C_x$.

The standard assumptions are $G(\overline{DE}-R)=\delta\cdot(\overline{DE}-R)$, $D_R<0$, $D_{RR}>0$, $C_R<0$ and $C_{RR}>0$. Following Goulder and Mathai, the integration of (13) yields

$$q_R(t)=\int_t^{T=\infty}-\left[C_R\left(\overline{DE}-R(s)\right)+D_R\left(\overline{DE}-R(s)\right)\right]\cdot e^{-(r+\delta)(s-t)}ds \tag{14}$$

where we have to assume additive separability in C between R and τ_E (i.e. $C_{R,q_R}=0$) to make C_R independent of q_R. This equation states that the shadow cost of a CO_2-unit equals the discounted sum of the two marginal damages to production and the environment that this unit would inflict over all future time. Substituting $q_R(t)$ into (13) yields

$$\dot{q}_R=$$

$$(r+\delta)e^{(r+\delta)t}\int_t^{\infty}-\left[C_R\left(R(s)\right)+D_R\left(R(s)\right)\right]\cdot e^{-(r+\delta)s}ds+C_R\left(R(t)\right)+D_R\left(R(t)\right) \tag{15}$$

If C_R and D_R were constant, then the first term on the right hand side would reduce to - [C_R+D_R], and the tax would also be constant since $\dot{q}_R=0$. If $R(s)$ decreases with $s>t$, i.e. CO_2 emissions will increase even under an optimal policy, as environmental economists expect, then the convex damage and cost functions ensure that the first term in the $\dot{q}_R$ equation would be larger than the second term and the slope of the tax path would be positive.

If we substitute $e\cdot q_R$ for τ_E in the two equations for $\dot{R}$ ((12)) and for $\dot{q}_R$ ((13)), we obtain a system of two differential equations in the two unknowns R and q_R. They are necessary conditions that govern the motion of an optimal program. The system must know, in addition, where it is going or where it should start. $R(0)$ is given by the state of the CO_2 fill capacity at the beginning. Yet the determination of $q_R(0)$ is obviously crucial. If the terminal values of $R(T)$ and $q_R(T)$ are known, then it would be possible to use the equations of motion to trace back through time to find the appropriate $\tau_E(0)$ $(=e\cdot q_R(0))$ to get the CO_2 reduction program off to the "right start". For this purpose transversality conditions are used.

We first assume that absorption can be neglected $(G\approx 0)$. Then we can distinguish two cases at the final time T:

(a) $R(T)=0$, $q_R(T)>0$.

The capacity of the resource is exhausted at T and the resource stock must have no value according to the transversality condition $q_R(T)\cdot R(T)=0$. From the second transversality condition, $H(T)=0$, we obtain

$$H(T)=\int_0^x p(\zeta)\,d\,\zeta-C[\cdot,q_R(T)]-D(\overline{DE})+q_R(T)\cdot G(\overline{DE})=0.$$

Consumer and producer surplus plus the value of the (minor) absorption capacity is equal to damage. From this condition $q_R(T)$ can be determined. Since no CO_2 emissions are permitted after period $T(R=0)$, fossil fuel is not an essential input anymore because a carbon-free technology is available at *T*, or energy can be produced from renewable resources (solar-, wind- or hydropower-energy). If we introduce the possibility of some absorption after *T*, then $q_R(T+1)$ follows from

$$e \cdot E\left[x(p,\tau_E), \overline{DE}, PE_0 + e \cdot q_R\right] = G(\overline{DE}). \tag{16}$$

Since supply is a function of q_R as $\tau_E = e \cdot q_R$, x is also determined from (16), given *p*. The tax restricts energy demand such that $\dot{R} = 0$.

(b) $R(T) > 0$, $\quad q_R(T) = 0$.

Now we exclude the case a) (a corner solution) that pollution will ever reach its critical limit, and assume $R(T) > 0$. This situation will occur if marginal cost and marginal damage become extremely high if $R \to 0$. When temperature increases up to a threshold value, derived from *DE*, damage increases sharply. Then a catastrophe occurs if an (uncertain) level of the stock variable (pollution or aggregated extraction of the CO_2-fill capacity) is exceeded. The specification of a damage function

$$D\left(\frac{\overline{DE} - R}{R}\right).$$

would reflect such a situation where $R \to 0$. If we consider the CO_2-fill capacity *R* as a non-renewable resource ($G(\cdot)$ is small or zero), then the transversality condition $q_R(T) \cdot R(T) = 0$ must be satisfied at a final time *T* if such a time exists. Since $R(T) > 0$, the resource price of the remaining stock has fallen to zero, i.e. $q_R(T) = 0$. The absorption capacity cannot be used anymore because a catastrophe might occur. The second transversality condition postulates that the current value of the performance indicator (H) must be zero at *T*; i.e. $H(T) = 0$. This implies according to (10) that consumer and producer surplus is equal to damage from global warming (it is $\tau_E(T) = e \cdot q_R(T) = 0$ at T) and the government terminates the program (no energy tax anymore). $H(T) = 0$ can be solved for *R*(*T*) and after *T* a backstop technology must be available. Permitting some absorption, a resource price q_R after *T* should control (16) (with $\overline{DE}$ - *R*(*T*) instead of $\overline{DE}$).

(c) A steady state with R^*, q_R^* if $G(\cdot) > 0$.

If we assume that absorption is not irrelevant then we can analyze a steady state for *t* > *T*, the time where CO_2 emissions are limited by natural absorption. The steady state tax τ_{E^*} and the resource stock R^* follow from (13) with $\dot{q}_R = 0$ and from (7) with $\dot{R} = 0$, i.e. (16) with $\overline{DE} - R^*$ instead of $\overline{DE}$. The tax is

$$\tau_E^* = e\, q_R^* = -\frac{C_R + D_R}{r - G_R} \tag{17}$$

where C_R, D_R and G_R are functions of R^*. Since G_R is very small (about -1%), the tax rate is the present value of all the increased production and damage costs arising from burning a bit more fossil energy today. The tax captures the present value of the future added costs from energy consumption not incident on a private energy intensive firm, but on the other energy consuming firms and the society.[12]

In models where the social planner maximizes the stream of utility from energy use by choosing an optimal path of fossil fuel extraction, he implicitly also determines the optimal path of a CO_2 tax because the co-state variable is equivalent to a CO_2 tax if markets are perfectly competitive. In our model the social planner maximizes the performance of the economy by choosing only the optimal path of a CO_2 tax. Energy consumption is determined indirectly by the agents' cost-minimizing behavior.

2.4 A Phase Diagram for the Steady State

If we believe that in the near future a high percentage of energy is produced from carbon-free resources, then a standard phase diagram could be used to show the dynamics of the stock R and the energy tax around the steady state. Figure 1 presents the result of the analysis.

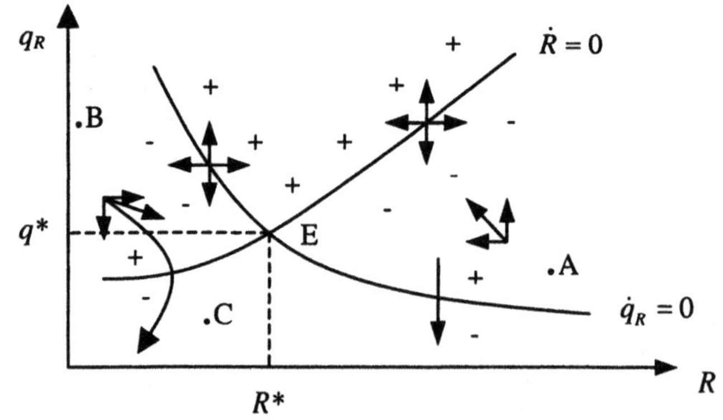

Figure 1: The Dynamics of the CO_2 Fill Capacity R and of the Resource Price or CO_2 Tax q_R.

12 If the two transversality conditions cannot be satisfied, then the program will have to run forever. This is typically the case for renewable resources that can be characterized in final steady state with (q_R, R) positive. For these, transversality is satisfied in the limit as $T \to \infty$: $e^{-rT}\, q_R(T) \cdot R(T) = 0$, $e^{-rT}\, H(T) = 0$.

To show the motions of R and q_R off the steady state (R^*, q_R^*), we first determine for the steady state $(\dot{q}_R = d\,\dot{q}_R = \dot{R} = d\,\dot{R} = 0)$ the slope of the $\dot{q}_R = 0$ equation and of the $\dot{R} = 0$ equation:[13]

$$\dot{q}_R = 0: \qquad \frac{d\,q_R}{d\,R} < 0, \tag{18}$$

$$\dot{R} = 0: \qquad \frac{d\,q_R}{d\,R} > 0. \tag{19}$$

For the motions off the steady state we obtain for the $\dot{q}_R = 0$ equation

$$\frac{\partial\,\dot{q}_R}{\partial\,q_R} > 0 \quad , \qquad \frac{\partial\,\dot{q}_R}{\partial\,R} > 0 \tag{20}$$

and for the $\dot{R} = 0$ equation

$$\frac{\partial\,\dot{R}}{\partial\,q_R} > 0 \quad , \qquad \frac{\partial\,\dot{R}}{\partial\,R} < 0. \tag{21}$$

As shown in Figure 1, the control variable τ_E, which corresponds to $e \cdot q_R$, is used to guide R from $R(0)$ (given exogenously) to R^* (solved for) in an optimal way. Let us assume that today there is enough capacity for CO_2 deposit, i.e. $R(0)$ is to the right of R^*. If society is in A (i.e. $q_R(0), R(0)$), this corresponds to case (III) in Section 2.3; R is high and D_R is negligible. Then a raising tax leads to the steady state. If society is in B, this corresponds to case (IV); R is low and D_R is high. A falling tax and an increasing R (i.e. less CO_2 emissions) are in line with Goulder and Mathai's (2000) analysis of a declining tax under less CO_2 emissions, and a steady state will be reached. If society is in C, the tax is too low and immediate command and control regulations are required to stabilize the situation.

2.5 Comparative Statics in the Neighborhood of the Steady State

We next want to analyze the system, consisting of the maximum principle, $\tau_E = e \cdot q_R$, the dynamic constraint (12) and the portfolio balance condition (13). For that purpose we have to assume functional forms for $G(\overline{DE} - R)$, $C(x, \overline{DE} - R, \tau_E)$ and $D(\overline{DE} - R)$:

$$G(\overline{DE} - R) = \delta(\overline{DE} - R), \qquad \delta > 0 \tag{22}$$

13 The signs are proven in the Appendix, assuming standard specifications of $C(\cdot)$, $G(\cdot)$ and $D(\cdot)$.

$$C(\cdot) = x^{1/h} \cdot c(PE_0 + \tau_E), \tag{23}$$

i.e. we omit the productivity externality $\overline{DE} - R$. All other input prices are held constant and are subpressed in $c(\cdot)$.

$$D(\overline{DE} - R) = \frac{1}{2}\gamma(\overline{DE} - R)^2. \tag{24}$$

δ in (22) is a parameter for natural mitigation, $h < 1$ in (23) is the degree of homogeneity of the production function and expresses decreasing returns to scale. The damage function in (24) is quadratic with $\gamma > 0$ as a damage parameter. From $p = MC(\cdot)$, we obtain

$$x = \left[\frac{h \cdot p}{c}\right]^{\frac{h}{1-h}}. \tag{25}$$

Energy demand E can be derived from Shephard's lemma, using the specification of the cost function in (23) and then replacing x by its expression, derived in (25). This yields:

$$E = \left[\frac{h \cdot p}{c(q_R)}\right]^{1/1-h} c_{PE}(q_R). \tag{26}$$

Since $C_R = 0$, (13) reduces to

$$\dot{q}_R = (r + \delta)\, q_R - \gamma \cdot (\overline{DE} - R) = q(q_R, R). \tag{27}$$

The second equation of motion (27 is the first one) is

$$\dot{R} = \delta(\overline{DE} - R) - e \cdot E(\cdot) = R(q_R, R) \tag{28}$$

with $E(\cdot)$ as given in (26). The steady state solution (R^*, q_R^*) can be obtained by setting $\dot{q}_R = R = 0$ in (27) and (28).

For comparative statics with respect to the exogenous variables $r, \delta,$ and γ we consider the two-equation system for the steady state, i.e. (27) and (28):

$$(r + \delta)\, q_R^* - \gamma \cdot (\overline{DE} - R^*) = 0 \tag{29}$$

$$-e \cdot E(\cdot, q_R^*) + \delta \cdot (\overline{DE} - R^*) = 0 \tag{30}$$

To find

$$\frac{d\,q^*}{d\,r} \text{ or } \frac{d\,R^*}{d\,r}$$

we have to solve these two equations simultaneously after having injected some $d\,r$. The differentiation of these equations with respect to the two endogenous variables (q^*, R^*) and the exogenous variable in question (r) yields:

$$(r+\delta)\,d\,q_R^* + \gamma\cdot d\,R^* = -q_R^*\,d\,r, \qquad -e\cdot E_{q_R}\,d\,q_R^* - \delta\cdot d\,R^* = 0\,. \tag{31}$$

The solution is:[14]

$$\frac{d\,q_R^*}{d\,r} < 0 \quad , \qquad \frac{d\,R^*}{d\,r} < 0\,. \tag{32}$$

A higher interest rate depresses the steady state capital value q_{R^*} of the CO_2-fill capacity and lowers its level. The emphasis is on production and profit now and not later, and therefore on a lower CO_2 tax and less CO_2-fill capacity.

In a similar way, we can find out how q_R^* and R^* respond to changes in δ:

$$\frac{d\,q_R^*}{d\,\delta} < 0 \quad , \qquad \frac{d\,R^*}{d\,\delta} > 0\,. \tag{33}$$

More absorption of CO_2 by the biosphere will increase the CO_2-fill capacity and permits to lower the tax rate.

If damage parameter γ changes, we obtain:

$$\frac{d\,q_R^*}{d\,\gamma} > 0 \quad , \qquad \frac{d\,R^*}{d\,\gamma} > 0\,. \tag{34}$$

This again confirms intuition. If damage from CO_2 emissions is higher, then R^* and the CO_2 tax should be higher, too.

---

14 For technical details see the Appendix.

3 Intertemporal Allocation of Fossil Fuel as a Non-Renewable Resource and as a Source of CO_2 Emissions

We next link the problem of controlling CO_2 emissions to the problem of controlling the extraction of fossil fuel, the exhaustible source of CO_2 emissions. Although the domestic planner has no real control over resource exhaustion and he cannot affect the price path for the exhaustible resource, he can raise the domestic energy price by a tax in order to remind the domestic economy that fossil fuel is an exhaustible resource. We state a dynamic optimization problem with two state variables: $R(t)$, the CO_2 fill capacity; $RF(t)$, the level of the stock of fossil fuel at time t (crude oil, coal, natural gas). The government selects values for the control variable $\tau_E(t)$ for $t = 0, 1, \ldots T$ so as to maximize

$$\max_{\{\tau_E(t)\}} W = \int_0^T e^{-rt} \left\{ \int_0^{x(\tau_E)} p(\zeta)d\,\zeta - C(x, \overline{DE} - R, PE_0 + \tau_E) + \tau_E \cdot E - D(\overline{DE} - R) \right\} d\,t$$

subject to the constraints

$$\dot{R} = G(\overline{DE} - R) - e \cdot E(\cdot) \tag{36}$$

$$\dot{RF} = -E(\cdot) \quad , \qquad RF(0) \geq \int_0^T E\,dt\,. \tag{37}$$

Ulph and Ulph (1994) construct such a model with a CO_2 stock externality which is related to the use of a non-renewable resource. But in their model an agency maximizes the present value of a stream of utility where utility depends on consumption of energy. It also exploits its own resource by choosing an optimal time path of extraction. In our model the government is concerned about the negative externalities from the derived demand for energy. Although it does not own the stock of the non-renewable resource, it knows that production of goods in its country contributes via energy use to the exhaustion of world wide fossil fuel reserves. The energy tax therefore addresses not only the problem of global warming but also the problem that fossil fuels will be exhausted in the near future.[15]

The current-value Hamiltonian is

---

15 As the rest of the world also consumes energy, we had to subtract also this energy use in (37). Since we are not interested in doing numerical simulation of a world model we assume this consumption to be constant. Since a constant does not affect FOCs or the portfolio balance condition we have dropped it.

$$H = \int_0^{x(\tau_E)} p(\zeta)\, d\,\zeta - C(\cdot) + \tau_E \cdot E - D(\overline{DE} - R) + q_R\, \dot{R} + q_{RF} \cdot \dot{RF}$$

and the maximum principle yields

$$H_{\tau_E} = \frac{d\,E}{d\,\tau_E}(\tau_E - q_R \cdot e - q_{RF}) = 0\,.$$

The energy tax consists of the shadow price of CO_2 pollution, $q_R{\cdot}e$, and of the shadow price of fossil fuel, q_{RF}, which in our model is received by the resource-concerned government but not by the owner of the resource:

$$\tau_E = q_R \cdot e + q_{RF}\,. \tag{38}$$

From the portfolio balance condition $\dot{q}_R = r\, q_R - H_R$ we obtain

$$\dot{q}_R = r\, q_R + C_R + D_R - q_R\, G_R + (-\tau_E + q_R \cdot e + q_{RF}) E_R \tag{39}$$

which reduces to the same condition as in (13). From the second portfolio condition, $\dot{q}_{RF} = r\, q_{RF} - H_{RF}$, we obtain the Hotelling rule:

$$\dot{q}_{RF} = r \cdot q_{RF}\,. \tag{40}$$

The optimal tax τ_E in (38) derived from our approach, is in principle the same as that obtained by Ulph and Ulph (1994). It is known that perfect competition on the resource market with price taking resource suppliers will result in the same price and quantity paths like those chosen by a planner. In our model, however, we consider firms under perfect competition which produce anything except of being a mining firm. They use energy at a price PE_0 which could be supplied under any kind of market structure.

The change of the tax over time is $\dot{\tau}_E = \dot{q}_R \cdot e + \dot{q}_{RF}$ or, if we use (39), (40), and (38):

$$\dot{\tau}_E = e\big[(r - G_R)\, q_R + C_R + D_R\big] + r\, q_{RF} = r \cdot \tau_E + e\big[-G_R\, q_R + C_R + D_R\big]\,. \tag{41}$$

As for the slope of the energy tax we could repeat the argument by Goulder and Mathai (2000), presented in (14) and (15), by saying that the tax path will be upward sloping because q_{RF} rises according to Hotelling's rule and q_R rises because we expect that accumulated CO_2 emissions will steadily increase. This sufficient condition for an upward sloping path of τ_E might not hold if a model takes into account that the source of CO_2 emissions is exhaustible fossil fuel.[16] Since it can be shown that $\dot{x} < 0$ and $\dot{E} < 0$, it is possible that finally $R(s) > R(t) \quad \forall \quad s > t$ and the tax path then becomes downward sloping, a possibility also described by Ulph and Ulph (1994). Since it is difficult to derive general conclusions for this kind of model, we therefore discuss only some cases that might occur.

(a) $R(T) = 0 \quad , \qquad q_R(T) > 0 \quad , \qquad G(\cdot) = 0.$

As the CO_2-fill capacity is exhausted at T, no fossil fuel can be used after T. Therefore, an optimal extraction path would require that $RF(T)$=0. However, as the country cannot sell fossil fuel because it does not own the stock, it is not in its interest that the stock of fossil fuel is exhausted. From (3), (6) and (37) follows that obviously $e \cdot RF(0) = P(T) = R(0)$ holds. The tax $\tau_E(T)$ solves the second transversality condition:

$$H(T) = \int_0^{x(\tau_E)} p(\zeta)\, d\,\zeta - C(\cdot, \tau_E) - D(\overline{DE}) = 0; \tag{42}$$

the sum of producer and consumer surplus at T is equal to damage. After T, energy is produced by a backstop technology.

(b) $R(T) > 0 \quad , \qquad q_R(T) = 0 \quad , \qquad G(\cdot) = 0 .$

In this case, damage becomes catastrophic and emissions are not permitted anymore. Again, fossil fuel should be exhausted at T, i.e. $RF(T) = 0$. This time, from (3), (7) and (37) follows

$$\int_0^T E(\cdot)dt = RF(0) \quad , \qquad \int_0^T e \cdot E(\cdot)dt = \overline{DE} - R(T) \tag{43}$$

and therefore $R(T) = \overline{DE} - e \cdot RF(0)$. Then the tax $\tau_E(T)$ solves $H(T) = 0$ with $D(\overline{DE} - R(T))$ in (42).

(c) $RF(T) = 0 \quad , \qquad q_{RF}(T) > 0 \quad , \qquad R(T) > 0 \quad , \qquad G(\cdot) = 0 .$

This case represents the aspect that the stock of fossil fuel is exhausted before global warming becomes a serious problem. $R(T)$ can be determined similarly as in case b), and so can $\tau_E(T)$.

---

16 In a dynamic game of CO_2 emissions Hoel (1993) studies the intertemporal properties of an international carbon tax. He shows that an optimal tax rate should be increasing over time, but at a rate lower than the interest rate, like our result in (41). But he did not add the aspect that the source of CO_2 emissions is a non-renewable resource.

(d) A steady state R^*, q_R^* with $G(\cdot) > 0$ and $RF > 0$.

With absorption, the CO_2 problem has reached a steady state before the resource RF is exhausted. Then

$$q_R^* = \frac{-D_R(\overline{DE} - R^*)}{r - G_R(\overline{DE} - R^*)}$$

from (39) ($\dot{q}_R = 0$) together with

$$G(\overline{DE} - R^*) = e \cdot E\left[x(q_R^* \cdot e + q_{RF}), q_R^* \cdot e + q_{RF}\right] \tag{44}$$

have to be solved for R^* and q_R^*.[17] It is $R^* = R(q_{RF})$ and q_{RF} must keep R^* constant. The tax component q_{RF} is therefore no longer on its optimal path.

We could also think of introducing two control variables, an energy tax τ_{EN} and a CO_2 tax τ_{CO}. We have to modify the welfare function by defining $PE = PE_0 + \tau_{EN} + e \cdot \tau_{CO}$ and by introducing two types of tax revenues. From $H_{\tau_{EN}} = 0$ we obtain $\tau_{EN} + \tau_{CO} \cdot e = q_R \cdot e + q_{RF}$, and from $H_{\tau_{CO}} = 0$ we obtain this condition again. Since q_R and q_{RF} follow the same path in each FOC, there is no need to have an energy tax as well as a CO_2-tax. Welfare cannot be improved by choosing two instruments.

It is important to note that our restriction to one industry is not a restriction of generality of the results. If we affix an index i for industry i to p, C, x and E, and introduce a sum over i in the welfare function at the beginning of this section and in (36) and (37), the tax formula (38) is the same, i.e. independent of the number and size of industries. If $C_{i,R} = 0$, then also the equation of motion of q_R is the same and so is (40). If $C_{i,R} < 0$, then

$$\dot{q}_R = q_R(r - G_R) + \sum_i C_{i,R} + D_R . \tag{45}$$

The stock externalities on the costs of production lower the slope of $\dot{q}_R / q_R$, that is, the initial level of the CO_2 tax should be higher.[18]

To our knowledge, all authors look at a mining firm or at a government which owns the resource. If all countries are interested in reducing CO_2 and in conserving fossil fuel, however, then it is irrelevant whether the use of fossil fuel is taxed or the extraction of it. The path of the tax is the same. Not the same is the

---

17 We assume here that $C(\cdot)$ does not depend on $\overline{DE} - R$ anymore.

18 See also the higher value of $q_R(t)$ in (14) if marginal damage to production increases.

distributional effect between the suppliers and users of fossil fuel.[19] If producers charge the CO_2 tax and the royalty, they obtain the revenues. If the user country charges a CO_2 tax, then the consumers of the user countries benefit from the recycling of the tax revenue. The resource supplier receives only the price he can receive on the resource market (PE_0 in our model). We have not modeled the first phase of a resource model, i.e. the policy of the country that owns the resource. If extraction costs do not depend on the stock RF, then the price path $PE_0=r\cdot PE_0$ follows the Hotelling rule. With mitigation policies in the user countries, the price PE_0 does not depend on τ_E. This can lead to the curious result that the governments of the user countries have higher revenues from taxing oil and gas than the producer countries earn from selling their resources. In our model the government of a resource importing country raises the tax on the stock externality, the revenue of it is recycled to the consumers. There is therefore a rent-transfer of $RF(0)\cdot\tau_E$ in favor of the importers. If the oil producer would internalize at the well the stock externality in their price and extraction policy of fossil fuel, this would change the distributional outcome under a CO_2 mitigation policy.

4 On the Improvement of Abatement Measures in an Intertemporal Setting

In this section we include also pollutants of energy use which can be abated such as SO_2 and NO_X. The firm can react to standards and/or emission taxes by factor substitution or by abatement activities or by both. It has an abatement function and determines the level of the abatement activity by equating marginal abatement costs to the uniform tax rate on emissions. The firm produces output x at cost C $[x, PE(a,\tau_E,\tau_{SO})]$where $PE(\cdot)$ consists of the basic price PE_0, the cost of abatement ca, the cost from taxing non-abated emissions (e.g. SO_2) and the energy tax τ_E

$$PE(a,\tau_E,\tau_{SO}) = PE_0 + \tau_E + ca\cdot a\cdot e_{SO} + \tau_{SO}(1-a)e_{SO} \tag{46}$$

$ca=ca(a)$ is the unit cost of abatement which depends on the degree of abatement activity, a, $(0\le a\le 1)$, e_{SO} is the emission coefficient of the input energy (e.g. tons of SO_2 per ton of input) and τ_{SO} is an SO_2 emission tax rate. We assume $ca_a > 0$ and $ca_{aa} > 0$. The firm maximizes profit choosing output x and the degree of abatement:

$$\max_{x,a} \quad \pi = p\cdot x - C\big(x, PE(a,\tau_E,\tau_{SO})\big). \tag{47}$$

19 See Blank and Ströbele (1995) on this point.

From the FOC $p = C_x$ output is a function of the tax rates $x = x(\tau_E, \tau_{SO})$. Maximizing (47) with respect to the degree a is equivalent to min $PE(a, \tau_E, \tau_{SO})$. The FOC is

$$\frac{d\,PE}{d\,a} = \left(ca_a \cdot a + ca - \tau_{SO}\right) \cdot e_{SO} = 0\,; \tag{48}$$

that is, marginal cost of abatement is equal to the tax rate. Comparative statics for (48) yields $da/d\tau_{SO}>0$; the government can stimulate abatement by raising the tax.

Similarly to the previous section we are interested in the structure and the optimal path of the energy tax τ_E. It affects substitution and lowers output (p is fixed) in order to cope with the CO_2 pollution problem and with the resource scarcity. In order to stimulate abatement, the government introduces another tax τ_{SO}. Its path has an impact on substitution, on output (marginal cost C_x will change) and on abatement. The objective function of the government is

$$\max_{\{\tau_E(t), \tau_{SO}(t)\}} \quad W = \int_0^T e^{-rt} \Bigg[\int_0^x p(\zeta) d\,\zeta - C\left(x, PE(a, \tau_E, \tau_{SO})\right) + \tau_E \cdot E$$
$$+ \tau_{SO} \cdot (1-a) \cdot e_{SO} \cdot E - D_1(\overline{DE} - R) - D_2(\overline{DE}_{SO} - R_{SO}) \Bigg] dt \tag{49}$$

where analogously to the CO_2 accumulation problem $\overline{DE}_{SO}$ denotes the critical deposit capacity of the environment for SO_2. If the SO_2-fill capacity R_{SO} is exhausted, the acidification of land, forest and lakes will become a very serious problem with huge damages to the economy. The dynamic constraints are (36), (37) and

$$\dot{R}_{SO} = G_2(DE_{SO} - R_{SO}) - (1-a) \cdot e_{SO} \cdot E\,. \tag{50}$$

We note that $E = C_{PE}$ and that the government knows that a, x and E are functions of the tax rates with the corresponding reaction of the firm on their levels ($a(\tau_{SO})$, $x(\tau_E, \tau_{SO})$ and $E(\tau_E, \tau_{SO})$). The Hamiltonian is

$$H = \int_0^x p(\zeta) d\,\zeta - C(x, PE(\cdot)) + \tau_E \cdot E + \tau_{SO} \cdot (1-a) \cdot e_{SO} \cdot E - D_1(\overline{DE} - R)$$
$$- D_2(\overline{DE}_{SO} - R_{SO}) + q_R \cdot \dot{R} + q_{R_{SO}} \cdot \dot{R}_{SO} + q_{RF} \cdot \dot{RF}$$

Using $p = C_x$ and (48), the FOC $H_{\tau_{SO}} = 0$ yields

$$\tau_E - q_R \cdot e_{CO} - (q_{R_{SO}} - \tau_{SO}) \cdot (1-a) e_{SO} - q_{RF} + \frac{\frac{d\,a}{d\,\tau_{SO}}}{\frac{d\,E}{d\,\tau_{SO}}} \left[q_{R_{SO}} - \tau_{SO} \right] e_{SO} \cdot E = 0 . \quad (51)$$

The second FOC $H_{\tau_E} = 0$ yields

$$\tau_E + (1-a) \cdot e_{SO} \cdot \left(\tau_{SO} - q_{R_{SO}} \right) - q_R \cdot e_{CO} - q_{RF} = 0 . \quad (52)$$

A solution of this system (51) and (52) in the two unknown variables τ_E and τ_{SO} is:

$$\tau_{SO} = q_{R_{SO}} \quad , \qquad \tau_E = q_R \cdot e_{CO} + q_{RF} . \quad (53)$$

As before, the energy tax takes care of the CO_2 problem and of the resource scarcity, whereas the SO_2 tax τ_{SO} addresses only the problem of acidification. As we have introduced this time two instruments instead of only one, we have assigned two problems of market failure to one tax and the other one to the other tax. Finally, there are three equations of motion for the resource prices. Since that for $q_{R_{SO}}$ is similar to the one for q_R, all three shadow prices rise over time if both stocks of pollutant increase.

Since $da/d\tau_{SO} > 0$, the degree of abatement will then rise over time. If energy consumption has finally dropped over time to a low level, and fossil fuel is close to exhaustion, the tax components q_R and $q_{R_{SO}}$ might become downward sloping as the stocks of pollutants do not increase anymore. This then mitigates the effort to abate.

5 Concluding Remarks

Our main intention was to use a basic micro model of an industry which maximizes profits, uses energy as one of its inputs and is confronted with a varying energy tax. The firm reacts by substitution, changing its output level or by purchasing abatement equipment. The government is well aware about firms' reaction on price signals. It maximizes a stream of social welfare by choosing an optimal path of its instrument – an energy tax. We find that our tax rate is the same as the one derived from the model of an energy extracting benevolent government.

In this paper we extend the Hotelling model by including stock externalities (CO_2, SO_2) and a flow externality like pollutants which can be abated (SO_2). We examine how an energy (CO_2/SO_2) tax can be used to address these problems in an optimal way. The concern about the time profile of an energy tax arises from the facts that fossil fuels are an exhaustible resource and that global warming, being a consequence of carbon accumulation in the atmosphere, is a stock externality problem. It has been in response to these concerns that several authors have addressed the question of the optimal time path of a tax to control the two stock problems. The objective in these papers has been to design a tax, both in level and in time profile, as to bring about socially desirable paths of fossil fuel consumption and carbon accumulation. Unfortunately – and economists are used to that – these studies differ in their conclusions about the shape of the time path of a carbon tax. The optimal time path is either increasing over time (Hoel, 1993) (but at a rate lower than the interest rate), or inverted U-shaped (e.g. Ulph and Ulph, 1994), or monotonically decreasing (e.g. Sinclair, 1994). In addition, by specifying a more complex decay of carbon, the tax may as well be constant through time, increase monotonically, or have a U-shape (Farzin and Tahvonen, 1996). The objective of this paper has been to emphasize not just the level of an energy tax but to think about the time path of such a tax. The evolution of the tax should reflect the problem that CO_2 emissions accumulate in the atmosphere over time. It should also incorporate the aspect that the source of CO_2 emissions, fossil fuel, is an exhaustible resource. We finally addressed the problem of CO_2 accumulation, SO_2 accumulation, the exhaustibility of fossil fuels and the need of abatement measures by introducing two taxes. The message turned out to be rather simple: do tax energy because of its exhaustibility and source of CO_2 emissions and use a separate tax on SO_2 emissions to enhance its abatement. We also explored the idea of a tax first rising and later falling over time, first favored by Ulph and Ulph (1994). With positive discounting, producers need to face a higher specific tax in the future than today to bring about the same reduction in energy use. If the CO_2-fill capacity is nearly exhausted and CO_2 emissions are not permitted in the years to come, a unit of CO_2 emitted at a later date contributes less to marginal damage and should be taxed less, especially as a backstop technology is close anyway.

Appendix

Proof of (18)-(21):

In order to evaluate the slopes of the two equations of motion (12) and (13) near the steady state we choose the specification (22)-(24) given in Section 2. We then differentiate each of the equations of motion (27) and (28) with respect to each of the included variables. For example, for the $q_R(q_R, R)$ equation,

$$d\,\dot{q}_R = (r+\delta)\,d\,q_R + \gamma\,d\,R$$

and for the $R(q_R, R)$ equation

$$d\,\dot{R} = -\delta \cdot d\,R - e \cdot E_{q_R} \cdot d\,q_R$$

where $E_{q_R} < 0$, using (26).

Next we verify (18) as the slope of the $\dot{q}_R = 0$ equation at the steady state:

$$\frac{d\,q_R}{d\,R} = -\frac{\gamma}{r+\delta} < 0 \tag{18'}$$

and of the $\dot{R} = 0$ equation

$$\frac{d\,q_R}{d\,R} = -\frac{\delta}{e \cdot E_{q_R}} > 0\,. \tag{19'}$$

In order to verify (20), i.e. the motions off the steady state, we obtain for the $\dot{q}_R = 0$ equation

$$\frac{\partial\,\dot{q}_R}{\partial\,q_R} = r + \delta > 0, \qquad \frac{\partial\,\dot{q}_R}{\partial\,R} = \gamma > 0 \tag{20'}$$

and for the $\dot{R} = 0$ equation,

$$\frac{\partial\,\dot{R}}{\partial\,q_R} = -e \cdot E_{q_R} > 0, \qquad \frac{\partial\,\dot{R}}{\partial\,R} = -\delta < 0\,. \tag{21'}$$

Proof of (32) to (35):

The system (31), but with total differentiation with respect to all three exogenous variables, is:

$$\begin{aligned} (r+\delta)\,d\,q_R^* + \gamma\,d\,R^* &= -q_R^*\,d\,r - q_R^*\,d\,\delta + (\overline{DE} - R)d\,\gamma \\ -E_{q_R}\,d\,q_R^* - \delta\,d\,R^* &= -\overline{DE}\,d\,\delta \qquad (e=1). \end{aligned} \tag{A1}$$

The determinant of the characteristic matrix of this system is

$$D = -(r+\delta)\cdot\delta + E_{q_R} \cdot \gamma < 0\,.$$

By Cramer's rule $(d\,\delta = d\,\gamma = 0)$:

$$\frac{d\,q_R^*}{d\,r} = \frac{1}{D}\begin{vmatrix} -q_R^* & \gamma \\ 0 & -\delta \end{vmatrix} = \frac{1}{D}\left(q_R^*\,\delta\right) < 0$$

and

$$\frac{d\,R^*}{d\,r} = \frac{1}{D} \begin{vmatrix} r+\delta & -q_R^* \\ -E_{q_R} & 0 \end{vmatrix} = \frac{1}{D}\left(-E_{q_R} \cdot q_R^*\right) < 0$$

which proves (32). Similarly $(d\,r = d\,\gamma = 0)$:

$$\frac{d\,q_R^*}{d\,\delta} = \frac{1}{D} \begin{vmatrix} -q_R^* & \gamma \\ -\overline{DE} & -\delta \end{vmatrix} = \frac{1}{D}\left(q_R^* \cdot \delta + \overline{DE} \cdot \gamma\right) < 0$$

because the positive term $\gamma \cdot \overline{DE}$ dominates the small negative term $q_R^*\,\delta$. Furthermore,

$$\frac{d\,R^*}{d\,\delta} = \frac{1}{D} \begin{vmatrix} r+\delta & -q_R^* \\ -E_{q_R} & -\overline{DE} \end{vmatrix} = \frac{-1}{D}\left((r+\delta)\,\overline{DE} + E_{q_R} \cdot q_R\right) > 0$$

because the positive term $(r+\delta)\,\overline{DE}$ dominates the small negative term $E_{q_R} \cdot q_R$. This proves (33). Next $(d\,r = d\,\delta = 0)$:

$$\frac{d\,q_R^*}{d\,\gamma} = \frac{1}{D} \begin{vmatrix} \overline{DE} - R & \gamma \\ 0 & -\delta \end{vmatrix} = \frac{1}{D}\left(-\delta(\overline{DE} - R)\right) > 0$$

and

$$\frac{d\,R^*}{d\,\gamma} = \frac{1}{D} \begin{vmatrix} r+\delta & \overline{DE} - R \\ -E_{q_R} & 0 \end{vmatrix} = \frac{1}{D}\left(E_{q_R}(\overline{DE} - R)\right) > 0$$

which proves (34).

References

Blank, J. and W. Ströbele (1995), Das CO_2-Problem aus ressourcenökonomischer Sicht, *WiSt* 6, 283-289.

Conrad, K. (1993), Taxes and Subsidies for Pollution-Intensive Industries as Trade Policy, *Journal of Environmental Economics and Management* 25, 121-135.

Conrad, J.M. and L.J. Olson (1992), The Economics of Stock Pollutant: Aldicarb on Long Island, *Environmental and Resource Economics* 3, 245-258.

Dasgupta, P. (1982), *The Control of Resources*, Oxford.

Farzin, Y. and O. Tahvonen (1996), The Global Carbon Cycle and the Optimal Time Path of a Carbon Tax, *Oxford Economic Papers* 48, 515-536.

Forster, B. (1977), On a One State Variable Control Problem – Consumption Pollution Trade-Offs, in: Pitchford, J.D. and S. Turnovsky (Eds.), *Applications of Control Theory to Economic Analysis*, Amsterdam.

Goulder, L.H. and K. Mathai (2000), Optimal CO_2 Abatement in the Presence of Induced Technological Change, *Journal of Environmental Economics and Management* 39 (1), 1-38.

Hoel, M. (1993), Intertemporal Properties of an International Carbon Tax, *Resource and Energy Economics* 15, 51-70.

Keeler, E., M. Spence, and R. Zeckhauser (1972), The Optimal Control of Pollution, *Journal of Economic Theory* 4, 19-34.

Van der Ploeg, F. and C. Withagen (1991), Pollution Control and the Ramsey Problem, *Environmental and Resource Economics* 1, 215-236.

Plourde, C.G. (1972), A Model of Waste Accumulation and Disposal, *Canadian Journal of Economics* 5, 199-225.

Sinclair, P. (1994), On the Trend of Fossil Fuel Taxation, *Oxford Economic Papers* 46, 869-877.

Tahvonen, O. (1995), Dynamics of Pollution Control when Damage is Sensitive to the Rate of Pollution Accumulation, *Environmental and Resource Economics* 5, 9-27.

Tahvonen, O. (1997), Fossil Fuels, Stock Externalities, and Backstop Technology, *Canadian Journal of Economics* 30, 855-874.

Toman, M. and K. Palmer (1997), How Should an Accumulative Toxic Substance be Banned? *Environmental and Resource Economics* 9 (1), 83-102.

Toman, M.A. and C. Withagen (2000), Accumulative Pollution, "Clean Technology", and Policy Design, *Resource and Energy Economics* 22, 367-384.

Ulph, A. and D. Ulph (1994), The Optimal Time Path of a Carbon Tax, *Oxford Economic Papers* 46, 857-868.

Balancing the Interests of the Present and the Future*

Wolfgang Buchholz

Department of Economics, University of Regensburg, 93040 Regensburg, Germany, *wolfgang.buchholz@wiwi.uni-regensburg.de*

Summary:
From an ethical perspective intergenerational discounting often is justified by the desire to avoid distributions of welfare that are immoderately unbalanced in favor of future generations. In this paper it is shown in a two-generation framework that for productive economies it is well possible to use undiscounted *utilitarianism instead in order to make ethically acceptable choices on intergenerational distributions. When intergenerational discounting is nevertheless used as a pragmatic device it also follows from our approach that time-inconsistency might no longer be an important problem.*

Keywords: Intergenerational justice, discounting

1 Ethical Problems with Discounting

In theoretical as well as empirical models of intertemporal allocation it is quite common to use a utilitarian objective function with a constant discount rate. This means that aggregate discounted utility

$$\sum_{t=1}^{\infty} \frac{u(c_t)}{(1+\rho)^t} \qquad (1)$$

is maximized among all paths that are feasible given a certain technology and some initial endowment. Discounting future costs and benefits is a characteristic of individual behavior and individual preferences – and, depending on the circumstances, it is judged quite differently. In business life discounting is obviously compatible with rational behavior when individuals use it to adapt to a positive market interest rate. According to a widespread view discounting,

* Acknowledgements: I have received many helpful comments from Geir Asheim and the seminar participants at ZEW Mannheim, especially from Stef Proost.

however, seems to be irrational when it stems from pure time preferences i.e. if people are impatient or "myopic". Then by some "weakness of will" or "lack of imagination" individuals attach too low a value to payments and utilities that will accrue to them only in future periods. This neglect of the future then counteracts the true long-run interests of the individuals: Eventually arriving in future periods individuals will regret their former decisions wishing e. g. that they had saved more when they were young. This provides the standard paternalistic argument in favor of compulsory pension schemes.

The devaluation of future needs that is implied by discounting which is thus deemed as imprudent from the individual perspective is often considered as being unethical from a collective perspective when different generations are involved. Giving future generations less weight in social evaluations seems to be questionable from an ethical viewpoint because it would entail an arbitrary and thus unwarranted distinction between different generations. It is only by pure chance that one generation happens to appear later on the time axis and not by effort or desert. Thus no convincing reason seems to exist why generations should not be treated equally and thus be discriminated.

The effects of discounting on future generations may be dramatic, in particular if the welfare of far distant generations is at stake. This is clearly shown by the famous Dasgupta-Heal-Solow model where an exhaustible natural resource can be substituted as a factor of production by man-made capital. In this model a utilitarian criterion with a constant discount rate inevitably will lead to a consumption path that converges to zero in the long run. (Dasgupta and Heal, 1979; Pezzey and Withagen, 1998). Therefore, discounting is in clear contradiction with the idea of economic progress and even with "sustainability", i.e. with having utility paths being non-declining over time, which might be considered as a minimal requirement for progress. When despite these ethical objections intergenerational discounting is still applied this could, in a naive interpretation, simply be attributed to some kind of moral insensitiveness. Then Harrod's (1948) famous dictum that pure time preference is a "polite expression for rapacity" would simply be true.

When examined more closely the problems turn out to be more intricate as intergenerational discounting is often defended with ethical reasons, too. In this paper, we will, as a first step, retrace the basic ideas that are underlying this justification of intergenerational discounting. In a second step, we will then discuss another possibility for implementing these norms that even though being familiar in other fields of economics has not yet been systematically applied to the intergenerational context. Finally, we will consider the implications of this approach for the time consistency problem that has accompanied the discounting question for a long time.

2 The Ethical Argument in Favor of Discounting

Giving numerical calculations of costs and benefits of different abatement strategies for greenhouse gas emissions in the framework of the DICE model Nordhaus (1999: 155-156) concludes that "...lowering the pure rate of social time preference form 3% to 0%" would imply "that the net global saving rate rises from 8.1% of net global net product ... to ... 24.3% ...". Nordhaus qualifies this outcome as being "seriously out of proportion". Alternatively, this means that discounting is required in order to avoid a distribution of welfare that is immoderately unbalanced not in favor of the present but in favor of the future generation. As undiscounted utilitarianism "imposes excessive hardships to future generation" (Rawls, 1999: 259-262) fairness towards the present calls for a remedy which in an easily comprehensible way is provided by a positive rate of time discount. The idea underlying this argument is in fact very old. More than 100 years ago the Austrian economist Böhm-Bawerk (1971: 269-270) clearly identified the circumstances under which discounting may be required for distributional reasons.

"...if every employment of goods for future periods is not only technically, but economically more remunerative than the employment of them for the present or near future, of course men would withdraw their stocks of goods, to a great extent, from the service of the present and direct them to the more remunerative service of the future. But this would immediately cause an ebb-tide in the provision for the present, and a flood in the provision for the future, ..."

Note that Böhm-Bawerk did not explicitly refer to different generations, i.e. to a distributional problem occurring between different individuals. Rather it seems that he had in mind a very general and intuitive idea of equilibrating interests at different points in time ("ebb-tide" vs. "flood") – and what he called "more remunerative" will nowadays be coined as "productive".

The following diagram gives a graphical illustration of Böhm-Bawerk's argument for the case of two generations. Let u_i denote the utility level of generation $i = 1,2$ where generation 2 is a successor of generation 1 on the time axis where "utility" is assumed to be cardinally measurable and level comparable. It would well be possible to make utility u explicitly dependent on bundles c of consumption goods such that utility would be given as $u(c)$. For the argument that is given here this additional assumption, however, is not required. In Figure 1 the line AB describes the transformation curve $u_2 = T(u_1)$ between utility of generation 1 and generation 2 which is assumed to be decreasing, concave, and differentiable everywhere.

Call $s(u_1, u_2)$ the marginal rate of transformation between utility of generation 1 and utility of generation 2. The technology underlying AB will be (short-term) productive if

$$s(u_1,u_2) > 1 \tag{2}$$

holds at every point along AB. This says that reducing generation 1's utility by one marginal unit makes possible to increase generation 2's utility by more than a marginal unit. This property will hold in many cases:

Just imagine that we are in the special case where "utility" can be identified with "consumption" where the consumption good can be stored without any loss by depreciation. This at least gives a weak form of productivity ($s(u_1,u_2) \geq 1$). If moreover the saved consumption good can be used as a capital good that has a yield in the next period (just like in the famous "corn"-example) then a reward for waiting is obtained and productivity as defined here will hold.

If the undiscounted utilitarian criterion is applied to this situation, i.e. aggregate utility u_1+u_2 is maximized among all feasible utility paths the optimal solution is in $B = (0, \bar{u}_2)$ where the transformation curve AB intersects the u_2-axis. In B utility of generation 2 is at its maximum and utility of generation 1 is at its minimum. Hence, distribution of utility is extremely skewed to the damage of the earlier generation.

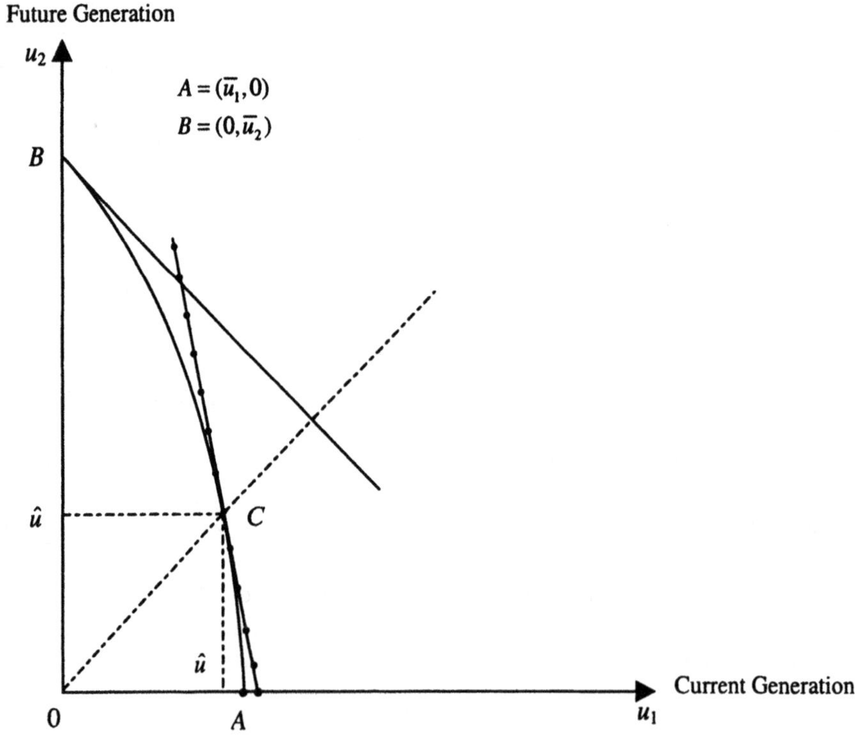

Figure 1: Utilitarian Optima in the Two Generations Case.

If we now introduce a time discount rate ρ into the utilitarian criterion this turns into

$$u_1 + \frac{u_2}{1+\rho}. \tag{3}$$

In particular, let us assume that ρ is chosen as

$$\hat{\rho} := s(\hat{u}, \hat{u}) - 1 \tag{4}$$

where $\hat{u}$ is utility in the point of intersection *C* between the transformation curve *AB* and the equal distribution line. The social indifference curves corresponding to the discounted utilitarian criterion (3) are straight lines with the slope $-(1+\rho)$ which is equal to $-s(\hat{u}, \hat{u})$ if $\rho = \hat{\rho}$. (See the dotted line in Figure 1). Then *C* is the unique optimal outcome of welfare maximization according to (3). Hence applying a utilitarian criterion with an adequately chosen discount rate may even yield a completely egalitarian distribution of utility as the unique welfare maximum.

Ethical intuition will object to this "maximin solution" as it does not allow exploitation of the existing technological possibilities for improving welfare of future generations. (See Solow, 1974, for treatment in the far more general framework of the Dasgupta-Heal-Solow model with an infinite number of periods.) Even though the economy is productive it possibly might stay in perpetual poverty which conflicts with the idea of progress. But this argument in the end has no bite: Discounted utilitarianism also shows considerable flexibility in making choices among the thus desired growth paths lying on the segment *CB* on the transformation curve. If the discount rate ρ varies between 0 and $\hat{\rho}$ this will give every utility distribution on *CB* as the best solution. "What we are doing is adjusting certain parameters so as to reach a conclusion more in line with our intuitive judgements" (Rawls, 1999: 262).

But there are also problems with this approach. On the one hand, as seen from the perspective of ethical principles, the use of discounting for protecting the interests of the future is not only ad hoc but also violates the equal treatment postulate that cannot easily be disregarded in the intergenerational context (see also the very clear discussion of this topic in Rawls, 1999: 262). On the other hand, a certain discount rate ρ that gives an ethically acceptable growth path within one technological environment may lead to economic decay within another technological environment where productivity is lower. With a "wrong" discount rate in Figure 1 an outcome on the segment *AC* below the equal distribution line would be attained. If $\rho > s(\overline{u}_1, 0) - 1$ the extreme solution *A* would be obtained such that the desire for preserving the living conditions of future generation would clearly not materialize.

Therefore, discounted utilitarianism even though being helpful in avoiding unappealing outcomes of intertemporal choices in many cases remains questionable as an ethical criterion. We now want to ask whether it is possible to avoid this dilemma between having ethically acceptable results on the one hand and having an a priori ethically acceptable criterion for getting these results, on the other.

3 Finding Ethically Acceptable Intergenerational Distributions in the Framework of Undiscounted Utilitarianism

The ethical argument in favor of discounting is based on the intention to make distributional decisions within a certain technological environment such that too much inequality in the final distribution of welfare is avoided. But instead of going back to Böhm-Bawerk one could as well refer to Pigou (1912) and Dalton (1920) who suggested a much more general way for incorporating normative judgements into social evaluations. By applying the same strictly increasing and concave evaluation function $h=h(u)$ to the utility levels of every generation before summing up the (modified) undiscounted utilitarian criterion for the two-generations-case becomes

$$h(u_1)+h(u_2) \tag{5}$$

In the theory of inequality measurement this idea has been used by Atkinson (1970) to conceive the famous Atkinson inequality indices. This approach explicitly allows for different degrees of inequality aversion by varying the valuation function h which in the central cases considered by Atkinson was assumed to be isoelastic (having the form $h(u) = h^{1-\varepsilon}$ if $0 \leq \varepsilon < 1$ and $h(u) = -u^{1-\varepsilon}$ if $\varepsilon > 1$). In the intergenerational context it was shown for the same kind of valuation functions (see e.g. Dasgupta and Heal, 1979) that a higher value of the elasticity parameter ε (i.e. a higher degree of concavity of the valuation function and thus a higher degree of inequality aversion) can be used to slow down growth and thus to generate a more equal intergenerational distribution.

So it is well-known that there is some flexibility for making intertemporal choices by adjusting the valuation function. But this approach is also all-encompassing as the degree of flexibility in all ethically relevant situations is not smaller than that which is given by varying the discount rate. We will show this for the case of two generations.

Adhering to the ideas of economic progress or sustainability choices have to be made among the paths on segment CB of the transformation curve where $u_2 \geq u_1$

holds. Let a certain utility profile $(\tilde{u}_1, \tilde{u}_2)$ with $\tilde{u}_2 > \tilde{u}_1$ be given. Then a marginal valuation function $h'(u)$ constructed in the following way: We only fix the values of $h'(u)$ at $\tilde{u}_1$ and $\tilde{u}_2$ by

$$h'(\tilde{u}_1) = 1 \qquad \text{and}$$

$$h'(\tilde{u}_2) = 1/s(\tilde{u}_1, \tilde{u}_2) \qquad (< 1).$$

otherwise we allow for any strictly non-decreasing, continuous and positive function $h'(u)$ on $[0, \infty[$ that is bounded from above. Then the integral

$$h(u) = \int_0^u h'(v)\ dv$$

seen as a function of u is differentiable, strictly increasing and concave (see Figure 2).

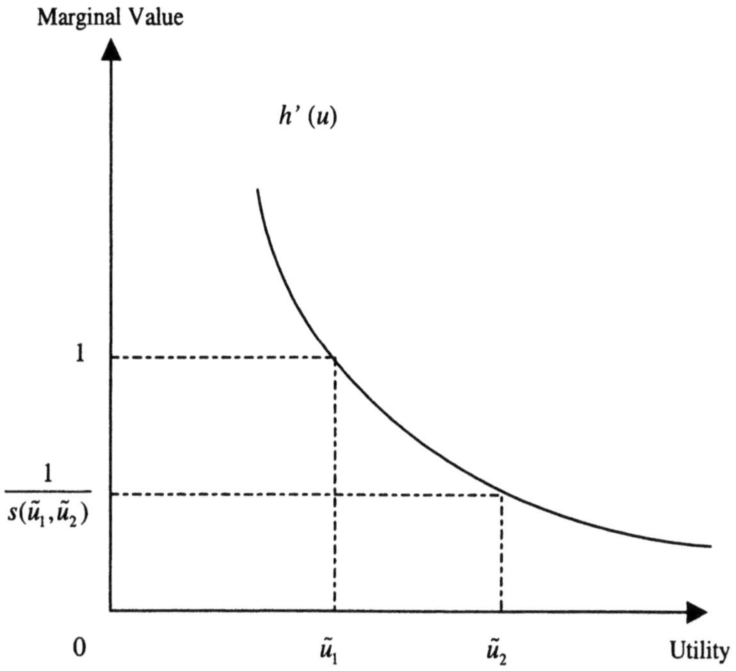

Figure 2: Constructing the Evaluation Function.

Maximizing $h(u_1)+h(u_2)$ along $u_2 = T(u_1)$ then gives

$$\frac{h'(u_2^*)}{h'(u_1^*)} = \frac{1}{s(u_1^*,u_2^*)} \tag{6}$$

as the marginal condition for the optimal solution (u_1^*, u_2^*). This marginal condition obviously is fulfilled if $u_1^* = \tilde{u}_1$ and $u_2^* = \tilde{u}_2$. From concavity of $T(u_1)$ and convexity of the corresponding social indifference curves if follows that $(\tilde{u}_1, \tilde{u}_2)$ is the unique solution to this optimization problem. If, however, $\tilde{u}_1 = \tilde{u}_2$ the maximin criterion has to be used as welfare function. In both cases a sustainable path can be selected without abandoning equal treatment of different generations in the welfare function which seems ethically appealing.

It is important to note that the construction of such welfare functions is definitely not feasible if $\tilde{u}_1 > \tilde{u}_2$ is assumed. It has been demonstrated for the general case of an infinite number of generations by Asheim et al. (2001) that non-sustainable utility path cannot be the best element for intergenerational social preferences that fulfil the Pareto criterion and give equal weight to every generation. For a graphical representation of the underlying argument in the two generations diagram see Buchholz (1997).

The local degree of (absolute) inequality aversion of the valuation function $h(u)$ is given by the elasticity of marginal utility $-h''(u)u/h'(u)$. If an initial valuation function is substituted by another one which shows a higher degree of inequality aversion for all utility levels u in the relevant range then the optimal solution moves closer to the equal distribution line. Therefore, a general fact is reflected in the intergenerational case considered here, which can, by extending a recent result by Fleurbaey and Michel (2001), easily be demonstrated.

Asheim and Buchholz (2000) have shown that the central argument on the "malleability" of undiscounted utilitarianism in finding out a unique solution essentially even carries over to the general case of infinitely many generations $t = 1,2....$ The result obtained in that paper heavily relies on using the 'catching up' criterion for comparing intergenerational allocations and assuming the condition of finite utility (or "consumption") value. This condition means that along a non-decreasing efficient path $(\tilde{u}_1, \tilde{u}_2,...)$ which should become the unique maximal element

$$\sum_{t=1}^{\infty} p_t \tilde{u}_t \quad < \infty \tag{7}$$

has to hold. Here the ratio p_{t+1} / p_t is the marginal rate of transformation between utility in period t and period t+1 along the given path. As can be shown finite consumption value is given for any non-decreasing efficient path in a linear technology and for all relevant cases in the Dasgupta-Heal-Solow model. In the one sector Ramsey model with only one type of man-made capital finite consumption value is obtained for all consumption paths that are bounded from above.

But the meaning of condition (7) might be even more fundamental than is suggested by these examples. Assume that the given utility path $(\tilde{u}_1, \tilde{u}_2, \ldots)$ is the maximal element w.r.t. to some discounted utilitarian criterion which is "pure" in the sense that no valuation function is applied to the initially given utility levels u_t i.e. w.r.t. to the objective function

$$\sum_{t=1}^{\infty} \lambda_t u_t \, .$$

Then by a simple arbitrage argument it follows from the efficiency requirement that the utility discount rates λ_t have to be proportional to p_t. Then without finite utility value the sum of discounted utility values

$$\sum_{t=1}^{\infty} \lambda_t \tilde{u}_t$$

would diverge which restricts the applicability of discounted utilitarianism in this case and takes away much of the attractiveness of this criterion. Therefore, in a certain sense, discounted utilitarianism is not more flexible in making the desired choices as its undiscounted counterpart would be.

4 The Possible Role of Discounting

The message of this paper, however, should not be that discounting is inappropriate in all cases. Discounting may well be justified

- as a description of existing individual preferences which have to be taken as given from the perspective of methodological individualism. These preferences may well be based on some intergenerational altruism. But if there is nothing else than such individual preferences no complaint on a possibly non-sustainable development would really be justified. If present people show enough sympathy with their descendants a sustainable development will be ensured. Otherwise, the world might perish which would seem all right on the basis of the given preferences.

- as a tool for taking uncertainty into account especially if some chance that the world will cease to exist at some time is given. Applying an ethical criterion will clearly not forbid some stochastic calculation.

The message of this paper rather is that some commonly maintained reasons for the inevitability of discounting in fact are not valid. If the problem is to find ethically acceptable distributions between different generations undiscounted utilitarianism is a more obvious starting point – and contrary to a widespread belief there is not so much restriction for using it.

This argument, however, does not exclude discounting as a purely pragmatic device for counteracting the detrimental effects productivity would otherwise have on the welfare of the present generation. People are not only accustomed to treat intertemporal distributional problems by using discount rates but discounting is also very easy to apply because it directly allows for comparisons with the given productivity rates. This is just what is called for by distributional reasons. In this way discounting as a "rule of thumb" may provide a transparent instrument for facilitating complex decisions. By the argument sketched in this paper even some reassurance is provided that such a discounting can be done without a bad conscience because no intrinsic conflict with the equal treatment principle would exist.

This idea for motivating the discounting procedure can also be related to Böhm-Bawerk's famous "third cause" for a positive interest rate which, in Böhm-Bawerk's own words, is to account for "why present goods are worth more than future" and which should thus give an "explanation and justification of interest" (Böhm-Bawerk, 1971: 260).

But if the purpose of discounting is perceived in this way as an easily tractable tool for achieving certain distributional objectives then it becomes clear that discount rates have to be endogenized and will depend on the given technological environment. Also Rawls (1999: 262) himself seemed to have had something like this in mind when saying that "the most suitable discount may depend upon how swiftly population is growing, upon the productivity of capital, and so on."

As an example for this general statement we may consider the Dasgupta-Heal-Solow model with a Cobb-Douglas-production function when a positive consumption level can be sustained forever. If the distributional objective is assumed to be strictly egalitarian then hyperbolic (e.g. Heal, 1998) or gamma discounting (Weitzman, 2001) with non-constant time-varying discount rates turns out to be the suitable form of discounting.

Weitzman (2001) gives a convincing foundation for this specific form of discounting by aggregating expressed opinions on reasonable constant discount rates. But he also touches the problem of time consistency that then seems to be implied by having non-constant discount rates. So Weitzman (2001: 266) concludes by putting the result of his paper into a broader context:

"I think it is fair to characterize this paper as suggesting empirically that long-run time inconsistency may run deeper and may be more generic than was previously suspected."

An implication of our interpretation, however, could be that there is not so much need to worry about the problem of time-inconsistency. If some distributional objective is a primitive then the pattern of discount rates has to be adjusted to changing productivities when time goes by. An illustration is again provided by the Dasgupta-Heal-Solow example. Here, in the long run the remaining stock of the natural resource converges to zero whereas the stock of man-made capital goes to infinity. In order to maintain a constant consumption level as the optimal outcome when a discounted utilitarian criterion is applied then the instantaneous discount rates have to be decreased more and more when we move along the time axis (see, e.g. Asheim, 2000).

In order to give an empirical vindication of this approach to having endogenously determined discount rates it might be helpful to conduct experiments in order to examine whether (and possibly how) people will vary their individual discount rates when productivities are changing. If this were the case it could be inferred that pure time preference as a part of ethical intergenerational social preferences might not be as prominent as it is often suggested.

References

Asheim, G. (2000), Green National Accounting: Why and How?, *Environment and Development Economics* 5, 25-48.

Asheim, G. and W. Buchholz (2000), *The Malleability of Undiscounted Utilitarianism as a Criterion of Intergenerational Justice*, CES ifo Working Paper, No. 392, München (To appear in a revised version in *Economica*).

Asheim, G., W. Buchholz, and B. Tungodden (2001), Justifying Sustainability, *Journal of Environmental Economics and Management* 41, 252-268.

Atkinson, A. (1970), On the Measurement of Inequality, *Journal of Economic Theory* 2, 244-263.

Böhm-Bawerk, E. von (1971), *The Positive Theory of Capital*, New York (Reprint of the First English Edition 1891).

Buchholz, W. (1997), Intergenerational Equity, in: Zylicz, T. (Ed.), *Ecological Economics. A Volume in the Series: A Sustainable Baltic Region*, Uppsala.

Dalton, H. (1920), Measurement of the Inequality of Incomes, *Economic Journal* 30, 348-361.

Dasgupta, P. and G. Heal (1979), *Economic Theory and Exhaustible Resources*, Cambridge, U.K.

Fleurbaey, M. and P. Michel (2001), Transfer Principles and Inequality Aversion, With an Application to Optimal Growth, *Mathematical Social Sciences* 42, 1-11.

Harrod, R.F. (1948), *Towards a Dynamic Economy*, London.

Heal, G. (1998), *Valuing the Future*, New York.

Nordhaus, W. (1999), Discounting and Public Policies that Affect the Distant Future, in: Portney, P.R. and J.P. Weyant (Eds.), *Discounting and Intergenerational Equity*, Washington, D.C.

Pezzey, J. and C. Withagen (1998), The Rise, Fall and Sustainability of Capital-Resource Economies, *Scandinavian Journal of Economics* 100, 513-527.

Pigou, A.C. (1912), *Wealth and Welfare*, London.

Pigou, A.C. (1920), *The Economics of Welfare*, London.

Rawls, J. (1999), *A Theory of Justice*, Revised Edition, Oxford (First Edition 1971).

Solow, R. (1974), Intergenerational Equity and Exhaustible Resources, *Review of Economic Studies*, Symposium.

Weitzman, M. (2001), Gamma Discounting, *American Economic Review* 91, 260-271.